THE

LOST

MEN

THE LOST MEN

---　✳　---

The Harrowing Story of
Shackleton's Ross Sea Party

KELLY TYLER–LEWIS

BLOOMSBURY

First published in Great Britain 2006
Copyright © 2006 by Kelly Tyler-Lewis
The moral right of the author has been asserted
Bloomsbury Publishing Plc, 36 Soho Square, London W1D 3QY

The maps on page 93 and 174 were prepared by Kelly Brunt and Elles Gianocostas using data from charts by Æneas Mackintosh (Scott Polar Research Institute, University of Cambridge) the Aurora 1916–17 log of John King Davis (State Library of Victoria) and the Antarctic Digital Database version 4.1 (copyright Scientific Committee on Antarctic Research, 1993–2003). The maps on pages 20 and 70 are courtesy of the La Trobe Picture Collection, State Library of Victoria. The map on page 147 is licensed with permission of the Scott Polar Research Institute, University of Cambridge, ITAE program.

A CIP catalogue record for this book is available from the British Library

Hardback ISBN 0 7475 6926 6
Hardback ISBN-13 9780747569268
10 9 8 7 6 5 4 3 2 1

Export paperback ISBN 0 7475 8414 1
Export paperback ISBN-13 9780747584148
10 9 8 7 6 5 4 3 2 1

Printed in Great Britain by Clays Limited, St Ives plc

The paper this book is printed on is certified by the © 1996 Forest Stewardship Council A. C. (FSC). It is ancient-forest friendly. The printer holds FSC chain of custody SGS-COC-2061

FSC
Mixed Sources
Product group from well-managed forests and other controlled sources
Cert no. SGS-COC-2061
www.fsc.org
© 1996 Forest Stewardship Council

FOR NICK

CONTENTS

LIST OF MAPS AND ILLUSTRATIONS

THE ROSS SEA PARTY OF THE IMPERIAL
TRANS-ANTARCTIC EXPEDITION, 1914–17

The Shore Party

Æneas Lionel Acton Mackintosh	*Commander*
Ernest Edward Joyce	*Sledging Equipment and Dogs*
Harry Ernest Wild	*Storekeeper*
Reverend Arnold Spencer-Smith	*Chaplain and Photographer*
John Lachlan Cope	*Biologist and Surgeon*
Alexander Stevens	*Chief Scientist*
Richard Walter Richards	*Physicist*
Andrew Keith Jack	*Physicist*
Irvine Owen Gaze	*General Assistant*
Victor George Hayward	*General Assistant*

The *Aurora's* Officers and Crew

Joseph Russell Stenhouse	*First Officer*
Leslie James Felix Thomson	*Second Officer*
Alfred Herbert Larkman	*Chief Engineer*
C. Adrian Donnelly	*Second Engineer*
Lionel Hooke	*Wireless Telegraph Operator*
Clarence Charles Mauger	*Carpenter*
James Paton	*Boatswain*
Aubrey Howard Ninnis	*Motor Tractor Specialist*
Sydney Atkin	*Able Seaman*
Arthur Downing	*Able Seaman*
William Kavanagh	*Able Seaman*
A. "Shorty" Warren	*Able Seaman*
Charles Glidden	*Ordinary Seaman*
S. Grady	*Fireman (Stoker)*
William Mugridge	*Fireman*
Harold Shaw	*Fireman*
Edward Wise	*Cook*
Emile Octave d'Anglade	*Steward*

THE ROSS SEA RELIEF EXPEDITION, 1916-17

John King Davis	*Captain and Commander*
Clarence P. de la Motte	*Chief Officer*
Austin le Gros	*Second Officer*
W. Aylward	*Third Officer*
Morton Moyes	*Fourth Officer*
Ernest Shackleton	*Supernumerary Officer*
Dr. Frederick G. Middleton	*Surgeon*
Aubrey Howard Ninnis	*Purser*
T. M. Ryan	*Wireless Operator*
Frederick J. Gillies	*Chief Engineer*
Arthur Dakin	*Second Engineer*
James Paton	*Boatswain*
M. Hannan	*Donkeyman*
Alexander Webster	*Chief Steward*
Baden P. Robertson	*Second Steward*
Henry Voegeli	*Cook*
C. Brock	*Able Seaman*
William Kavanagh	*Able Seaman*
Ewan McDonald	*Able Seaman*
Alasdair MacKinnon	*Able Seaman*
Malcolm MacNeill	*Able Seaman*
William Peacock	*Able Seaman*
E. Murphy	*Fireman*
J. Rafferty	*Fireman*
T. Smith	*Fireman*

THE

LOST

MEN

Preface

By March 20, 1916, the earliest days of the Great War were fading from memory. The hopes of swift victory on both sides had long since evaporated. On that day, the *Times* of London advertised "History of the War, Part 83," revisiting the beginnings of the bewildering descent into "the war to end all wars" after a century of relative peace in Europe. As the battles dragged on into a new year, submarine attacks, aerial bombardment, and mass slaughter on the western front dominated the headlines. With the number of British men mobilized climbing into the millions, page upon page listed the losses in Europe, Mesopotamia, East Africa, the Mediterranean, and Egypt: killed, accidentally killed, missing believed killed, wounded, died of wounds. Dozens of charities entreated readers for help, among them "Friendless Belgian Prisoners of War," "Polish Victims Relief Fund," and the "Waifs and Strays Society," along with personal pleas: "Wanted, home for boy of two years."

That same day, on page 7 of the newspaper, a short item announced:

News of the expedition which Sir Ernest Shackleton led to the Antarctic with the object of crossing the South Polar continent is expected to reach England at any moment (says Reuter's Agency). Sir Ernest Shackleton left for the Antarctic in 1914. Nothing has since been heard of the expedition. He estimated that it would take him four months to cross the Antarctic continent, and on the other side he expected to join hands with the party which, sailing in the *Aurora* from Tasmania in December 1914, were to make a base at the Ross Sea and go to meet him early this year.

The article was the first mention of Shackleton in the pages of the *Times* in well over a year. It had not always been so. In the early stages of his expedition, bulletins about the celebrated explorer appeared every day and were eagerly absorbed by a fascinated public. Then, on August 1, 1914, German forces invaded Belgium, and in a matter of days, the nations of Europe were at war. In December, Shackleton's ships, the *Endurance* and the *Aurora*, sailed south into the Antarctic and vanished from the public eye, their adventure eclipsed by a war whose devastation surpassed all others in the history of humankind.

On Monday, March 20, the story excited little attention in London, and escaped the notice of the *Southland Times*, the newspaper of Bluff, New Zealand. Bluff was one of the farthest corners of the British Empire, in another hemisphere and the next day on the calendar. Sailing due south from the port across the Southern Ocean, the next landfall as the crow flew was Antarctica. The *Southland Times* was preoccupied with shipping news, patent remedies, racing results, and reports on the distant battles. There was little threat of the war reaching New Zealand soil, but its effects were apparent in the unusual hush that muted everyday life in Bluff. So many of the town's eligible men had enlisted that the farms and sheep stations were short-handed. The coastal wireless station in nearby Awarua, too, was hard pressed to man the continuous listening service for distress calls. Shipping traffic had dwindled as well, but still Awarua listened, the station's Amalgamated Wireless tower standing sentinel over the barren landscape, waiting for the swarming signals to return.

In the autumnal stillness of Thursday, March 23, telegraphist Alfred Goodwin strode to the Awarua receiving hut for the night shift. Goodwin swung the dial of the Telefunken receiver to either side of the five-hundred-kilocycle band dedicated to distress calls. The high-powered thirty-kilowatt masts could detect an electromagnetic rustle in the atmosphere from as far away as China. But on this particular night, the only sound was atmospherics.

It was almost midnight before a trace of barely detectable chatter cut through the static. The incoherent taps resolved into snatches of Morse code: CT VLB VLB VLB. *Seeking Awarua Station*, over and over again. Then, the transmission changed: CT VIH VIH VIH. The sender was also calling VIH, casting his net wide and trying to reach Hobart, the southernmost station in Australia. It was likely a ship, plying the waters of the Tasman Sea between Australia and New Zealand. Then, a snippet of another call came on the air for the Chatham Islands station, indicating that the vessel could be somewhere in the high latitudes of the Southern Ocean.

Goodwin switched to transmit and rapped out a reply: CQ, *seeking you*, which was the routine call to an unidentified vessel. The ship failed to respond. The feeble stream continued, the ship's operator apparently deaf to the responses from Awarua, Hobart, and Chatham. Then, after two hours, the baffling signals faded and stopped. The operator may have given up, or the ship may have drifted beyond the range of its low-powered wireless set—or she might have been in distress. The only course of action was to keep monitoring the band, but the signal failed to materialize. Goodwin settled in for a long, slow night.

Then, just before two in the morning, the signal came on the air again, instantly recognizable by the handwriting, as veterans called a telegraphist's signature style. Goodwin acknowledged, and the ship finally responded: CT VLB VLB VLB de WVSQ WVSQ WVSQ *Awarua Station, this is British ship call sign WVSQ calling.* After an exchange of acknowledgments, the staccato Morse continued. To Goodwin's astonishment, it was a message for King George V:

> *Sire,*
> *Aurora driven from Winter Quarters Cape Evans Blizzard May 6th and set north frozen in pack ice. Rudder smashed ship disabled present position lat 65°00 S long 155° E. Prospects of relative safety of Southern party is doubtful. Little provisions and clothing at Ross Sea Base. I pray Your Majesty will permit ship proceed with all haste to Cape Evans McMurdo Sound with provisions and clothing.*
> > *Your Majesty's Humble & Devoted Subject*
> > > *Stenhouse*
> > > *Master Aurora*

The cryptic transmission was the first indication that something had gone terribly wrong with Shackleton's bold endeavor, but the full implications of the message were beyond Goodwin's comprehension. The party at the Ross Sea base, responsible for building a chain of supply depots along his intended route, had been stranded with scant supplies when their ship was swept away with the pack ice in May 1915. It was questionable whether they could save themselves, much less lay the depots to sustain Shackleton's transcontinental crossing. Goodwin responded immediately to offer assistance to the disabled *Aurora*, six hundred miles south of Bluff. But there was little he, or even the king, could do for the castaways at the Ross Sea base. Beyond wireless contact, they were marooned with no hope of imminent rescue, as remote from humanity as if they had been on the moon.

Goodwin could not be faulted for being unaware of Stenhouse and the *Aurora*. Neither was a household name. It was Shackleton, the national hero, who captured the headlines and the public imagination as he and the company of the *Endurance* headed south to blaze a glorious path through undiscovered lands in the name of the British Empire. The men of the Ross Sea party were the foot soldiers of a historic expedition, charged with manual labor rather than discovery.

Little changed in the intervening decades. In popular accounts of the Imperial Trans-Antarctic Expedition, then and now, they are overshadowed by Shackleton's heroic exploits or excluded entirely, though it was not his intention to consign them to obscurity. Shackleton devoted a quarter of his 1919 chronicle of the expedition, *South*, to their story, which is how I first became aware of the Ross Sea party in 1994. The astonishing efforts of Shackleton and his men to save their own lives after the loss of the *Endurance* in the Weddell Sea seemed the stuff of myth, but I found something even more compelling about the men of the *Aurora*. For in the face of calamity, they persevered contrary to the very instinct of survival. When the *Aurora* disappeared with most of their clothing, food, and equipment aboard, the stranded men chose to risk their own lives for the sake of Shackleton, a man most of them had met only briefly, or not at all. Believing that his life and those of his companions depended upon them, the Ross Sea party made do with homemade clothing, scavenged food, and salvaged gear, marching 1,300 miles to build the lifeline of depots for Shackleton's party. It was an extraordinarily selfless act, carried out with singular fortitude.

Intrigued, I set out to learn more about them. They were, on the face of it, ten ordinary men. None had made his mark on the grand scale of Shackleton by the time he joined the expedition in 1914. Two of the men had sailed with him to Antarctica before and the rest were amateurs: two teachers, a clergyman, a geologist, a medical student, a clerk, a seaman, and a college athlete. Most were unconcerned with fanfare and posterity, although two members, Ernest Joyce and Richard Richards, had published books. In the first wide-ranging biography of Shackleton, published in 1957, James and Margery Fisher noted "the need for a further book in which the attitudes and recollections of the other members of the party could be expressed." Nearly forty years later, the book had yet to be written.

The common wisdom was that little had been preserved; an account published in 1982 stated that only two complete diaries of the Ross Sea expedition members had survived. The full names of some of the men were never even recorded, and little was known about their origins. But as I delved into research, I discovered that many of the survivors had indeed left their records in the care of archives and descendants. In time, I located the diaries and logs of sixteen men, and for some of the diarists, multiple versions. One member had deposited a journal and memoirs in an archive; yet another diary, kept especially for the sweetheart he left in England in 1914, remained with her family. Other members recopied their diaries for Shackleton and kept the unabridged originals. Joyce left no fewer than three handwritten versions of his diary, a bounty for a historian—and then a collector presented me with Joyce's original diary from the Antarctic, bound in the same green tent canvas as the Ross Sea party's makeshift trousers. As the chorus of voices swelled, a rich tone poem of their experiences and thoughts began to emerge. Ringing out above the rest, Joyce contributed a bit of wisdom that stayed with me as I wrote the book. "Talking is easy and everyone is wise after the event," he declared, daring historians and armchair explorers to judge too hastily.

I found documents as well, public records and private papers. The Rosetta Stone was a collection of Shackleton's documents relating to the Ross Sea party, entrusted by the family to the Scott Polar Research Institute in 2002, which I was fortunate to examine in its entirety soon after its arrival. Carted back from Antarctica in boxes in 1917, this trove included the original orders of Shackleton and Mackintosh, the party's logs, letters, reports, and notes, all permeated with a strange musky smell and smudged with soot. The party's charts and plans were there too, creased from folding to pocket-size for the epic journey into the heart of the Antarctic continent.

Also at the Scott Polar Research Institute were the papers of the ship's chief officer, Joseph Stenhouse, recently opened to historians by his family. When Captain Æneas Mackintosh left the ship to oversee the sledging operations, Stenhouse took command of the ship. In elegant copperplate script, he described his battles to defend the *Aurora* from the pack ice of the Ross Sea, drawing the first comprehensive picture of those events and the remarkable men behind them. Stenhouse also kept meticulous records of the expedition's preparation stage, every last invoice and receipt, allowing me to reconstruct in detail the days before the *Aurora* sailed in 1914. Also among his papers was a curious untitled journal. As I began to read, I realized it was not part of Stenhouse's log, but a missing fragment of Mackintosh's diary, presumed lost since 1919, which detailed events during a decisive period of the expedition.

This diary, and others, included tantalizing references to a motion picture camera and scenes captured on film. I learned that the official photographer's belongings had been entrusted to the chief scientist, Alexander Stevens, when he sailed home from New Zealand in 1917. In sight of England, a German submarine torpedoed the steamship. Stevens lost everything but his life. I persisted with the search, and in 2001 I identified some of the missing footage, watching in disbelief at the British Film Institute as the men flickered to life on the screen.

Knowing the places was a vital part of my research as well. I tracked down the men's former homes, from a smart London townhouse near Buckingham Palace to a basement flat, and visited the disused Awarua wireless station, now surrounded by a sheep station. During the Antarctic summer of 2002, as a participant in the National Science Foundation's Antarctic Artists and Writers Program, I spent two months in the Ross Sea region and traveled to many of the locales frequented by the Ross Sea party. At Hut Point, where the men lived in the squalor of a drafty wooden shelter battered by blizzards, a seal carcass still lay next to the door, ready for butchering. Inside, the air was thick with the smell of burned blubber, instantly familiar as the odor clinging to the expedition papers in England. Crudely stitched clothing hung from lines overhead. A pan of seal meat sat on the stove, remnants of their last meal. As I examined the soot-grimed walls with a flashlight, the beam caught a trace of looping curved lines—the penciled graffiti of the first party to enter the hut. Stenhouse's upright script recorded the date, January 18, 1915.

There was graffiti in a hut thirteen miles north at Cape Evans as well, scrawled by scientists Keith Jack and Richard Richards. This hut, too, seemed suspended in time since the Ross Sea party left, with a toothbrush hanging on a wall and clothing piled on the bunks. Pinned to one wall was a handmade collage of crudely pasted magazine pictures of dogs. Outside, the partly preserved remains of a sledge dog lay curled in a melting snowdrift.

On a balmy fifteen-degree summer day, I joined a research group flying out to the Beardmore Glacier, the end of the epic depot-laying journey, 360 miles south of Hut Point. I have never felt so dwarfed by any landscape. The Ross Sea party had hoped to meet Shackleton at the Beardmore and march by his side for the last leg of his transcontinental crossing, but he never arrived. As months passed, it became clear that either Shackleton had not embarked on the crossing or he had perished in the attempt. Their sacrifices had been for naught.

Or so it might seem to a modern observer, from the vantage point of another time, on the far side of the broad divide of a tumultuous century. But the meaning of these extraordinary events for these ordinary men is best expressed in their deeds and words. This, then, is their story.

— 1 —

"That Restless Spirit"

"Men go out into the void spaces of the world for various reasons," Sir Ernest Shackleton declared. For some, it was science. For others, it was the siren song of "little voices," as he put it, calling them forth to adventure. For Shackleton, the rationale was crystal clear. "War in the old days made *men*. We have not these same stirring times to live in and must look for other outlets for our energy and for that restless spirit that fame alone can satisfy. The exploration of the Polar regions is one of these outlets where the value of the work done is second only to the endurance and courage of the explorer. It is because we are what we are that it is necessary for us to become the leader in all such work as this." So in 1913, he announced his second expedition to Antarctica, this time to unlock the secrets of the interior of the great continent by becoming the first man to cross its breadth from coast to coast.

Winston Churchill was having none of it. "What is the use of another expedition?" was his response to Shackleton's request for government support for the venture. With those words, the first lord of the Admiralty dismissed Shackleton's grand scheme. By 1913, the North and South Poles had both been claimed, glory attained, discoveries achieved—but no economic, strategic, or political value had been revealed in the icy wastes that he could discern. Further expense in pounds sterling and human life was, to his mind, pointless, if merely to chart a frozen hinterland. In his words, "Enough life and money has been spent on this sterile quest."

Others saw it differently. "Bold & Napoleonic" pronounced Lord Curzon of Kedleston, former viceroy of India and president of the Royal Geographical Society, rendering his view no less imperiously than Churchill. "One of the most courageous and splendid ventures of modern times." It was, in his estimation, "one of the few great achieve-ments in exploration that are still open to the human race—a greater achievement, indeed, than any yet recorded in the history of the Antarctic or the Arctic."

When Shackleton conceived his glorious endeavor, the era of Antarctic land exploration spanned just twenty years. For most of hu-man history, the seventh continent's very existence was a pipe dream of mapmakers. To the ancient Greeks, the undiscovered *Antarktikos* was a figment of logic and symmetry: Surely, reasoned Aristotle, a southern continent must exist to balance the lands of the opposite pole. The de-finitive proof waited two thousand years. The great explorers of the Age of Discovery, Magellan and Drake, circled the globe but never strayed far enough south to sight the so-called *Terra Australis Incognita*, the un-known southern land depicted by fanciful cartographers.

During the eighteenth century, stories abounded of ice islands glimpsed through the mists by southerly navigators. In 1768, Captain James Cook was secretly charged to find the rumored lands when he sailed from England, ostensibly on a scientific expedition. He, too, was confounded by the quest, even as he sailed within eighty miles of the coast during a series of epic voyages, the first ever south of the Antarctic Circle.* Finally, in 1820, Antarctica was sighted for the first time. Months later, American seal hunter Captain John Davis set foot on the frozen terra firma of the Antarctic Peninsula. The first known landing on the mainland of the continent was not until seventy-five years later. In the same year, the Sixth International Geographical Congress convened in London and resolved that "the exploration of the Antarctic regions is the greatest piece of geographical exploration still to be undertaken."

In 1895, Ernest Shackleton was a twenty-year-old ship's officer, con-sumed by his career in the merchant marine. His attention shifted southward as polar aspirations ignited in Europe, and French, Belgian, and Swedish expeditions scrambled to launch assaults into the "star-ing white blank" of Antarctica. The British were determined to lead the charge. In 1901, Britain's national academy of science, the Royal Society, and the Royal Geographical Society (RGS) appointed a naval lieutenant,

---

*Distance is expressed in British nautical miles unless otherwise indicated. (See ap-pendix, "Units of Measurement.")

Robert Falcon Scott, to lead the nation's first foray into the remotest southern latitudes since the 1840s. The RGS was a preeminent force in British exploration, contributing funds to generations of explorers who etched a tracery of mountains, valleys, rivers, and lakes onto the voids of Victorian maps. Its fellows and honorees included the likes of Alfred Russel Wallace, Charles Darwin, David Livingstone, Henry Morton Stanley, and Richard Francis Burton. But it was the prospect of vaulting beyond a career impasse, not adventure, that inspired Scott to pursue command of the *Discovery*. "I may as well confess at once that I had no predilection for Polar exploration," he admitted, though he was determined to fulfill his mandate as the commander of the British National Antarctic Expedition and oversee the most comprehensive scientific program yet attempted in the Antarctic.

That was the ostensible priority. Privately, RGS president Sir Clements Markham disdained science and shared Scott's determination to blaze a trail toward the South Pole in the name of British prestige. Resplendent in muttonchops and starched wing collar, poring over geographical tomes by candlelight, Markham harkened back to the heyday of Victorian exploration. He cherished the notion of recapturing the bygone days of the nineteenth-century British Arctic expeditions, extravagant affairs driven by the conviction that nature was no match for the might of the Royal Navy. In 1850, Markham had sailed as a midshipman to the Arctic and reveled in the camaraderie of amateur theatricals and lecture evenings. Markham's memory, however, was gilded with nostalgia. More often than not, these grandiose expeditions degenerated into unspeakable misery, as the officers shivered in unsuitable clothing and starved on inadequate rations served with engraved flatware—reduced, in one pathetic instance, to eating leather boots and, more than once, to cannibalism. Nonetheless, Markham held the reins and succeeded in running the Antarctic expedition on his lines.

After some lobbying, Shackleton's name was placed before Markham, and the charismatic young merchant officer was accepted. Like Scott, his chief interest was career advancement. As a friend put it, Shackleton regarded the expedition "as an opportunity and nothing more. He would have tried to join just as eagerly a ship bound to seek buried treasure on the Spanish Main." Once in Antarctica, however, he fell under its spell and was overjoyed when Scott selected him to join the attempt to unfurl the Union Jack farther south than any man had gone before.

And so on November 2, 1902, Scott set out from his base on the shores of McMurdo Sound accompanied by nineteen dogs and two companions: Dr. Edward Wilson, a courtly zoologist, and the exuberant Shackleton. Unmatched in background and outlook, the three men had a common, notable trait: their utter lack of experience with the terrain that lay ahead, a hallmark of Markham's philosophy that good breeding counted more than a man's practical experience. With heraldic banners fluttering on the sledges, they embarked on their southern journey.

Scott and his contemporaries ushered in what historians later christened the Heroic Age of Antarctic exploration, although a knowing veteran of the era acerbically called it "the Stone Age of polar travel." Snow and ice travel did not come naturally to these natives of a temperate land. Relying on some desultory practice, Scott, Shackleton, and Wilson struggled awkwardly on skis, finally ditching the vexing equipment to slog through the drifts on foot. The dogs scrapped and mutinied against their novice handlers, forcing the ill-clothed party to step into the harness and stagger along at a crawling pace, less than ten miles each arduous day. Realizing their provisions were severely stretched, Scott reluctantly elected to turn for home at latitude 82°17' south, nearly five hundred miles short of the South Pole.

The homeward journey became a desperate race for life. There was scant margin for error, no allowance for delays due to foul weather, no time to search for buried supply depots, no surplus food for their overworked bodies. One by one, the dogs perished and the trio man-hauled through the agony of snowblindness and malnutrition. Their gums turned swollen and spongy, a telltale sign of scurvy, the dreaded bane of explorers. Unaccountably, Shackleton suffered the most, his once-robust frame wracked by coughing fits that spattered bright blood on the snow. He overheard Wilson tell Scott that he was likely to die. In the end, all three survived their grim ordeal, although their shipmates barely recognized them, their bodies wasted and faces disfigured by frostbite.

When a relief ship arrived, Shackleton was one of the few to be sent home, quietly discharged for reasons of health, even though, as the ship's doctor attested, Shackleton was rapidly on the mend. Prodded from the blind of discretion, Scott cryptically stated, "If he does not go

back sick, he will go back in disgrace." There had been hints of acri-
mony on the southern journey: querulous bullying, exchanged words,
a defiant challenge to authority, recriminations. The three men said little, but speculation was rife. To some observers, the note of discord
bore the hallmarks of an insecure superior discomfited by a promising
subordinate. Scott and Shackleton were united in ambition but divided
by opposing temperaments. Scott, the reserved disciplinarian, commanded respect but not the easy rapport with men of all stripes that
came so naturally to Shackleton, an informal man of genuine warmth.

Branded as an invalid, Shackleton was no doubt mortified by his pre-
mature dispatch home. Yet he arrived a new man, no longer the hesitant young merchant officer who once quoted Robert Browning in a
tumult of doubt to his fiancée, Emily Dorman: "That man should strive
to the uttermost for his life's set prize. I feel the truth of it: but how can
I do enough if my uttermost falls short in my opinion of what should
take the prize?"* Back in Britain, Shackleton channeled his boundless
energy into assorted business ventures and dalliances with politics
and journalism, all the while with an eye to someday leading his own
Antarctic expedition, one that would make his name.

It would be a private enterprise, that much he knew. He was not a
favorite son of the RGS or the Admiralty. Without official patronage,
Shackleton's time would be monopolized by the business of fundraising. Yet there was a silver lining. A private venture would free him
from the kind of institutional control that had allowed Markham's
prejudices and outmoded ideas to stifle innovation and talent on the
Discovery expedition.

The campaign was not easy. Scott was plotting another voyage to the
Antarctic and claiming establishment support, while Shackleton struggled to sway financiers to back his own cause. At long last, in 1907, with
a clutch of hard-won donations and loans just sufficient to leave the
pier, he boarded the *Nimrod* and sailed south, leading the British
Antarctic Expedition.

Numbered among the faithful were two seamen Shackleton had befriended on the *Discovery,* Frank Wild and Ernest Joyce, who gladly
abandoned their careers as Royal Navy petty officers to follow him, and

* Spelling, punctuation, and grammar are reproduced verbatim in quotations.

a first-rate scientific staff led by the esteemed geologist T. W. Edgeworth David. But the ultimate object of Shackleton's desire was the South Pole. In deference to Scott's wishes, he would attempt to blaze a new trail across the Ross Ice Shelf several hundred miles east of the established route. Once in the Ross Sea, however, the *Nimrod* cruised the coast for a new landing place in vain, ultimately dropping anchor in the familiar waters of McMurdo Sound, where the crew built a hut at Cape Royds. Just nineteen miles north of Scott's old quarters, Shackleton readied for his fresh attempt on the Pole with Wild, Dr. Eric Marshall, and Jameson Adams.

On October 29, 1908, Shackleton and his comrades struck out with a team of Manchurian ponies and a customized motorcar, the first ever in Antarctica. Training had been haphazard before departure, since financing travails left just seven months for planning and preparation before the *Nimrod* sailed. While several men had undergone an intensive course in typesetting in order to print a book, none had been instructed in skiing and ice travel.

The car soon bogged down in the snow, and the ponies proved wretchedly ill suited to life on the ice shelf. The journey degenerated into a reprise of Shackleton's miserable march with Scott. "Surely, among all the nationalities of men who have made journeys along the edges or into the interior of the last continent, few have done so more uncomfortably or with greater hardship than the British," the *Nimrod*'s first officer later wrote. Nevertheless, by January 9, 1909, the party had bested Scott's southern record by 366 miles. The forlorn ponies were long since dead, the men pulling the sledges like beasts of burden, supplementing their dwindling rations with the fodder and flesh of the slaughtered ponies. Shackleton faced a decision: to claim the grail and perish, or turn back with a fighting chance for survival. He chose retreat. According to the navigational calculations, they were just ninety-seven miles from the South Pole, the farthest south ever attained. "The last day out we have shot our bolt and the tale is 88.23 S. 162 E.," he recorded impassively. "Whatever regrets may be we have done our best."

Because the party was so ominously late in returning to the *Nimrod*, the captain counted Shackleton's party as dead and prepared to sail. Then, just hours before departure, figures appeared onshore, waving a banner. All four men had survived.

As well as claiming the southern record, the achievements of the expedition were formidable. Members of the scientific staff reached the South Magnetic Pole for the first time ever, and, with three others, made the first ascent of volcanic Mount Erebus. Their studies in geology, biology, and terrestrial magnetism would fill four published volumes. Shackleton opened the way to the interior for future expeditions by revealing its geography and weather conditions. But perhaps his most celebrated achievement was the simple fact that he was alive. As he said to his wife Emily, "A live donkey is better than a dead lion, isn't it?" His difficult choice of life over glory—his own and his companions'—won the admiration of renowned Norwegian explorers Fridtjof Nansen and Roald Amundsen, men who had revolutionized polar travel. Amundsen called it "the most brilliant incident in the history of Antarctic exploration." In his words, Shackleton had shown "what the will and energy of a single man can perform."

By sheer force of will, it seemed, Shackleton had achieved a signal vindication. He emerged as a national hero, embraced by the populace and Fleet Street with patriotic fervor. At the Crystal Palace, his portrait was blazoned in fireworks for a gala in his honor. A jubilant throng unhitched the horses from Shackleton's family carriage and bore it triumphantly through the streets, to the terror of his young nieces inside. Readers thrilled to his account of the expedition, *The Heart of the Antarctic,* and it became an instant sensation. He was living proof, as the broadsheets and tabloids alike had it, that the doomsayers proclaiming the decadence of the British people and the decline of the Empire were half-cocked. Royal acclaim followed, as King Edward VII knighted this son of a middle-class Anglo-Irish doctor at Buckingham Palace. Brilliantined and assured, Shackleton moved with ease from London salons to country shooting parties. His breezy repartee charmed baronets and old-age pensioners alike, contacts he adeptly cultivated with an eye to someday courting the main chance.

Fame came easily, yet fortune and satisfaction eluded him. Shackleton labored to pay off the staggering debts incurred by the *Nimrod* expedition with lecture tours and business schemes, biding his time impatiently until he could launch a new expedition. There were high-flown speculations—Hungarian gold mines, cigarette factories, Mexican timber, and real estate—seemingly everything ventured and nothing gained. He fretted on the sidelines as Japanese, German, and British expeditions prepared to go south. Former *Nimrod* scientist Douglas Mawson mounted an Australasian expedition to seek out his "Land of Hope and Glory" in unexplored regions of the Antarctic, and rival Scott put to sea once again.

Since Shackleton's return, Scott had fulminated with moral indignation as his former subordinate gathered laurels on ground he claimed as his own preserve. In 1910, Scott set forth in the *Terra Nova*, with a righteous determination that the South Pole, at last, would be his. On the eve of his departure from Australia, he was stunned by an enigmatic telegram: "Beg leave to inform you *Fram* proceeding Antarctic Amundsen." The gauntlet was thrown down by Roald Amundsen, who had departed on an Arctic expedition with secret plans to change course for the Antarctic and capture the Pole. The Norwegian challenge, so unexpected, was scarcely believable—until months later, when the men of the *Terra Nova* beheld the unwelcome specter of Amundsen's ship *Fram* at anchor in the ice of the Ross Sea. The two camps staked out their positions: Scott following the trodden route across the Ross Ice Shelf, Amundsen heading out across virgin territory to the east.

The competitors were woefully mismatched. For Amundsen and his well-drilled men, the journey was a sportive *langlauf,* a Nordic cross-country ski race transposed to a novel, albeit harsher, locale. A seasoned Arctic traveler, his methods were rooted in Scandinavian and Inuit traditions, with his clothing, shelter, transport, and rations all proven in the polar climate and topography. The British seemed at odds with the environment, best characterized by junior member Apsley Cherry-Garrard's mournful observation, "Polar exploration is at once the cleanest and most isolated way of having a bad time which has been devised." With the wisdom of hindsight, the outcome now seems preordained. Amundsen and his team reached the South Pole on the fourteenth day of December 1911 and returned safely. Scott and his four colleagues did not arrive at the Pole until January 17, where they were greeted by the Norwegian flag whipping stiffly in the chill air. "Great God!" Scott anguished in his diary, "this is an awful place and terrible enough for us to have laboured to it without the reward of priority."

All that remained was the indignity of retreat, some eight hundred miles to the coast. With supplies calculated to the bone and scant allowance for poor weather, the return journey deteriorated into a slow death march, the men's bodies racked with hypothermia, frostbite, and malnutrition. On March 21, little more than a hundred miles from safe haven at the base, a storm halted their progress. Tentbound and starving for days, the men exhausted their supplies along with the last vestiges of hope. "Had we lived, I should have had a tale to tell of the hardihood, endurance, and courage of my companions which would have stirred the heart of every Englishman," Scott scrawled, laboring to compose his last words before perishing with his companions. His story of suffering and failed promise, eloquently told, resonated deeply with the British public as a heroic ideal, and in death Scott and his companions assumed the mantle of martyrdom.

Shackleton was doubly stunned by the loss of his chief rival as well as his cherished aim of being first at the Pole. He hailed the news of Amundsen's triumph with public congratulations, deftly concealing agitation over his own unfulfilled polar ambitions. He was rounding on forty and yoked restively to the obligations of Edwardian bourgeois life, with two children in school and a third in the cradle. With the obvious geographic prizes already claimed, there was little else to inspire him—except for one breathtaking idea. "I feel that another Expedition unless it crosses the continent is not much," he confided to his wife Emily. It was a supremely daunting notion, one that had lingered in his mind for some years. In the ashes of his unrealized prospects, he conceived an audacious plan.

Shackleton dubbed the venture, with characteristic grandeur, the Imperial Trans-Antarctic Expedition. His strategy for the journey borrowed liberally from a scheme worked out by Scottish explorer William Speirs Bruce, who gladly supported Shackleton's bid when he was unable to obtain his own funding. The proposed route bisected the continent at its narrowest span on a doglegged course through the Pole, a distance he accurately estimated to be about 1,500 miles. At first, Shackleton envisioned equipping a single ship, which would sail from South America into the Weddell Sea to disembark a party of six as far south as possible to begin the crossing. The ship would then sail around to the opposite coast, to the Ross Sea, to await the party's arrival and then sail north for home. On reflection, Shackleton decided that two ships, one sailing to the Weddell Sea and another to the Ross Sea, might be necessary. The Ross Sea party would lay supply depots along the last quarter of his route. The deciding factor would be money. Outfitting both vessels for two years in the Antarctic, crewed with up to forty-two men, would cost an estimated £50,000.*

Confident that cash would be forthcoming, Shackleton opened negotiations on a vessel called the *Aurora*. Cousin to the legendary polar warhorses *Nimrod* and *Terra Nova*, she was built in the late nineteenth century for the Arctic whaling and sealing trade. In her heyday, the vessel returned from a single season with a prized cargo of 28,000 seal pelts and 400 tons of whale and seal oil. In 1911, Douglas Mawson bargained hard and paid £6,000 for the three-masted barquentine for his expedition. Pending Mawson's return from the Antarctic, Shackleton tendered an offer for the vessel. In the end, he paid £3,200 for whatever life, gear, and stores were left in her after thirty-eight brutal years in the ice.

The moment had arrived. On Monday, December 29, 1913, Shackleton unveiled his bold new endeavor in the *Times* of London—though not, as legend has it, in the form of an advertisement reading "Men wanted for hazardous journey. Small wages, bitter cold, long months of complete darkness, constant danger, safe return doubtful. Honour and recognition in case of success." No trace of that apocryphal ad has ever been found, though the tale suits the epic undertaking. Instead, the archives yield a terse letter to the editor, penned by Shackleton:

> *Sir,—It has been an open secret for some time past that I have been desirous of leading another expedition to the South Polar regions. I am glad now to be able to state that, through the generosity of a friend, I can announce that an expedition will start next year with the object of crossing the South Polar continent from sea to sea. I have taken the liberty of calling the expedition "The Imperial Trans-Antarctic Expedition," because I feel that not only the people of these islands, but our kinsmen in all the lands under the Union Jack will be willing to assist towards the carrying out of the full programme of exploration to which my comrades and myself are pledged.*

* See Notes and Appendix for currency conversions.

– 2 –

The Imperial Trans-Antarctic
Expedition

DECEMBER 31, 1913–SEPTEMBER 8, 1914

In Shackleton's announcement, the press found a rallying cry for king and country. "While there is yet a blank space upon the map of the world," London's *Evening News* opined, "you will find an Englishman ready to go forth to gain the knowledge which shall be the means of filling it." Shackleton played to the imperialist mood. "We'll have some British flags and we'll put them up. There will be islands and mountains and lands to name with British names," he enthused, adding cautiously, "It is possible that we may make a considerable contribution to Antarctic exploration." On New Year's Eve, Sir Ernest and his wife went to the Coliseum Theatre in St. Martin's Lane. As they made their way through the orchestra seats to the stalls, a murmur rippled through the house. Suddenly, the audience erupted in applause and a chorus of "For He's a Jolly Good Fellow." Shackleton stepped forward and acknowledged a firestorm of cheers. A popular hero, he would go forth in the spotlight under the fascinated gaze of the public. "This Expedition calls for more preparation and also for more responsibility than when I went last without the eyes of the British nation upon me," he wrote to an old shipmate. "I am in a great rush as you can imagine."

Shackleton proposed to leave the London docks in just eight months. Amundsen had obsessed over the organization of his most recent expedition for nearly two years. Shackleton was unperturbed: Intuition, adaptability, and improvisation had served him well in the past. Regular bulletins crowded the newspapers, charting every aspect of the expedi-

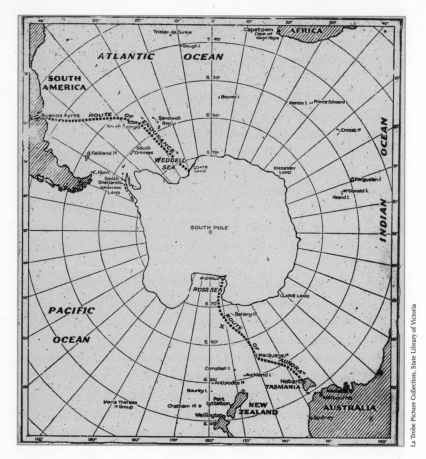

THE IMPERIAL TRANS-ANTARCTIC EXPEDITION

tion's progress. With a showman's flair, Shackleton captivated the press with "little romantic touches of Jules Verne flavour." Items about propeller-driven motorized sledges, scientifically formulated rations, and newfangled "igloo" tents fueled public fascination.

Mail inundated the expedition's London offices at 4 New Burlington Street. Envelopes filled with notes and silver, unsolicited donations from the public, arrived daily. Applications from scientists and adventurers streamed in for positions on Shackleton's two ships. Four hundred men had applied for berths on the *Nimrod*; now thousands scrambled for a place in the new venture, the applications finding their way into three file drawers labeled "Mad, Hopeless, and Possible." One earnest applicant offered the services of "three sporty girls," keen "to undergo any hardships." Shackleton politely declined.

Choosing the right men mattered above all else to Shackleton. A few trusted compatriots from his *Nimrod* days already formed the inner circle. He was unequivocal about his choice to captain the Weddell Sea ship, a man he called "the best navigator and sailor that ever went into the Antarctic." John King Davis had embarked as *Nimrod*'s chief officer and finished the expedition as master. By the age of thirty, he had made five Antarctic voyages. But when Shackleton announced his appointment, Davis was battling an Antarctic gale in the Southern Ocean at the end of a toilsome two-year expedition with Douglas Mawson. The formalities would have to wait.

Nimrodonian Frank Wild was appointed second-in-command. A tough, quiet Yorkshireman, Wild was not easily impressed, and he regarded Shackleton with near reverence. On their South Pole attempt in 1908–9, Wild was overawed by Shackleton's selflessness. "Poor S. works away like ten devils," he observed as Shackleton strained to compensate for his lagging companions. As the rations dwindled, Shackleton urged his paltry share, a single biscuit, on the ailing Wild. "I do not suppose that anyone else in the world can thoroughly realise how much generosity & sympathy was shown by this; I do, & by God I shall never forget it."

Wild's boon companion from the *Nimrod,* Ernest Joyce, was also one of the chosen. Joyce had since emigrated to Australia, where Shackleton sent a letter offering him a berth with the Ross Sea contingent. If funding failed to materialize for the Ross Sea ship, Shackleton offered Joyce a position in the group manning the Weddell Sea base through the winter after the transcontinental party's departure.

Shackleton had known Joyce as a brash able seaman on the *Discovery* voyage. During the *Nimrod* expedition, he had been Wild's bunkmate at Cape Royds, sharing digs they called "Rogues' Retreat" and working on the first book printed in the Antarctic, *Aurora Australis*. Before leaving Cape Royds to set out on his polar trek, Shackleton had charged Joyce with laying a supply depot on the homeward trail. Joyce taught himself dog driving and drilled the scrapping animals into a crack team, then led two journeys south to cache the supplies, staking bamboo poles with flags en route to guide Shackleton's party. By the time Shackleton and his companions spotted the first flag, their supply of raw horsemeat and dry biscuits had long since been exhausted. "After months of want and hunger," Shackleton wrote, "we suddenly found ourselves able to have meals fit for the gods." Joyce's depot was stocked with cases of mutton, plum puddings, gingerbread, glacé fruit, and a Fortnum & Mason birthday cake. He saved the party from starvation in flamboyant style, and ever after reveled in telling the tale.

Joyce's swagger may have rankled some, but it had been a stunning achievement. Until then, Shackleton saw sledge dogs as troublesome and inefficient. He believed Manchurian ponies could haul heavy loads twenty to thirty miles per day in frigid temperatures. Shackleton was not alone in his attitude. Most British explorers of the era held a "deep-rooted conviction that the use of dogs as transport animals was cruel," as one British explorer observed. "Gentlemen used horses." Traditionally, the only acceptable alternative was raw manpower. When he embarked on his southern journey, Shackleton left the dogs behind. Joyce started training the animals, and gradually worked the team up to forty-five miles per day, nearly triple the ponies' best pace. In the wake of Joyce's and, indeed, Amundsen's successes with dogs, Shackleton was persuaded to count on them for his new venture.

Also destined for the Ross Sea party was Æneas Lionel Acton Mackintosh, a lieutenant in the Royal Naval Reserve. With no previous polar experience, he had sailed as second officer aboard the *Nimrod* in July 1907. The former merchant officer rapidly earned Shackleton's confidence and impressed his fellow officers as "gay, dashing, and with a will of iron" as they sailed south. On January 31, 1908, the ship moored in McMurdo Sound. As the crew unloaded stores, a cargo hook swung wildly and struck Mackintosh's right eye. Ship's surgeon Eric Marshall removed the damaged eye, and Mackintosh lost his chance of a place in Shackleton's party for the South Pole bid. Marshall was deeply impressed by Mackintosh's fortitude, observing, "Poor fellow, it hit him very hard, but no man could have taken it better. He seems to feel having to go back to New Zealand very much more than the loss of his eye."

Mackintosh shunned discharge and joined the support work. His unswerving devotion to duty won him a place in the Ross Sea party in charge of navigational and meteorological instruments.

For command of the Ross Sea party, Shackleton approached Cambridge-educated Marshall, a stalwart of the farthest south journey. He had handled the brutish aspects of expedition life—from extracting a tooth on the march to butchering ponies for food—with dispassionate cool. His insistence on an experimental regimen to prevent the mysterious and incapacitating malady of scurvy, including bottled fruits and vegetables and fresh seal meat, protected the entire crew from its ravages. On Marshall's return from the Antarctic, he joined a 1909 British expedition into the unexplored interior of Dutch New Guinea, where the pistol-packing physician handily fended off reticulated pythons, beriberi, starvation, malaria, and a murderous tribesman to emerge as the sole survivor. In a crisis, as one acquaintance later put it, "He is a tough nut if ever there was one."

But the question of leadership of the Ross Sea party was entirely hypothetical until Shackleton raised the necessary funds. Like the *Nimrod* expedition, the ITAE would be forced to rely on private financing for the most part. The name "Imperial Trans-Antarctic Expedition" was not without a touch of presumption, since the expedition was neither sponsored nor endorsed by His Majesty's Government, save for £10,000 in seed money awarded grudgingly after a year of deliberation. If Shackleton could not find the balance from outside sources, the grant would be withdrawn. And it was sorely needed; his budget had escalated from £60,000 to £100,000 for his full program.

From the outset, he refused to solicit donations from the public, saying it was "undignified to take money from poor people." Contributions from wealthy patrons were a surer means of reaching his daunting target than farthings from social clubs and schoolchildren. Shackleton also sought discounts and gifts from suppliers in exchange for product endorsements. In his boldest entrepreneurial stroke, he pitched a newspaper exclusive, a future book, and a motion picture. "The mercenary side of a Polar 'stunt' is absorbing," the *London Mail* tattled on January 17. "Any day you may see Sir Ernest—always alone—taxi-ing from one newspaper office to another. He is trying to arrange the best terms and it is going to be a battle royal both for the news and pictorial rights." The *Daily Chronicle* won the day, and the paper's representative, Ernest Perris, joined the company in the expedition offices.

However great his need, Shackleton was loath to turn to the "hidebound and narrow" Royal Geographical Society for support. Notwithstanding Lord Curzon's praise of his "Napoleonic" enterprise, Shackleton was well aware of the politics behind the scenes. Many of the society's fellows were Scott partisans, notably Sir Clements Markham, and hostile to Shackleton. Markham's Machiavellian favoritism did him little credit, but his assertion that Shackleton placed "geography and every other science a secondary object" struck a chord with members of the RGS Council. Even Shackleton's longtime friend and supporter, former RGS librarian Hugh Robert Mill, was concerned that the risks of the expedition far outweighed the potential for new geographical work. Still, Shackleton knew the imprimatur of this bastion of the exploration establishment was essential to leverage more financing. With the society's goodwill at stake, Shackleton appeared before a confidential RGS committee chaired by Curzon on March 4, braced for questioning.

He outlined his plan. The main ship would sail for the Weddell Sea in November 1914 to disembark Shackleton's party and return to South America before winter. After landing at Vahsel Bay in the Filchner Ice Shelf, Shackleton and five men would set out with one hundred dogs, bound for the Ross Sea coast, about 1,500 miles across the continent. Motor vehicles would accompany them early in the journey to carry additional supplies. Even with that support, Shackleton's team could not carry enough rations and fuel for the entire journey, so he also planned to employ a support party in the Ross Sea region.

In December, a second ship would sail to the opposite coast of Antarctica, and disembark men and supplies. The Ross Sea party would then establish a lifeline of supply depots, 360 miles long, over the final quarter of Shackleton's route, ending at the foot of the Beardmore Glacier. Shackleton anticipated that his journey would take about four months, enabling his party to reach the safety of the McMurdo Sound base before late April, when the continent would be plunged into darkness and bitterly cold winter temperatures.

The committee questioned the logistics in detail. A key objection was that no explorer had yet made a secure landing on the Filchner Ice Shelf at Vahsel Bay. Thus, the terrain in the first half of the journey was unknown. Shackleton conceded the point, but claimed that previous difficulties in the Weddell Sea occurred because explorers went south too late in the season, sailing directly into the maelstrom of ice at its worst. Of equal moment was whether Shackleton's party would be able to carry sufficient food to last through the proposed journey, one of the longest ever attempted in the history of polar exploration. Shackleton estimated that traveling fifteen miles daily, on average, he could cross the continent in 115 to 139 days. On his South Pole attempt, he had averaged 11.8 miles per day, but Amundsen had achieved sixteen miles per day with dogs. Shackleton planned to stretch his supplies by slaughtering seventy-five of the dogs to feed the team en route. Still, the depots would be critical and Lord Curzon insisted it was imperative that the Ross Sea party be "well commanded, properly equipped, and sufficiently organized, much in the same way as if it were an independent expedition." Shackleton confidently rejoined, "Except for the fact that it is a known route." The Ross Sea party's route would follow in the footsteps of the previous Shackleton and Scott expeditions.

That the expedition would, in the main, retrace well-trodden footsteps was a major objection in itself. The journey across the unmapped lands in the Weddell Sea quadrant would be too hurried for significant scientific work. Critics asked whether the crossing was merely a competitive stunt, if it did not entail extensive new geographic and scientific discovery. According to the prospectus, the expedition would resolve questions that had long puzzled scientists in the fields of terrestrial magnetism, glaciology, meteorology, and geology. The committee was not entirely convinced. "Geology, I suppose, means in the main picking up stones and putting them on a sledge?" commented a skeptical Curzon. Shackleton's sketchy answers and his flippant remark that to trim the budget, he "might have to cut off a geologist here and a geologist there," did not help his case. In the end, he was forthright about his aim: "The journey across is the thing that I want to do." The meeting ended in a polite standoff, with Shackleton closing testily, "I am perfectly aware that a lot of people do not take me very seriously. We do the work, anyhow." Despite their reservations, the committee knew well enough to take Shackleton's determination and resourcefulness seriously. They had seen the power of his vision, judgment, and formidable will, and it boded well for a future success. Shackleton left with a tacit endorsement of the expedition and a promise of £1,000 that sealed the bargain.

Not long after, Shackleton secured his ship. For £14,000 he bought the 144-foot wooden barquentine *Polaris*, newly built and fully outfitted in Norway for tourist cruises in the Arctic seas. Like the *Aurora*, the *Polaris* was a steam yacht, equipped with coal-powered engines and square sails on her foremast. Sail was essential for Antarctic exploring vessels, since there were no ports south of the Antarctic Circle for refueling. He renamed her the *Endurance*, a nod to his family motto, *Fortitudine vincimus* (By endurance we conquer). The warhorse *Aurora*, which awaited refitting in Hobart, Tasmania, would transport the Ross Sea party.

With cash in hand, Shackleton could at last proceed with the development of innovations he hoped would revolutionize polar travel. Though Shackleton's previous experiment with a motorcar had failed, he was determined to perfect a vehicle that would perform far more work with less fuel than ponies or dogs. A dome-shaped tent design promised easier pitching and less resistance in the windy Antarctic. His keenest interest was diet. Etched indelibly in Shackleton's memory was the paradoxical torment of dragging sledges overladen with food while slowly starving. For the answer to the perennial dilemma of polar explorers—how to fill the belly yet carry less weight—he turned to Colonel Wilfred Beveridge of the Royal Army Medical College. Beveridge devised a special high-carbohydrate, high-fat diet for easy digestibility and energy production. Palatability was decidedly secondary. Beveridge recommended sausages made of lard and powdered beef, sugar, oats, and patent milk products called Glidine and Trumilk as the staple foods, supplemented by lime juice to prevent scurvy.

Shackleton was obsessed with warding off the crippling disease that had once ravaged him and had plagued mariners since the epic voyages of da Gama and Magellan. The cause was a puzzle, but the symptoms evolved in a familiar, dreaded pattern: ulcerated gums, loose teeth, sore joints, and blackened limbs, all too often ending in death. The culprit was vitamin C deficiency, but by 1914, scientists had not yet isolated the mysterious elements they theorized to be essential to good health. Captain Cook noted that citrus juice and fresh greens somehow healed the affliction, and Royal Navy ships henceforth carried lime juice as a preventative. Yet the cure was not always effective—a switch to cheaper West Indian limes saved money but not lives, since the variety contained much less of the vitamin, leading later generations to seek new remedies. Theories abounded as to scurvy's cause; some thought it was a kind of ptomaine poisoning or high acidity of the blood. In 1907, Norwegian scientists Axel Holst and Theodore Frölich theorized that it was a nutritional disease that could be cured by diet. Shackleton's experience seemed proof. His *Nimrod* attempt on the Pole was scurvy-free, he suspected, because his party had consumed pony meat on the trail. Amundsen's party, too, had returned from the Pole in excellent health with the aid of seal meat depots en route.

In the days that followed, however, Shackleton was distracted from his research by a series of reversals, the gravest being the loss of John King Davis. Davis was wary of the staffing. In his view, an expedition demanded utmost professionalism, and he worried that Shackleton was recruiting adventure seekers. "I have learnt something of men and the difference between talking before the event and acting when you are up against it in the Antarctic," he wrote to Shackleton, refusing the offer of command of the *Endurance*. He was also skeptical of the vessel. Deluxe as the fittings of the *Endurance* may have been, in his view, "modern engines, fine cabins, and good laboratories for the scientists could not compensate for lighter scantlings if she met with ice conditions similar to or worse than those recently encountered by the *Aurora*" in the Ross Sea. Yet another setback came when Marshall refused to consider Shackleton's invitation to command the Ross Sea party unless he could approve the personnel. Privately, he also questioned Shackleton's fitness to withstand the physical exertion of the proposed crossing. Marshall had detected a heart murmur when he examined Shackleton in 1908, a diagnosis that Shackleton ignored. Also rejected was Shackleton's request to the Admiralty for twenty-three Royal Navy personnel to man the *Aurora* at government expense. "These polar expeditions are becoming an industry," jotted Churchill on an internal memorandum.

Æneas Mackintosh, Shackleton finally decided, would command the Ross Sea party and captain the *Aurora*, with Ernest Joyce reporting to him in charge of dogs, stores, and sledging equipment, though not second-in-command. The appointments were apparently not without reservations. Staunchly loyal, Mackintosh was an impeccable executive officer who commanded the respect of the men serving under him, but he had little polar travel experience. Joyce, on the other hand, had acquitted himself brilliantly in the field, but could be a loose cannon, a petty officer whose feet were firmly planted in the forecastle with the boys. He was not of the same breed as Shackleton, Mackintosh, and John King Davis, the cultured merchant naval officers. Neither was he akin to Frank Wild, the educated Royal Navy petty officer who recited Shakespeare on the trail. To many, Joyce was a "jolly good sort," but not suited for command. Shackleton's hope was that Joyce would complement Mackintosh. In any case, Shackleton held the reins firmly in hand for both parties. Even during the preparation stage, Shackleton chose not to delegate management of the Ross Sea side of the expedition to either man.

Other than this unlikely pair, the *Aurora* remained unstaffed, and time was running short. Historically, expeditions departed England for the Antarctic in June or July to allow for the final loading of supplies and coaling at southern hemisphere ports en route. As the summer wore on, the window of opportunity narrowed for favorable pack ice conditions in the Antarctic. The auspicious moment for fundraising also seemed on the wane. With no major benefactors on the horizon, Shackleton wrote to hundreds of individuals, hoping to drum up £50 from each of two hundred prospects. Neville Chamberlain, then a local politician in the Midlands, responded with a check for £5 and the crisp remark that he had more pressing demands on his wallet "in present political conditions."

Currents of unease rippled through Europe, as fractious relations between Serbia and Austria escalated tensions in the Balkans. At home, expanded social welfare programs demanded higher income tax and death duties. The mood was perceptibly shifting inward, the purse strings tightening. Then a Scottish manufacturing magnate, Sir James Caird, stepped into the breach, contributing £24,000. This windfall, in addition to substantial contributions from two other philanthropists and the fruits of Shackleton's dogged letter-writing campaign, rendered the expedition solvent at last. News of the assassination of Austrian Archduke Franz Ferdinand and his wife in Sarajevo by a radical Bosnian nationalist seemed too distant to intrude on Shackleton's jubilation. The *Endurance* would set out in August.

On July 14, a steamship arrived from Canada bearing a rambunctious cargo of ninety-nine sledge dogs. Playing the moment for fresh publicity, Shackleton arranged to have the animals paraded to quarantine the next day in a fleet of wagons touting "Spratt's Dog Cakes," one of several suppliers providing goods in exchange for endorsements. Tactfully omitted in the major news accounts, which featured portraits of the grinning canines, was their probable fate. The public had accepted the use of draught animals for food as a last resort on other expeditions, but Shackleton guessed that British sensibilities would be offended by calculated slaughter. The pageantry continued the following day, when Queen Alexandra and her sister, the dowager empress of Russia, arrived for an official visit, squired smartly about the ship by Frank Wild. Two weeks before sailing, the *Endurance* had a nearly complete crew, but staffing of the *Aurora* had reached the crisis stage.

Shackleton had planned the expedition as a single entity, intending to order equipment, provisions, and dogs for both parties at once. With finances uncertain, however, he postponed as many orders as possible for the Ross Sea party and focused on staffing and provisioning the *Endurance*. In July, his request to the Royal Australian Navy to borrow personnel was officially denied. Aside from Mackintosh and Joyce, only three men were pegged for the Ross Sea party.

After much deliberation, Frank Wild's brother, Ernest, signed on to assist Joyce with stores. Aubrey Howard Ninnis, called Howard, enlisted as a specialist motor engineer for the sledging vehicle. He also boasted impressive family connections. His uncle, Belgrave Ninnis Senior, had been surgeon to a renowned British Arctic expedition, and his cousin, Belgrave Junior, had sailed with Mawson. The newspapers also touted the addition of Alexander Macklin, a Lancashire doctor, as the *Aurora*'s surgeon.

On August 1, however, Macklin was waving at spectators from the deck of the *Endurance* as she prepared to depart London's West India Docks, having impressed Shackleton with his indispensability to the Weddell Sea contingent. Mackintosh also strode the decks, but only to join photographs for the press and attend to final details with Shackleton. After a few more weeks of preparation, Mackintosh would depart

London aboard a steamer for Hobart, Australia, where he would join the *Aurora* and assume command of the Ross Sea party. In September, the crew of the *Aurora*—such as it was—would follow to Hobart.

As the *Endurance* eased from the wharf, a woman yelled above the din of a Highland bagpiper, "Are you going near Chile, Sir Ernest?" To the delight of reporters, he shouted back, "No; but it's chilly enough where we're going." The story would be a welcome distraction from the front-page news of the mobilization of troops in Russia, Austria-Hungary, Germany, France, and Belgium, the consequences of events in Bosnia. The web of alliances that had guaranteed a semblance of peace for the last century in Europe now drew the major states into the conflict. But direct British involvement appeared avoidable for the moment, and in less than a week, the *Endurance* would be bound for Buenos Aires.

On Monday, August 3, the *Endurance* lay at anchor off Margate as the sunrise gilded the shallows of the estuary. Shackleton learned that the Thames had been closed to traffic. With German troops moving into Luxembourg, Britain had begun to mobilize forces, drawing closer to war. Among the expedition members were eight Royal Naval Reserve and Army officers all liable to be called up. Shackleton went ashore and wired Churchill, placing "ship staff stores and provisions" at his disposal. In the tense hours that followed, he felt sure the "excursion will be postponed." The unanticipated response from Churchill came quickly—"Proceed"—affirmed later by a note and an audience for Shackleton with King George V.

Before the day was out, Berlin's ambassador had delivered a declaration of war to the French premier and German troops were surging into Belgium. Crowds thronged around Whitehall and Buckingham Palace, chanting exuberantly, "We want war." On August 4, Britain issued an ultimatum: Either Germany guaranteed Belgian neutrality, or Britain would declare war. Amid the euphoria, the somber foreign secretary, Sir Edward Grey, watched the lamplighters on the streets of London that evening and said, "The lamps are going out all over Europe; we shall not see them lit again in our lifetime." By midnight, Britain was at war with Germany.

Despite Churchill's dispensation, some members obeyed the mobilization order for the Royal Naval Reserve and left the expedition.

Shackleton wondered if he could retain a full enough complement to sail. But by Wednesday, the last of the conscientious members had left the ship, now lying at Plymouth. Finally, on Saturday, August 8, the *Endurance* slipped the bonds of shore. Shackleton, who remained behind to attend to loose ends, intended to follow later by steamer.

Among a host of other problems, the Ross Sea party was still incomplete. Mackintosh's staff—first, second, and third officers—had yet to be appointed. A boatswain was needed to ride herd on the able seamen, four or five in all, and take charge of rigging, sails, anchors, and cables. A chief engineer and second engineer were essential to maintain the ship's compound two-cylinder steam engines, with three firemen to stoke the boilers. The carpenter, too, had yet to be selected. And apart from Joyce, Ninnis, and Wild, there were no new shore party recruits to lay the depots and perform scientific work.

For the *Aurora*'s scientific and technical staff, Shackleton pinned his hopes on an eleventh-hour recruitment effort at Cambridge University, aided by venerable zoologist Sir Arthur Shipley, master of Christ's College. In the end, four men joined. John Lachlan Cope was a former biology student under Shipley. He exaggerated his age, from twenty-one to twenty-five, to improve his chances. Reverend Arnold Spencer-Smith, engaged as expedition photographer, was a Queens' College man soon to be ordained as an Anglican priest. Geologist James Wordie and physicist Reginald James also joined, though Shackleton was undecided whether to assign them to the *Endurance* instead.

Emily Shackleton, for one, was concerned with the level of experience within the Ross Sea party. "I cannot afford to risk anything with that ship," Shackleton wrote to ease her fears. "You can rest assured that I will have sufficient trained men besides the University men: I have already 3 sailors and there is Wild's brother and the Engineers and Carpenter will be experts." The experts were chief engineer Donald Seth Mason, thirty-two, on leave from the Peninsular and Oriental Steam Navigation Company (P&O); second engineer Alfred Larkman, whose father ran a navigation school; and Clarence Mauger, twenty-two, an Irish-born ship's carpenter. Second officer Henry Goldsmith Leonard came from the White Star line. Victor Hayward was the only one of the new members with a modicum of wilderness experience. A clerk for a London financial firm, he had spent the better part of a year working on a ranch in Canada. There were no other candidates in the wings. Shackleton wrote wearily to Emily, "I am rather tired to go into all this now but I am sure to get it eventually fixed up: I have the complement of the ship *Aurora* practically complete." He decided to leave the rest of the recruitment to Mackintosh in Australia.

With the nation at war, Shackleton's financial worries reared up once again. The delay in settling affairs for the Ross Sea party proved costly, as he told a backer, "I am having a bad time now as all my Ross Sea stores have gone up in price and expenses increased enormously." The steamship lines that had offered free passage to expedition members before the war withdrew their offers. Shackleton sprinted to Scotland to lay his case before contributor James Caird. "I get very tired of this rush and anxiety," he confessed to his wife. "I am sure the Expedition will not be as arduous as all this." He considered suspending the expedition and volunteering, if Caird would only make the ends meet for a postponement. Caird declined and Shackleton banished doubt from his mind. He would seize the day before it passed.

On the day of departure, September 18, the Ross Sea party numbered twelve men. Two weeks before, Mackintosh had sailed. In his absence, newly engaged chief officer Joseph Russell Stenhouse, a merchant mariner with experience in sail, stood in for him. At 10:30 AM, the entire Ross Sea party gathered in the vaulted neo-Gothic hall of St. Pancras station. A haze of blue locomotive smoke hung in the filigreed iron trusswork as the morning rush crowds eddied around the clutch of family and well-wishers. Shackleton arrived for the farewells in a gust of optimistic bonhomie. His presence buoyed the already high spirits of the members and soothed the nerves of loved ones. "I'm not going to give your son much hard work to do," Shackleton reassured Hayward's mother, to which the ebullient Hayward bantered, "But surely, Sir Ernest, this isn't going to fizzle out into a picnic—I could get that at home." Shackleton had drafted instructions for Mackintosh in the early hours of the morning, intending perhaps to hand them to Stenhouse, but at the last moment held the letter back, trusting to the Royal Mail. Then, the Ross Sea party boarded an eastbound train for Tilbury, where the passenger steamship *Ionic* awaited in the Lower Hope Reaches of the Thames.

By evening, the *Ionic* was stealing down the English Channel under cover of darkness, cabin lights doused to avoid the notice of enemy ships. The accustomed shimmer of the coastal villages was dimmed by curfew orders, save for the searchlights beaming from Dover. With somber pageantry, cruisers flanked the ship, alert to the approaching threat. Within days, the escorts peeled off and the *Ionic* was alone, plowing her way south.

—3—

Aurora

OCTOBER 8–DECEMBER 23, 1914

When Mackintosh arrived in Australia in the second week of October to take command of the *Aurora,* he was shocked to see the unseaworthy vessel moored at the Hobart docks. "The ship was leaking all over; water poured from the deck into the cabins and hold," he wrote to the expedition's organizers in London. Though the *Aurora* had just returned from two and a half years' hard service with Mawson's Antarctic expedition, Shackleton had purchased the ship sight unseen, without an inspection by a marine surveyor. The armored prow and propeller were damaged from ramming pack ice, the engines and winches were rusted, the bulwarks and bridge damaged and in dire need of repair. She was utterly unfit for the rigors of a new polar voyage.

As Mackintosh knew, the ice exacted a punishing toll from wooden ships. To rebuild her defenses before the sailing date of December 1, the refit of the *Aurora* would have to start immediately. He also knew it would be a costly enterprise, but had received only £1,000 from the expedition's organizers in London, even though Shackleton had told him £2,000 would be available, with the possibility of another thousand if necessary. Mackintosh turned to Professor T. W. Edgeworth David of the University of Sydney, whom Shackleton had tapped as scientific adviser to the expedition, for help.

Welshman Edgeworth David had been the scientific director on the *Nimrod* expedition, impressing his companions with his generous spirit and boundless energy. "He had that rare gift," as one colleague wrote, "of ennobling all he touched." Mackintosh and David had enjoyed a warm camaraderie on the expedition, a bond that deepened

34

when Mackintosh convalesced after the loss of his eye at David's family retreat. Once again, David came to his old friend's aid. As a prestigious academic and president of the Australasian Association for the Advancement of Science, David commanded a wide sphere of influence. He arranged a meeting with the prime minister of Australia, Andrew Fisher, on October 16. Fisher frowned on Mackintosh's appeal for government funds, looking askance at the fact that Shackleton had sent out the Ross Sea party with inadequate funding. Reluctantly, he approved a grant of £500 for docking and refit of the *Aurora* at the Cockatoo Island Commonwealth Dockyard in Sydney.

It was not the first time Mackintosh had triumphed over seemingly immovable opposition. He was known for his determination, which he credited to his fierce forebears. Æneas Mackintosh was descended from the ancient line of Scottish chieftains of Clan Chattan and had been named to inherit the title of thirty-first chief and the ancestral manor near Inverness. But he was born far from the Highlands in Tirhoot, India, on July 1, 1879, the eldest son of indigo planter Alexander "Alistair" Mackintosh and his wife Annie. The family enjoyed a prosperous life in expatriate society, but the verities of their Raj plantation existence crumbled when Annie Mackintosh abruptly left her husband and returned to Bedfordshire with her children. The secrets and sorrows of her failed marriage she placed firmly behind her. Independent of mind, if not of means, she instilled in her five sons and only daughter her own unconventional and adventuresome spirit. Of his father, Mackintosh heard no more.

Like Shackleton, Mackintosh left home in his midteens to join the mercantile marine. In the days of sail, a boy apprentice suffered a hard initiation. He endured a life of grinding toil, sleeping less than four hours at a stretch between watches and subsisting on short rations of salt meat and hardtack biscuits. Boys were known to filch food in port to quell their constant hunger. And that was a good berth. A boy might land on a ship at the mercy of a tyrannical bucko mate who heaped verbal abuse and draconian punishments on his charges. The unsparing apprenticeship under sail soon weeded out all but the toughest; with luck and persistence, talent was rewarded. As a young officer, Mackintosh cut a debonair figure in his merchant whites, and he rose through the ranks of the P&O line out of India, relishing his roving life. His cut-glass diction and refined manner belied his grit and gusto for hard labor. In the summer of 1907, he was third officer on a steamer and had just passed the master's examination when Shackleton offered him the appointment as second officer on the *Nimrod*. The expedition held the promise of adventure and a chance to break ahead of the pack in his career. But his polar experience counted for little when he returned to England in 1909. Calling at the P&O offices on his thirtieth birthday, he was informed that, owing to the loss of his eye, his name would be stricken from the books. His career with the company ended after nine years.

In the ensuing years, assorted endeavors failed to bear fruit. He joined Shackleton's Hungarian gold mining venture, and when it came to naught, launched his own expedition in search of pirate treasure on Cocos Island in the South Pacific. Like so many fortune-seekers before him, Mackintosh returned home empty-handed. Back in England, he resisted domesticity and could be "a little haphazard in his ways," as one friend described him, a conscientious officer at sea who "lacked foresight and a sense of responsibility" ashore.

Then in February 1912, at the age of thirty-two, Mackintosh turned over a new leaf, marrying a serene Scottish beauty named Gladys Campbell. A year later, with a wife and a baby daughter at home, he settled into a desk job as junior assistant secretary of the Imperial Merchant Service Guild. It was a steady living, but a modest one, and the chance of advancement limited. Privately, he yearned for his wayfaring days. "I am still existing at this job: stuck in a dirty office," he wrote to an old mate from the *Nimrod*. "I always feel, I never completed my first initiation—so would like to have one final wallow, for good or bad!" In early 1914, he leaped at the chance to join Shackleton again. The expedition offered the best of both worlds: adventure and brighter prospects for his family. But his leave-taking was an uneasy one; Gladys was pregnant with their second child.

She never strayed far from Mackintosh's mind as he scrambled to arrange for the refitting of the *Aurora*. He rounded up a scratch crew to sail up the coast to the dockyard in Sydney. At least one man would remain by his side for the expedition: James "Scotty" Paton, who had sailed on the *Nimrod* as a seaman. They pumped the water from the bilges and weighed anchor on October 26, praying for fair weather and smooth seas.

———

As Mackintosh raced through the hectic days in Hobart, the rest of the Ross Sea party was aboard the *Ionic* in the South Atlantic, cruising south to Cape Town, their final port before the last leg of the journey to Australia. To the wonderment of the landlubbers among them, the steamer cruised through tropical waters aglow with bioluminescent plankton, flanked by flying fish and dolphins streaking through the ship's wake.

These sights were commonplace to Chief Officer Joseph Stenhouse, a Scotsman with the square-jawed handsomeness of a silent film idol. Before joining the expedition, he had been an officer with the British India Steam Navigation Company, where he acquired a reputation as a sharp navigator and meticulous officer. He had apprenticed on the tall ships rounding Cape Horn for trade with America's west coast, an equally miserable experience for boys of humble origins and those, like Stenhouse, of distinguished heritage. His family's shipbuilding firm in Dumbarton had built the sleek clipper ships that raced China tea to market. Though he was an infant when the last of the elegant clippers left the shipyard of Birrell, Stenhouse and Company, he grew up with a passion for sailing ships. Stenhouse's upbringing also ingrained in him an enduring sense of propriety and principle.

His eleventh-hour appointment meant that he had spent little time with Shackleton and had never met Mackintosh. In the absence of explicit orders, he set about organizing a routine of training and duties to occupy the men during the six-week voyage to Hobart. The newcomers embraced the impromptu classes in seamanship, signaling, and first aid with enthusiasm. But their main work responsibility—the care and feeding of the sledge dogs—they approached cautiously.

Ranging in size from sixty to over one hundred pounds, the dogs were powerful animals who lunged, snarling, at all comers. "Thrashing seems a cruel method to employ as a reproof," wrote Stenhouse in exasperation, "but as kindness has had no effect, it is the only one." One obstacle to imposing discipline was the fact that they had been delivered without nametags. On board the *Ionic,* they were rechristened, with the result that mayhem ensued when more than one dog responded to the same name. Their new masters compounded the problem by changing the names. Christabel's antics earned her the nickname Maddo, later changed to Bitchie, then modified to Beechey by the men of finer sensibility.

Bitchie's ill temper was understandable. Since leaving Canada's Lake Winnipeg region in mid-June, the dogs had been cooped up in wooden crates for months as they were trundled from train to ship to quarantine station on three continents. Six had died before the expedition left England. Of the remainder, Shackleton had picked sixty-nine dogs for his transcontinental crossing, allocating twenty-six to the Ross Sea party. They were hardly prime specimens; many were old and most were plagued by parasites. As the *Ionic* sailed through the tropics, the dogs sweltered in their kennels on deck.

Howard Ninnis had some dog driving experience and, along with Ernest Wild and Victor Hayward, took a firm hand with the "sporty dogs," as he called the worst offenders. A chief petty officer with the Royal Naval Volunteer Reserve until 1909, the thirty-one-year-old Englishman had indulged his wanderlust in a roustabout lifestyle in recent years, winding up as a hired hand at a hunting lodge in Manitoba. Along the way, he picked up dog driving, motorcycling, shooting, photography, sketching, Morse code, Japanese, and the rudiments of jujitsu. Officially, he joined the expedition as a motor mechanic, but talked his way into contributing photography and illustration as well. His skills were self-taught and hard won, his slight build and pinched, wary expression giving little clue to his tenacity.

Raised by his widowed mother in impoverished circumstances, Ninnis had enjoyed none of the privileges of his illustrious relations. While his cousin, Belgrave Ninnis, attended Sandhurst Royal Military College and joined the Royal Fusiliers, Howard drifted, falling in with the Thameside dockers and frequenting opium dens. Yet, it was Howard who won a place on Scott's expedition by pitching in to load cargo on the *Terra Nova*, when Belgrave's father's connections could not secure a position for his son. Howard's polar ambitions were dashed, however, when he was suspected of having tuberculosis and was forced to leave the *Terra Nova* in Cape Town en route to Antarctica. The rivalry with his cousin intensified when Belgrave found favor with Shackleton, then sailed south with Mawson's expedition. Howard's lifelong sense of grievance did not fade when his cousin fell to his death in a crevasse in December 1912. Howard's long-awaited chance to prove himself came with the Ross Sea party.

Family connections were undoubtedly a factor in Ernest Wild's appointment, but whether elder brother Frank played the role of persuader or persuaded is unknown. The sons of Yorkshire school headmaster Benjamin Wild and his wife Mary, they were raised in a home of starch-collared propriety with six other brothers and five sisters. Both had followed the sea since their midteens. But where Frank was given to peripatetic wanderings and flights of fancy, claiming he was descended from Captain James Cook, Ernest was a steady Royal Navy petty officer. At thirty-five, Ernest took leave from the service, and his steady habits, to join the Ross Sea party.

Londoner Victor Hayward divided his time between corralling the dogs and mucking in with the *Ionic*'s crew to practice stoking coal, but his dapper uniform of white flannel trousers and spats placed him squarely with the first-class passengers. The son of a London train official who was the local vice president of the nationalistic British Empire Union Club, also known as the Anti-German League, Hayward was educated at an Essex boarding school. As a boy, he was an avid reader of Robert Ballantyne's rollicking adventure tales of stalwart lads braving the wilderness and high seas. They were coveted prizes for diligent Sunday school pupils, which is how Hayward came to possess Ballantyne's *The World of Ice; or, Adventures in the Polar Regions* at an impressionable age. After attending Saint Mark's College in Chelsea, he joined a financial firm in London. Twenty-six and newly engaged to be married, he nonetheless resisted a predictable future in the leafy contentment of the outer boroughs. He begged leave and went to northern Canada, where he spent seven months working on a ranch. Back in London he found motorcycling to the office a poor substitute for high adventure. "You are looking great in the photographs, you appear pretty fat. Eh! Think you must be living pretty well," needled his brother Stan in a letter, knowing full well how Victor dreaded bourgeois complacency. In August 1914, Hayward hurried to Shackleton's office. By the end of the month, he had received a letter of acceptance. His offer to "do anything" earned him the job of general assistant.

The struggles with the dogs proved an irresistible subject for photographer Arnold Spencer-Smith, who brought out the still and motion picture cameras to film the daily ritual. Prior to the expedition, Spencer-Smith had shown little inclination toward things polar or creative. Moving from the Westminster City Schools, where his father was an official, to King's College, London, Spencer-Smith wavered between law and the civil service. Still undecided, he followed his elder brother to Queens' College, Cambridge, to read history, where he soon became a man about town. Tall, lean, and genially handsome, smoking the "inevitable Woodbine," he steered through a mad whirl of punting parties on the River Cam and May Balls. His hectic social and sporting agenda—tennis, cricket, swimming, rowing, debating, and theological society—soon overtook his academic pursuits. The college magazine, which Spencer-Smith founded in a rare spell of leisure, waggishly observed, "For some unknown reason he was presented with a round dozen of books, which he hopes to read some day." Devoted to his parents and six siblings, he spent holidays at the family's Sussex country home or their London townhouse around the corner from Buckingham Palace. In 1907, he earned a pass degree in history, despite missing his final exams.

Spencer-Smith went on to teach at the Merchiston Castle School for boys in Edinburgh, where his interest in religion evolved from an intellectual pursuit to a vocation. In 1910, he was ordained as a deacon in the Anglican Church of Scotland, but he had no intention of becoming a tea-and-sympathy vicar. His was the robust faith of Anglican missionaries like David Livingstone, saving souls as he canoed down the Zambezi. When Britain declared war, Spencer-Smith was eager to serve his country in a noncombatant role befitting his calling and volunteered to join Shackleton's expedition. He was ordained five days before the Ross Sea party departed London and hoped the crew would view him as an unofficial chaplain, offering spiritual support as well as his physical strength to the sledging effort.

Once the duties were in hand, Stenhouse volunteered the party for a series of lectures to edify and entertain the *Ionic*'s passengers. Wild yarned about navy life, Hayward related his travels in Canada, Ninnis spun tales of his brief sojourn on *Terra Nova*, and biologist John Lachlan Cope held forth on a string of topics, including nutrition, evolution, the central nervous system, and penguins. Cope was a buoyant raconteur with an endless store of amusing anecdotes and "a habit of wandering about the room while talking, thoughts as to where he would wander & what he would say next kept everyone interested." At twenty-one, Cope was at loose ends. The son of the proprietor of Cope's Bank, a private bank serving a rarefied clientele, Cope attended Tonbridge, an elite boarding school in Kent, and matriculated at Christ's College, Cambridge, intent on becoming a physician. His father's astute financial dealings kept the family in comfortable circumstances with servants in residence as he earned a master's degree in biology. Then, in 1914, William Cope abandoned the family, leaving his wife as the sole support for their two sons and two daughters. Derailed from his medical studies, Cope deftly concealed any trace of bitterness or disappointment beneath his eccentricities and voluble banter.

Alexander Stevens was nearly his antithesis. Tall and angular, his speech peppered with Gaelic, the Scottish chief scientist was baffled by the reactions he sometimes evoked. At twenty-eight, he was a quiet, socially naïve loner. His sharp intelligence and shyness could be misinterpreted as abrasiveness. One sympathetic colleague described him as "not always striving to avoid treading on corns, yet essentially kindly." Stevens was solitary from a young age. Born to a domestic servant and his wife, he was raised in Kilmarnock near Glasgow but felt a stronger kinship with his grandparents' home on the gale-swept north coast of Scotland. Influenced by his father, a church officer, Stevens studied comparative religion at the University of Glasgow with the intention of becoming a minister. Struck by a sudden crisis of faith, he left and became a schoolteacher in the Outer Hebrides, spending his free time roaming the stark hills of Stornoway. Later, he returned to Glasgow University with a new zeal to study science and earned his degree with distinction. In 1913, he assumed a lecture post under the eminent geologist J. W. Gregory, who had been involved in the early stages of the *Discovery* expedition. Shackleton visited Glasgow in August 1914 and snared Gregory's former student for the Ross Sea party.

Like Stevens, most of the party had been signed late, so the time en route to Australia was their first opportunity to take each other's measure. The streetwise Ninnis cast a cynical eye on "the scientist brigade," scorning their clumsy efforts at manual labor. Ninnis often found

Stevens "lavishing attention on a seedy old wreck called Captain, no more fit to pull in a team than a rabbit." Stevens groomed the dog for hours and fed him hot beef bouillon. In the end, Ninnis euthanized and skinned the aged Captain to make a pair of fur mitts. In spite of his inexperience, Spencer-Smith, dubbed the Padre, universally endeared himself, even to the sardonic Ninnis, with his good cheer and willingness to work. Hayward, too, was a tireless worker. The notable exceptions were Second Officer Leonard and Chief Engineer Mason, who proved susceptible to the distractions of a passenger liner crowded with holidaymakers. "Our two 'Launcelots' are still deeply engrossed with the cloud of ladies," Ninnis observed acidly after they shirked their duties yet again.

Anticipation onboard heightened as the *Ionic* neared Tasmania, the island state lying farthest south in Australia and the last Australian port en route to Antarctica. At midnight on October 30, the Hobart pilot boarded to guide the *Ionic* into the harbor. To universal disappointment, they found that Mackintosh had sailed for Sydney a few days before to put the *Aurora* in dry dock. Stenhouse received Mackintosh's written instructions and set the crew to work. The first task would be to shift the tons of expedition cargo to a warehouse. Sorting freight consumed the next day, a job complicated by the fact that there was no inventory of the contents. Stenhouse delegated the training of the dogs to Ninnis and Wild, who were the most adept at coping with their antics. The two men joined their charges at the Taroona quarantine station, where they would remain until the *Aurora* returned on her voyage south in December. On November 3, Stenhouse and eight men boarded a steamer for the six-hundred-mile passage to Sydney.

They arrived on Thursday morning to find Mackintosh, smartly dressed and wearing a gold monocle oddly matched by the fixed gaze of his glass right eye, waiting on the quayside to greet them. His warmth set the crew immediately at ease. "He's an absolute 'dear,'" Spencer-Smith wrote in a letter home. "We are all, staff & men, absolutely in love with him!" Ernest Joyce's late arrival had been eagerly anticipated, as his reputation preceded him. A stout figure with extravagant mustachios, goatee, and a conspiratorial smile, he looked more buccaneer than polar explorer. Mackintosh was surprised to find such a small contingent. In particular, two scientists were missing, geologist James Wordie and physicist Reginald James. Shackleton had spoken of assigning the two scientists to the *Aurora,* but had evidently changed his mind after Mackintosh's departure. Mackintosh squired the entourage to the Cockatoo Island naval dockyard to see the *Aurora* in dry dock. Dwarfed by a destroyer alongside, she was not an impressive sight. "She's in a filthy state, inside & out," wrote Spencer-Smith. Stenhouse, surveying the work ahead, kept his composure and observed that after languishing for nine months, she was far from being trim.

Arguably, it was the ship's thirty-eight-year career in the Arctic and Antarctic telling on her. The *Aurora* was a relic of another age, built in 1876 by Scottish shipwrights Alexander Stephen and Sons, who made their sterling reputation in staunch sailing ships for Arctic whalers and sealers. Stephen's ships were masterworks of craftsmanship, with joinery as intricately crafted as fine cabinetry. By the turn of the century, when steel steamships edged out square-riggers on the high seas, the *Aurora* remained in demand. In encounters with unyielding masses of pack ice, wooden ships were unrivaled. The reinforced bows heaved up onto the encroaching floes and crushed them beneath the full weight of the ship, and her resilient timbers flexed under battering that would gash a steel hull asunder. To a steamer captain, a ship beset in pack ice was catastrophic; to an ice master aboard a Newfoundland whaler, she was merely "in the nips" for a spell, soon to be liberated by navigational skill and brute force. The *Aurora* had neither the grace nor the speed of Shackleton's *Endurance,* which handily made ten knots while the *Aurora* normally wallowed along at six: Instead, sheer strength was her forte. She was 165 feet from stem to stern and thirty feet across, double the gross tonnage of the *Endurance,* and had survived three previous voyages to Antarctica.

To set the *Aurora* to rights, Mackintosh put Stenhouse in charge. He immediately began tallying "the hundred & one things which require attention" in a refit and delegating duties to the men. The refit was just the beginning of making the *Aurora* ready for the expedition. A staggering task awaited after she was refloated from dry dock on November 14. Mackintosh was confounded to learn that much of the equipment and stores on board had been reclaimed by Mawson. Apparently, Shackleton

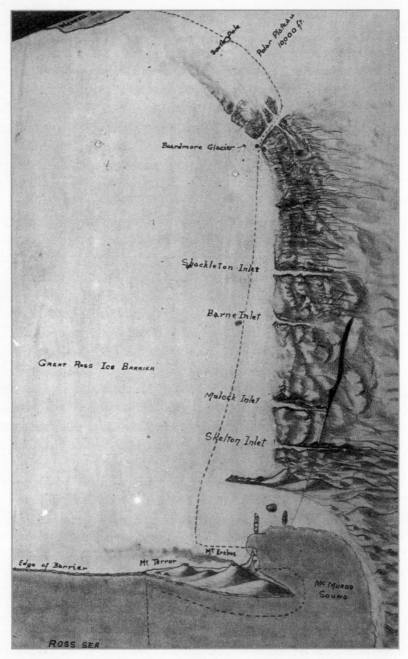

SHACKLETON'S PROPOSED ROUTE ACROSS ANTARCTICA

had misunderstood the terms of the *Aurora*'s sale, and the ship would have to be outfitted with the most basic items. Marine supplies, from paints to tools to ensigns, were needed. The bridge had to be supplied with nautical almanacs, charts, chronometers, compasses, and log books. The living quarters lacked mattresses, pillows, bed linens, blankets, pans, crockery, matches, and toilet paper.

Mackintosh would be provisioning the ship for a longer voyage than originally planned. Just before they parted in London, Shackleton had told Mackintosh to winter the ship in Antarctica to economize on coal. Instead of disembarking the shore party and returning to New Zealand to sit out the brutal Antarctic winter, expending fuel on two round-trip voyages instead of one, Mackintosh would moor the *Aurora* in McMurdo Sound until the job was done. Since Shackleton had not decided whether he would make his crossing in early or late 1915, the ship would have to be supplied to sustain them for as long as two years.

Yet the stores sent from England were grossly inadequate. The sledging gear—clothing, skis, sleeping bags—was just sufficient to outfit the six members of the depot-laying party for one season without spares. Some equipment was in short supply for even a small party— only two sledges were included—and other essential items, such as oilskins and ski boots, were missing entirely. Fur items had been damaged by rats and moths during shipping. There was not enough food to sustain six men for four months of sledging, let alone twenty-odd men for two years in the Antarctic. To make matters worse, the expedition's storage shed in Hobart was plundered by thieves.

Mackintosh was at a complete loss. He had been under the impression from Shackleton that all of the necessities would be shipped on the *Ionic* with Stenhouse, save for some supplies to be purchased in Sydney. There was no explanation for the shortfall; Shackleton had promised to send detailed final instructions in early October, but the letter had never arrived. And now Shackleton was beyond reach, already en route to the Antarctic. The *Endurance* was out of communication unless Shackleton entrusted a message to a northbound whaling ship.

By Mackintosh's reckoning, costs for Australian labor and materials were running double that in England. The £1,000 advanced to Mackintosh had already been depleted by expenses in port and lodging for personnel. On November 17, he cabled the expedition's secretary in

London, Frederick White, asking him to release the second £1,000 that Shackleton had earmarked for the Ross Sea party. White responded two days later. He was unable to send further funds, he explained, and suggested that Mackintosh mortgage the ship for up to £3,000, keep the minimum needed, and send the balance to him in London. Mackintosh was stunned, replying that without money to meet commitments, the *Aurora* would be detained. Though the schedule had slipped, Mackintosh had still planned to leave by the end of the first week of December, now a practical impossibility. White did not respond. On November 22, Mackintosh called a meeting with Stenhouse, Spencer-Smith, Cope, and Stevens to show them White's cable. He confessed that he was "astounded" and asked their advice. Stenhouse suggested that they raise the funds on their own and offered to approach a prosperous uncle in Melbourne about a loan. He boarded a train to meet with him and came back unsuccessful. Two days later, White finally replied to Mackintosh. No further money was forthcoming, he stated crisply, forwarding the ship's papers as collateral for a mortgage. He also enclosed the missing letter of instructions from Shackleton, dated September 18, with no explanation for the delay in delivery.

"My dear Mackintosh," Shackleton wrote, apparently on the morning of the Ross Sea party's departure from London. "The following are my final instructions before leaving, and, unless there is mail from Buenos Aires, the last you will hear from me until I meet you at the Ross Sea side." He ordered Mackintosh to sail the *Aurora* on December 1 at the latest, to reach Antarctica by January 1, 1915, and "send out a party to the South" immediately. He included a list of directives for the expedition, many of the provisos reversing his original understanding with Mackintosh. He also reduced his commitment of funds to the Ross Sea party to £1,000 in total, deciding that this would be enough to perform "the minimum amount of repairs compatible with safety" to the *Aurora*. He advised Mackintosh to obtain supplies "free as gifts: this especially refers to the coal," and to search the old expedition huts in Antarctica for extra food and clothing.

Mackintosh was appalled, not least by Shackleton's last-minute decision to cut his funding. He showed the correspondence to Edgeworth David. "[David] is not given to extravagant expression. The present situation of this party of the Expedition dumbfounded him," Mackintosh wired to White. David and local solicitors advised him that, as he was not the registered owner of the ship, no reputable bank would mortgage the vessel. "How coal and equipment are to be obtained, the very large deficiency of stores made up, and wages and advances to the crew paid is not indicated," he wrote, indignant at being forced into fundraising when time was of the essence. "Why should I be concerned in such business matters?" He demanded to know what had become of the extra funds Shackleton had said he would reserve for the Ross Sea party. Because of the cash problems, Mackintosh concluded, "The ship is held up, valuable time is being lost, and I have very grave fears for the result."

With White either unable or unwilling to come to grips with the crisis, Mackintosh decided to appeal directly to Shackleton's expedition attorneys in London, Hutchison and Cuff. He cabled them that he was legally unable to obtain a mortgage. Faced with spiraling expenses, he pressed for an additional £1,500 to complete the preparations. In the interim, Mackintosh turned to his chief benefactor, the Commonwealth Naval Dockyard. The yard managers understood the gravity of the situation. In addition to the refit, countless other items were essential to make the ship equal to the rigors of polar conditions. Mackintosh had impressed the dockyard staff as "extremely capable and a grand fellow" as he grappled with one calamity after another, the latest being a dockyard strike. The sympathetic yard manager approved a flurry of orders, including much-needed work clothing for the *Aurora*'s crew, but a request for a portable expedition hut and other unusual items raised eyebrows. Costs soon climbed over £3,000 for docking, overhauling, stores, and fittings.

Meanwhile, Mackintosh had no source of ready cash to cover food, clothing, wages, or living expenses for the party in port. The government of New South Wales had declined his plea for a grant. As a last resort, Mackintosh opened the ship to the general public and collected admission. As spirits reached a low ebb, David organized an informal committee of civic and business leaders who marshaled donations of supplies for the expedition.

John King Davis, too, rallied support. After refusing command of the *Endurance,* Davis still maintained a fraternal interest in the expedition, out of affection for his longtime friend, Mackintosh, and his old ship, the *Aurora*. While in Australia, Davis dined with the Ross Sea party officers and introduced them to seasoned polar explorers and scientists who freely offered their expertise. As word spread of the expedition's plight, donations flooded in: chronometers, a typewriter, an electric lighting system, motors, and concrete slabs for a hut. The general public responded with countless small gifts, including eggs, jam, books, and five cases of coveted Johnnie Walker Red scotch whiskey. A group of "very energetic" ladies "took fright lest anything should happen to the *Aurora* in the south and she should have no means of communicating with civilisation," and campaigned successfully for a wireless set for the ship.

Some of the donors were outraged that Shackleton had dispatched the expedition poorly prepared and dependent upon charity. "During the stay in Sydney an unpleasant feeling with regard to the Expedition grew and became strong," Stevens observed. "Unfortunately the feeling seemed to be much directed against headquarters, and to have a counterpart in sympathy for the Ross Sea party." Mackintosh tried to deflect the resentment, laying the "deplorable circumstances" at the door of the attorneys in London.

For the most part, the *Aurora*'s company rose to the occasion. Mackintosh sang the praises of the officers and men for "doing their utmost in the trying, unusual and unexpected circumstances and making sacrifices in all manner of ways." The crew worked long hours and agreed to defer their wages. Stevens, Cope, and Spencer-Smith begged and borrowed scientific and photographic equipment, even signing for personal loans. Spencer-Smith bought and donated books. Some members, though, kept an eye peeled for brighter prospects. "Looking extremely happy & smoking a vile cigar," the cook announced to Stenhouse that he "got tied up to a nice bit of stuff, only 38, with six houses." For the time being, the newlywed cook stayed, but three stewards in succession came and went, leaving him unassisted in the galley.

Even with the staffing at a crisis, two members were forced to resign. Leonard, as Stevens discreetly put it, "was not considered to be keeping himself sufficiently aloof from the other Passengers" on the *Ionic*, attributed by Stenhouse to a lack of experience on "better class ships." The chief engineer, Donald Mason, drank heavily and behaved badly. The swaggering Welshman had left his last ship after a reprimand for insobriety.

"Drink is a curse," Stenhouse complained as the sailors staggered in from late nights on the town. A more serious confrontation loomed. Ernest Joyce, by long habit, was a hard drinker, and soon found himself at odds with Stenhouse after a rumbustious drinking bout at a local hotel. "Jollification," in Stenhouse's book, was tolerable in the common lower-deck sailor, but his patrician backbone stiffened at like conduct in a man of Joyce's position. He insisted to Mackintosh that Joyce be sacked.

"A rotter," one of Joyce's detractors supposedly called him, not the first to regard him with distaste. More often than not, the well-born element of British Antarctic exploration—the former public schoolboys and Oxbridge men—sniffed at the likes of Joyce. His south London working man's accent and untutored grammar made his humble origins all too plain. His career strivings were undisguised and there was little of the gentleman in his braggadocio and sometimes boorish behavior. Those who regarded the Antarctic as a preserve for serious science also had little room for his carousing ways. They drew a sharp distinction between polar exploring and polar adventuring, and many placed him disapprovingly in the latter camp. As John King Davis had told Shackleton, "I absolutely decline to be associated with any enterprise with which people of the Joyce type are connected." Shackleton, fortunately for Joyce, saw the raw talent and grit behind the swagger and forgave his excesses.

Mackintosh told Stenhouse that he had neither the authority nor the funds to remove a key member of the expedition. Dismissing a second mate was one thing; a longtime associate of Shackleton was an entirely different matter. Firing Joyce would also mean dispensing with the only man with Antarctic dog driving and sledging experience. Mackintosh advised Stenhouse to hold his peace.

Mackintosh could ill afford to lose anyone. The *Aurora* was not fully staffed and recruiting was proving difficult. Word had spread on the Hobart docks, and most seafaring men found the idea of joining a blighted ship distinctly unappealing. To replace Mason, Stenhouse promoted Larkman to chief and began searching for his replacement. "Many engineers came on board with a view to sailing as Second; they took a look at the engines and promptly went ashore," Larkman re-counted. "One, with a sardonic grin, remarked how nice they would look in a museum." Eventually, Adrian Donnelly, a railway engineer with no sea experience, signed on.

Finding candidates for the scientific staff proved difficult as well. After an unsuccessful search for a physician, Cope's medical aspirations were suddenly fulfilled. He was named ship's surgeon without further ado. Richard Walter Richards answered an advertisement for an expedition physicist in a Melbourne newspaper and accepted the post at a nominal salary of £1 per week. Spencer-Smith's Australian cousin, Irvine Owen Gaze, enlisted as a general assistant at the same wage. Melbourne scientist Andrew Keith Jack, a friend of Gaze, also signed on. A former schoolmate of Jack, Lionel Hooke, was hired as the wireless telegraph operator and, at age eighteen, became the youngest member of the expedition.

On December 9, with the paint on the hull still wet and the crew shorthanded, Mackintosh prepared to weigh anchor and sail for Hobart. After weeks of stalemate, Shackleton's attorneys wired £700 with a message blaming the wartime economy for the financial troubles. "Ship must leave according to Shackleton's instructions," the wire insisted. "Have confidence you personally, rely your honour get ship away on minimum cash." Hutchison and Perris suggested that Mackintosh obtain goods and services without payment and mail the bills to London. David and his committee stepped into the breach, pooling donations totaling £700. In return, they insisted that Shackleton's London organizers sign a mortgage to guarantee repayment by Shackleton. In the few days remaining in Sydney, Mackintosh and Stenhouse plunged headlong into buying the essential food and clothing. For a two-year voyage, the quantities were immense. Just one of several orders of meat weighed in at 1,800 pounds.

On December 10, Mackintosh summoned the crew to the shipping office to sign ship's articles, the contract of employment for merchant mariners. Boatswain Scotty Paton, four experienced able seaman, one ordinary seaman, a cook, and a steward signed on. After weeks of ceaseless labor, Stenhouse gave them leave for twenty-four hours, and they set off without delay. "I sighted our crew in a taxi and all bound for the land of non-remembrance," Stenhouse wrote, not entirely certain they would all return. He hired day laborers to finish the rigging and enlisted the scientists in stowing cargo.

Stenhouse found it hard to shake the feeling that the *Aurora* was not a happy ship. In seaman's parlance, a happy ship meant harmony and efficiency, an elusive alchemy that had had all too little time to evolve during the chaotic days in Sydney. Most of the crew was green and the ship and equipment untested. The burden would be on his shoulders, since officially Stenhouse was in command. At the mercantile marine office on December 15, he was signed as master of the ship. Officials likely insisted on the change in command when they saw that Mackintosh's sight was impaired. The crew was not informed and Stenhouse carried on as if Mackintosh was in command.

Later the same day, after shipwrights scrambled to seal a leak near the sternpost, the ship started south for Hobart under clear skies. Even on a fine day, she proved to be a "champion roller," in the words of Spencer-Smith, who was laid low by seasickness as the ship rolled thirty degrees to port and starboard by turns. Belowdecks, water streamed through the gaps in the hull's planking and flooded the stokehold. Larkman and Donnelly manned the engines and pumps. The usually stoic Larkman veered to the bridge to see Stenhouse "with a very lugubrious countenance" and declared, "This isn't a ship, she's a bally pendulum."

When the *Aurora* arrived in Hobart just after dawn on December 20, the crew launched into feverish activity. Mackintosh had originally intended to berth there just long enough to load the stored cargo and the dogs, but the short voyage had revealed serious problems. The engineers went to work on the pumps and bilge pipes as the last recruits for the shore party climbed aboard, including physicist Richards, who jostled up the gangway behind dockers hefting cargo. Dazed, he found it "difficult to imagine a state of greater confusion."

Consumed by work on the ship, Mackintosh had delayed replying to a sheaf of cables from London, demanding an accounting of his expenditures. Their tone seemed at best obtuse, and at worst, disingenuously so: "Cannot understand extra expenditure . . . You have given no particulars gravest anxiety . . . Cannot explain troubles here nor why money wanted." He had equipped the ship on £1,700 cash from the London organizers, and at least triple that amount in Australian contributions.

After all his trials, one high-handed line in particular rankled Mackintosh: "Obligation yours to see starts proper time." He sat down on December 21 to explain the problems and costs to White, adding:

> My obligations as commander are to get the ship equipped and ready for sea, in such manner that she will be fit to withstand the rigours of an Antarctic voyage and wintering in McMurdo Sound. These I recognize fully and will carry out. There are, on the other hand, correlative obligations resting upon you and the others who hold Sir Ernest Shackleton's power of attorney, to see that I am supplied with the necessary money to fulfil my obligations. If you have great difficulty in raising the money, how much more have I, who have no status and no power? Nevertheless money I must have. There are lives of men on my hands.

Mackintosh had been placed in an untenable position. Still, his sense of honor and duty was steadfast. Shackleton had been emphatic about his party's utter dependence on the depots, stressing that it was "of supreme importance, as ensuring an absolutely safe return to the T. C. [transcontinental] party."

From his cabin adjoining the chart room on the *Aurora,* Mackintosh commanded a bird's-eye view of the pandemonium. Fifty sheep and a flock of ducks and chickens milled about the crates and heaps of coal as the motor tractor was hoisted aloft by crane. Hundreds of tons of cargo and nearly five hundred tons of coal had been loaded into the ship's holds and bunkers, with more still to come, though the ship was already heavily laden and low in the water. On December 22, "the disorder and general mess reached its climax," as Richards put it, when the Governor of Tasmania, Sir William Ellison-Macartney, and Lady Macartney visited the ship. The governor's wife, who was the sister of the late Captain Scott, presented the ship with a portrait of her brother; though an odd icon for a Shackleton vessel, it was nonetheless accepted with grace.

The following day, December 23, Mackintosh felt he could wait no longer. It was three weeks later than the departure date Shackleton had specified. Although the ship was not quite ready, Mackintosh gave the order to prepare for departure. There was no sign of most of the *Aurora*'s crew. The previous evening, advanced wages in pockets, the sailors and stokers had hightailed it for a last carouse on shore. The ship was undermanned, even if all the hands turned up. The boiler room needed a third stoker, the steward lay in a Hobart hospital, and most crucial, a second officer candidate was still undecided.

Improbably, three new lower-deck recruits appeared within hours of casting off the hawsers, two of them vague about their recent history. Stenhouse asked few questions. Brawny ex-policeman and boxer Harry Shaw looked capable of stoking the boilers. On a whim, diminutive Frenchman Emile Octave d'Anglade quit his kitchen job at a hotel to join the voyage as steward, his second into the Southern Ocean. To Stenhouse's relief, second mate Leslie Thomson strode up the gangway as the mooring ropes were thrown off onto the quayside, but, unlike the others, he was no impulsive pierhead jumper. A merchant officer who had trained in sail early in his career, the cool-tempered Australian craved a change from the predictable routine aboard passenger steamers. At 4:10 PM, on his twenty-eighth birthday, he cast the die, signing ship's articles with less than an hour to say his farewells.

Following in Thomson's wake were the forecastle hands, dodging stevedores and shambling up the gangway in morning-after disarray. Under the wire, twenty-six of the Ross Sea party's full complement of twenty-eight were aboard, but the stores were still not fully loaded; the ship would be forced to anchor in the stream overnight while the remaining crates were ferried out in launches. Wild, Ninnis, and the dogs waited to be embarked at the quarantine station. At five o'clock, the *Aurora* backed awkwardly and "nearly took a piece off the end of the New Wharf." A handful of passersby milled about, taking little notice of her inelegant maneuvers. "No one cares a 'hang' about this Expedition," Stenhouse brooded in his diary. "A few loafers & others were the only people on the wharf when we left."

— 4 —

Southing

DECEMBER 24, 1914–JANUARY 24, 1915

Mackintosh fixed a course for Ross Island, Antarctica, 2,100 miles distant. With the wind and sea rising, the *Aurora* bowled along the swell, driven south toward the Antarctic Circle by brisk northeasterly winds. The falling barometer promised a rough crossing of the Southern Ocean, a fierce navigational challenge even in fair weather. Yet Mackintosh set forth with full confidence in his vessel. "She rides along the waves in fine style," he recorded in his diary. "The old *Aurora* is behaving admirably and proving her worth as a ship."

His view was not unanimous. "She's a damned oscillating farmyard," groused Larkman. "What with the dogs, sheep & poultry & oil on deck it's the queerest vessel I was ever on." The dogs, agitated into fits by the livestock, strained at the end of their tethers. Green seas surged across the decks, flooding the galley and washing rivulets of animal urine, feces, and coal dust through the deck planking. The crew grappled with unlashed stores sliding from port to starboard with each roll.

The neophytes of the shore party were soon seasick. The seamen regarded their miseries with a jaundiced eye. "Many of our after crowd are 'hors de combat,' incapable of putting one leg before the other," veteran boatswain Scotty Paton commented dourly. "We see little of them, only when they are making a bee-line for the rail, as if their food did not agree with them." Too delighted to take offense, Reverend Arnold Spencer-Smith decided the Glaswegian bo'sun was the living image of a gruff sea salt "straight out of *Treasure Island,*" singing sea shanties as he hauled sail. Mastering his stomach, Spencer-Smith held an Anglican Christmas service and reported for duty.

Struggling to gain their sea legs, the shore party shoveled coal, greased the engine bearings, took in sail, and emptied buckets of ash overboard—unceasing work, as the engines consumed some five tons of coal each day. Hayward renounced his uniform of white flannel trousers and straw boater for coarse woolens and shaven head. But their exertions did little to improve their standing with the crew, who dubbed the shore party "the FuFu gang." For his part, Mackintosh was none too impressed with the seamen and firemen. "The only disagreeable people on the ship are the crew," he wrote. "They are brutes."

The tension was rooted in social rank. The shore party members were, for the most part, from educated middle- or upper-class families and could afford to take flight from their ordinary lives for a year or two for the sake of adventure or ambition. For the crew, the expedition meant a guaranteed billet, decent meals, and a steady living wage, by no means a certainty for working-class men on a merchant ship or at home in England.

Ernest Edward Joyce had a foot in both worlds. With a Polar Medal and bars to his name for his stints aboard the *Discovery* and the *Nimrod*, Joyce commanded respect as the most experienced Antarctic traveler in the entire company, an "old penguin," as he liked to call himself. His jovial manner coaxed more work out of belligerent seamen than iron discipline. He rarely pulled rank, choosing instead to muck in on the meanest chores and enlist by example. An inveterate storyteller with a touch of Barnum about him, Joyce habitually spun yarns that were always entertaining, if not always entirely factual. Joyce was instantly popular with the young Australian scientists of the shore party as well, who he gladly helped with advice and encouragement.

Joyce was, like his father and grandfather before him, a seaman, with the best part of twenty-four years at sea. Joyce's widowed mother Frances, left alone with three children to eke out a living as a seamstress, sent him to the Royal Hospital School for Navy Orphans in Greenwich when he was a boy. The grim Dickensian institution had undergone reforms, but he still slept in a fifty-bed dormitory and ate with a thousand boys in a vast dining hall. Their existence was defined by a strict regimen of parade drills, rudimentary academic work, and vocational instruction. The most promising boys were sent to the Upper School for advanced training, where the brightest could hope for careers as commissioned officers; the school even numbered admirals among its alumni. But Joyce, like most of his fellow pupils in the Lower School, was steered into the bottom ranks of the Royal Navy. At fifteen, he left his formal education behind and rose through the ranks from boy seaman to ordinary seaman to able seaman.

In 1901, the twenty-nine-year-old Joyce was serving as a leading seaman aboard the cruiser *Gibraltar* in South Africa when a call came for volunteers on Scott's first Antarctic expedition. Four hundred applied, and Joyce was one of only four chosen to sail aboard the *Discovery*. "An excellent, trustworthy man," was the appraisal of expedition organizer Clements Markham. "Well set up and handsome. A capital letter writer." Joyce lived up to expectations, impressing Scott as "sober, honest, loyal and intelligent." In spite of his susceptibility to frostbite, Joyce soldiered on with sledging and a failed attempt to reach the summit of Mount Erebus.

Returning to the Royal Navy, Joyce was promoted to petty officer, first class, on Scott's recommendation. At the end of his twelve-year engagement in 1905, he took discharge. Freedom was not all he had anticipated. Trained to the clockwork routine of the navy since childhood, he felt cast adrift in civilian life. He rejoined the navy in 1906, seeking a secure pension and the sense of purpose he so desperately craved. Fate intervened in 1907, as Joyce yarned, when Shackleton spotted him riding topside on a double-decker bus in London and shouted an invitation to his old *Discovery* chum to join him at the South Pole. Joyce left the navy, once and for all, and signed articles for the *Nimrod*.

Although he was not chosen for a coveted place in Shackleton's South Pole attempt, he threw himself unstintingly into any task set before him. His hard-driving work ethic and loyalty won him favor with Shackleton. "Joycey" returned home in high spirits, joining the inner circle when Shackleton celebrated his knighthood in December 1909. Shackleton prided himself in repaying the faithful in kind. He chose Joyce to run the *Nimrod* as a floating tourist attraction on the Thames and set him up with a job at £250 a year, a respectable sum when the average teacher earned £175.

Shackleton doubtless put in a good word when Douglas Mawson considered Joyce for his Australasian Antarctic Expedition in 1911. At Mawson's behest, Joyce sailed to Denmark for the vital task of choosing and training dogs, then joined the expedition in Australia. But on the eve of the expedition's departure for Antarctica, they fell out and parted ways, neither man anxious to air their differences after the fact. The problem, it was rumored, was drink. After his break with Mawson, Joyce stalled in the doldrums, drifting from job to job in Australia as he dreamed of reclaiming his glory days in the Antarctic. When the summons from Shackleton came in March 1914, Joyce was overjoyed to give notice on his shore job with the Sydney Harbour Trust and head south once again.

———

On December 29, a hazy silhouette of land appeared in the mist: Macquarie Island was the last outpost of human habitation en route. Three men operated a weather and wireless telegraph station on the island, established by Mawson in 1911. His was the first Antarctic expedition to be equipped with wireless, meeting with intermittent success in relaying Morse messages from the coast of Antarctica back to Australia via the Macquarie station.

The wireless set aboard the *Aurora* was regarded as an experiment, primarily for ship-to-shore messages in McMurdo Sound. Shackleton doubted the value of wireless. Technical limitations rendered it useless for communications between the Weddell Sea and Ross Sea parties across Antarctica. Though the parties could have been equipped to reach ship and shore stations during periods when they were in range, Shackleton believed that contact with the outside world might compromise his command. Then there was the expense when the expedition was severely strapped for cash. So the *Endurance* sailed from England without a wireless set, although a South American benefactor later presented Shackleton with a receiver. The *Aurora*'s set was capable of sending and receiving, but its limited range ensured that the two vessels would not communicate after leaving civilization.

The *Aurora*'s wireless operator, Lionel Hooke, matched Shackleton's skepticism with his unbounded enthusiasm for the invention. The eighteen-year-old Melbourne native was so anxious to join the expedition that he claimed to be twenty. Hooke attended the same grammar school as physicist Keith Jack, but despite his intense scientific curiosity, he decided to bypass university. He was captivated by electricity. At seventeen, he apprenticed with the Melbourne Tramways, the department superintending the electrification of the city. The following year, he trained as a marine wireless operator. The *Aurora* was Hooke's third shipboard assignment for Amalgamated Wireless Australasia.

Hooke signaled Macquarie before going ashore on the thirtieth. There were no messages waiting from Shackleton or his London organizers. Mackintosh could not be sure of the *Endurance*'s progress. Of one thing he was certain: The *Aurora* was behind schedule. He hurried the crew to unload the cargo and livestock intended for the station in a twenty-hour stint, and the men wrote last letters home before turning in. Some sent final wireless messages to loved ones. "Everything OK just leaving Macquarie. Do not forget," Hayward assured his fiancée. It would be at least a year, perhaps two, before the *Aurora*'s return. At sunset on the thirty-first, Mackintosh ordered steam and anchors raised. Jack, taking a turn at the wheel, watched the island recede on the horizon and mused, "We have severed our last link with civilisation."

Shackleton's instructions had stipulated that the *Aurora* arrive in McMurdo Sound by January 1. Anxious to win back the time lost in Australia, Mackintosh ordered all sail set. With steam, the *Aurora* was making nine knots, but it was "rather uncomfortable," as Stenhouse put it. "The old ship is beginning now to take seas on board & she occasionally stands on her head & threatens to somersault." With each passing day, the window of opportunity for laying the depots narrowed. "Worked out the programme for sledging, find we have a hard task in front of us," Mackintosh wrote in his diary on January 2. The warmest days of the fleeting Antarctic summer last from November until mid-February, when days of twenty-four-hour sunlight raise average temperatures above zero. After the solstice in December, the warm summer days begin to wane. As daylight fades in late February, the average temperature plummets below zero, and by late April, the sun dips below the horizon for four months, plunging the continent into winter. Mackintosh knew that survival would hang by a very slender thread in the bitter cold of March.

In the first week of 1915, Mackintosh's tactics paid dividends. "Ice blink," the steely glare of sea ice reflected on the clouds, appeared on the horizon. The first berg was sighted on January 3, towering higher than a twenty-story building. The next day, the *Aurora* entered the pack ice. Orcas breached alongside the ship as Mackintosh guided the

Aurora cautiously into the pack. Physicist Richards dangled on a rope over the stern to pole ice away from the propeller. As the ship neared the Antarctic Circle, squalls were supplanted by a steady westerly gale with furious winds and heavy confused seas that dispersed the ice and tossed the ship mercilessly. The seamen scrambled aloft to shorten sail on the pitching ship as they grappled with the ice-coated canvas, as stiff as sheet metal.

Off duty, six of the shore party crammed into crawl spaces flanking the rudder, sleeping side by side in slots as narrow as coffins. Even the officers' cabins were cramped, with four men assigned to a space six feet by nine feet and stacked with crates. The sailors in the forecastle slept in bunks crosswise to the hull, which seesawed the occupant from a headstand to his feet with each roll. Mackintosh fared little better in the captain's cabin, where the temperature hovered just above freezing and seawater flooded under the door. The calls from Thomson on the voice pipe, summoning him to the bridge, became a welcome escape. The latrines—"heads" in sailors' lingo—opened over the side of the ship and exposed the unlucky user to icy geysers in a heavy swell.

Chained on deck in kennels and lashed by wind and heavy seas, the twenty-six dogs were wretched. "They look objects of abject pity, and look appealingly at us for consolation in their discomfort," Mackintosh wrote. "But we are not able to do anything." Joyce was troubled by the condition of the animals. One dog died and another had convulsions. Recognizing that the teams would make or break the sledging efforts, Joyce saw to it that they were unchained for exercise and cooked pots of hot seal meat mash to nurse the animals. The sheep were too far gone; two had died. Joyce gave them a last hot meal before they were slaughtered and hung in the rigging, the carcasses swinging macabrely overhead as the ship pitched.

When a respite came, it confirmed the Antarctic's reputation as a realm of extremes. The barometer rose steadily and the storm-tossed voyage became, in Mackintosh's words, "a yachting cruise in the Mediterranean in summertime." On January 7, the cragged peaks of the Admiralty Range appeared, their first glimpse of Antarctica, "standing out in prominence like an immense sentinel to the South; an untrodden piece of the Earth," Stenhouse marveled. "I think most of us felt the wanderlust magnet, which takes men into the Wild, more strongly after viewing this noble work of nature." The *Aurora* ghosted along, her company basking in the sun with a gramophone playing opera as they readied the sledging gear and sorted provisions on deck. Ski boots were found to be missing, so Joyce devised a makeshift boot from canvas stitched to a sole of braided rope called sennit and the shore party set to work sewing.

By the light of the midnight sun, the *Aurora* ran the curve of the coast, heading south into the Ross Sea. James Clark Ross was the first explorer to sail these waters. No navigator had ventured as far south when Ross braved the pack in 1841 with two naval warships, *Erebus* and *Terror*. As hitherto unknown lands appeared, the explorer etched the names of a legion of royalty, patrons, mentors, faithful lieutenants, and scientists on the chart. After a run of nearly five hundred miles, Ross's southern dash had been checked by a strange spectacle, sheer cliffs of ice rising two hundred feet from the sea. "We might with equal chance of success try to sail through the Cliffs of Dover, as penetrate such a mass," he wrote, awestruck. It was the face of a floating slab of ice the size of France he called the "Perpendicular Barrier of Ice," a vast vertical wall stretching as far as the eye could perceive, which would later be named the Ross Ice Shelf. Ross's ships skimmed east along the edge of the Barrier, hoping to find a way south. They sighted an island held fast in the margin of the Barrier, surmounted by four lofty peaks. Convinced, rightly, that he could sail no farther south, Ross steered north.

On January 9, the snow-clad mountains of Ross Island loomed over the starboard bow of the *Aurora*. They told of the turbulent geological origins of the landscape. The quartet of volcanic peaks—Mount Bird, Mount Terra Nova, Mount Erebus, and Mount Terror—had risen from the sea four million years before and fused to form the island when the rivers of lava flowed together. Only Mount Erebus still fumed gently, the southernmost active volcano on Earth. The foothills of Mount Bird and Mount Erebus bounded the eastern shore of McMurdo Sound, with one long volcanic tendril snaking south to form the Hut Point Peninsula. The steep coasts were blanketed with glaciers, barring passage to the interior. Only lowlands on the fringes of the peninsula and capes offered a foothold to explorers. On the east side of the island, the lower reaches of Mount Terror fell steeply into the Ross Sea.

It was here, at a rocky promontory called Cape Crozier, that Mackintosh prepared for a landing. Belatedly, Shackleton had decided to add emperor penguin studies to the Ross Sea party's program. He instructed Mackintosh to land a prefabricated hut and supplies at Cape Crozier on the way to McMurdo Sound. Later in the winter, biologist John Lachlan Cope and two others would journey overland from the main base to live in the hut and study the penguins that, it was hoped, would provide some clues to the evolutionary origins of birds. Spencer-Smith and Stenhouse had volunteered. Ominously, a member of a previous expedition to Cape Crozier, Apsley Cherry-Garrard, called his own treacherous excursion "the worst journey in the world."

Led by Stenhouse, a party rowed to shore with the cargo. They searched in vain for a route up the sheer black cliffs, hundreds of feet high, which might lead to the penguin rookery on the ice shelf. Adjoining the land, the perpendicular face of the Barrier rose some sixty feet high. Admitting defeat, they rowed back to the *Aurora* in a dense fog. In their absence, the *Aurora* collided with the Barrier in the mist, snapping the jib boom as tons of ice crashed onto the deck. "This was regrettable, but the responsibility lay on my shoulders," Mackintosh candidly admitted, though none of the damage was irreparable. He declared landing impossible and headed west.

"We must bustle now, and have everything prepared for sledging to commence, as soon as we can get an opportunity," Mackintosh wrote as the *Aurora* rounded Cape Bird and neared the entrance of McMurdo Sound on January 10. Here, the Ross Sea channels into an open bay, forty-one miles across at its widest point and forty-nine miles long, bounded to the west by coastal mountains and to the east by Ross Island. At the southernmost end of the sound, the sea met the Ross Ice Shelf. Dotted along the western shore of Ross Island were the abandoned bases of the earlier pioneering British expeditions: the Cape Royds hut, Shackleton's former base, and the *Discovery* and Cape Evans huts, built by Scott. McMurdo Sound was the ideal staging area for inland forays. In the summer season, the sound was usually navigable, allowing ships to unload supplies not far from the edge of the Barrier. The explorers could build their wooden huts on the solid ground of Ross Island rather than risk a precarious existence on the ice shelf, with the constant threat that the ice could collapse under their feet if the shelf broke away. From the huts it was a relatively short march to the Barrier. Unlike the imposing cliffs at Cape Crozier, the face of the ice shelf dipped lower at the head of the sound and the frozen sea formed an upward-sloping ramp.

Ice conditions varied from year to year, however, and on January 10, McMurdo Sound was crammed with pack ice as far north as Cape Bird. Navigating the pack was an art, requiring a combination of skill and intuition. The challenge was sensing weakness and rushing into the breach at the right moment. If Mackintosh gained the upper hand, he could use the ship as a battering ram, riding up on the floes and crushing them. Somewhere in the maze of ice lay the "secret seas"—polynyas in the argot of modern science: great expanses of ice-free water created by winds and localized warmer currents—if only he could find them. If he steamed ahead too boldly, there was the danger of smashing the propeller or rudder, rendering the ship helpless; if he hesitated, the ship could be trapped and crushed. On the bridge, the officers flinched at each glancing blow, hour after hour, as the ship pushed deeper into the ice. "Down below the noise is deafening," Paton wrote, unnerved by the battering of the ice on the hull. "Aloft, our masts and spars remind one of some tall stately tree as it sways to and fro." Icemanship, as one of Mackintosh's former superior officers asserted, was above all a game of nerve and patience.

In the lulls between assaults when the ship was stalled by the pack, the men piled onto the floe for impromptu skiing lessons. Joyce went hunting and presented the cook with penguins for dinner. He was a firm believer in eating fresh meat to stave off scurvy. John Lachlan Cope, as medical officer, wholeheartedly agreed. "Scurvy is not due to any germ or bacteria and cannot be cured or prevented by tinned or dried foods. There is in the blood of man (and as far as is known at present in some other mammals) a substance known generally as a 'vitamine.' This substance is found *in* the blood but its origin is quite unknown," declared Cope, echoing the current ideas of Casimir Funk. They were right: undercooked seal and penguin meat is a rich source of vitamin C. The problem was the taste. One polar explorer described the flavor of penguin as a mélange of "beef, an odoriferous cod fish and a canvas-backed duck, roasted in a pot, with blood and cod-liver oil

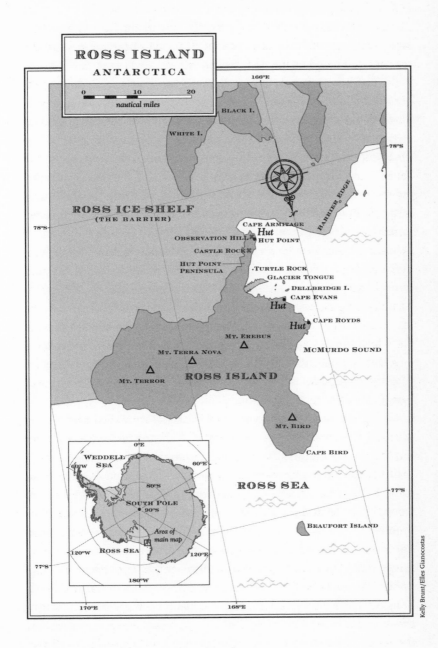

ROSS ISLAND
ANTARCTICA

0   10   20
nautical miles

166°E

BLACK I.

WHITE I.

78°S

ROSS ICE SHELF
(THE BARRIER)

78°S

CAPE ARMITAGE
Hut
OBSERVATION HILL HUT POINT
CASTLE ROCK

BARRIER EDGE

HUT POINT
PENINSULA

TURTLE ROCK
GLACIER TONGUE
DELLBRIDGE I.
CAPE EVANS

Hut

MT. EREBUS
MT. TERRA NOVA

MT. TERROR

Hut CAPE ROYDS

McMURDO SOUND

ROSS ISLAND

MT. BIRD

CAPE BIRD

ROSS SEA

77°S

WEDDELL
SEA
60°W

0°E

60°E

80°S

SOUTH POLE
90°S

Area of
main map

120°W ROSS SEA

120°E

BEAUFORT ISLAND

77°S

180°W

170°E

168°E

Kelly Brunt/Elles Gianocostas

for sauce," not an easy sell to men raised on mutton and Yorkshire pudding. Jack diplomatically claimed it tasted like steak.

For four days, the *Aurora* battled the ice, gaining a few hundred yards' southing one day, only to be pushed miles north of Ross Island the next. Encircled by the pack, ice-free waters always seemed frustratingly in sight. "Open water haunts us in all directions," Mackintosh observed. "It's like the position of a dog within a chain's length from his tin of water just outside his reach." Then, in the early morning hours of January 14, the ship broke the stalemate. The *Aurora* passed Cape Bird and reentered McMurdo Sound. Mount Erebus appeared, its lower flanks cascading into blue-tinted icefalls, then sloping moraine, pulverized into gravel by the glacial flows. Just offshore, the Dellbridge Islands—Big Razorback, Little Razorback, Tent, and Inaccessible—rose from the sea, the remains of an ancient volcano now mainly hidden beneath the sea. The *Aurora* cruised unopposed past Cape Royds, making seventeen miles of southing. The snowdrifts were peppered with hundreds of Adélie penguins. Mackintosh had decided to make their base at Cape Evans, about six miles southeast of Cape Royds. Their progress slowed, but by the sixteenth, the ship was in sight of the low black foreland of Cape Evans. They had reached the end of the voyage.

Named for Captain Scott's second-in-command, Edward "Teddy" Evans, the volcanic hills of the cape sheltered a hut built in 1911 as a base for Scott's bid for the South Pole. The shore party would spend most of their time either laying depots or at Cape Evans performing scientific work. Mackintosh and Joyce, with Alexander Stevens and Spencer-Smith in tow, skied to the hut to see if it was habitable.

One year before, Scott's expedition, still mourning the death of his party, had left the hut as though expecting to return momentarily. A stout oak table dominated the center of the high-eaved space, surrounded by bunks neatly stacked with clothing, some monogrammed. The names of the fallen marked the bunks. Secluded in one corner was Scott's own cubicle, lined with books. Beakers and test tubes crowded a laboratory bench. Shelves behind a cast-iron range were crammed with provisions and neat rows of crockery. Open crates spilled flour, sugar, and biscuits, as well as exotic delicacies such as canned gooseberries, liver pâté, chocolates, and smoked haddock. According to a detailed inventory left on the table by Scott's men, hundreds of pounds of foodstuffs were stowed there. Mackintosh assigned several men to haul surplus clothing and food aboard the ship to make up for shortages while the crew unloaded ten tons of coal and dozens of cases of fuel from the ship onto the gravel beach near the hut.

More supplies would be offloaded at Hut Point, thirteen miles south. Mackintosh was determined to deposit the gear and food for the sledging parties as far south as possible to shorten the upcoming journey. He hoped to use Scott's *Discovery* hut, built in 1902 and chosen by subsequent expeditions as a starting point for sledging journeys. Mackintosh sent Stenhouse and Joyce to the hut to report on its condition as he waited for an opening to push the ship through the pack.

On January 18, Stenhouse and Joyce disembarked on the sea ice in high spirits with a sledge and four volunteers: Stevens, Gaze, Hayward, and Wild. The coast between Cape Evans and Hut Point was a tumult of crevassed icefalls, making it impassable to men dragging heavily laden sledges. Previous expeditions had trekked parallel to the coast on the sea ice, though this route was not without risk. In the warmth of summer, the ice of the sound deteriorated into a patchwork of rotting ice and open water. The journey took twelve hours as they picked out a circuitous route to avoid cracks on the ice. Near the hut, Gaze plunged up to his neck in the frigid water. His companions dragged him out and rushed him to the shelter. The door had been blown open and two-thirds of the interior was filled with drifted snow. Strewn with crates and debris, the unfurnished hut was unsuitable for habitation, but Stenhouse and Joyce decided it would be adequate for temporary shelter and storage. A pile of old clothes and a bottle of crème de menthe surfaced for the shivering Gaze. Further rummaging only produced two sleeping bags, so the men took turns sleeping. When they emerged in the morning, open water surrounded Hut Point and a blizzard was gathering force. It was not until January 21 that the weather slackened and allowed them to travel. The crossing was treacherous and both Joyce and Stenhouse broke through the ice to their waists.

In their absence, the *Aurora* had only made a few miles of headway south. Mackintosh skied out with Jack to meet them. He was, Joyce recorded, "somewhat perturbed at our late arrival," and it was not entirely anxiety for their welfare. Mackintosh had decided to commence depot-laying operations immediately.

Joyce was taken aback. In his view, it was far too soon for sledging. Still penned on the ship, the dogs were in poor physical condition after the voyage. They had suffered intensely with constant exposure to the elements. Two were seriously ill, and many still suffered from parasites and unhealed bite wounds.

Purchased from the Hudson's Bay Company, the dogs were not huskies but mixed breeds, and few showed any trace of husky lineage. Newfoundland, collie, Labrador retriever, English setter, Scottish deerhound, and lurcher were all in evidence. Though not truly adapted to the polar environment, other breeds had lived and worked effectively as sledge dogs. The Hudson's Bay Company traditionally used mongrels for sledging, but they were well-conditioned dogs running shorter distances. Acclimatization was even more critical for the non-Northern breeds, especially for endurance in extreme conditions. It had been nearly a year since the dogs had pulled sledges in winter, and Joyce believed that they needed weeks of good nutrition, gradual exercise, and team training to become fit.

Joyce ushered Mackintosh aside for a private word, telling the captain that he had intended to start the dogs out with light loads and short days before beginning the depot-laying in earnest. "It will take some time before any of them will feel fit for hard pulling," he advised, mindful of his own hard-won experience with the dogs on *Nimrod.* Nor was the dogs' health and training the only issue; their novice masters, too, were untried. Dogs and men needed rigorous training together in the field to perform well. Joyce knew that, even matched with an experienced driver, a team required time to adapt to the personalities and voice of a new master.

Mackintosh rejected Joyce's argument and refused to alter his timetable. They were three weeks behind schedule, and the lives of Shackleton and his party were at stake. According to Shackleton's instructions, a party of six was to build the lifeline of food and fuel depots. From the opposite side of the continent, Shackleton's party would journey about seven hundred miles to reach the South Pole. After crossing the Pole, they would still have to cover about another 350 miles to reach the terminus of the Beardmore Glacier. By that time, their supplies would likely be exhausted, and their lives dependent on picking up the first of the Ross Sea party's depots as they trekked to the coast.

Mackintosh plotted a line of depots along longitude 169° east, about sixty nautical miles apart—at roughly 79°, 80°, 81°, 82°, and 83° south latitude—ending with a final depot at Mount Hope, a striking landmark at the foot of the Beardmore, about thirty miles beyond the 83° depot. The route was known, as it followed in the footsteps of Shackleton and Scott, but the accounts of those journeys made the dangers clear.

At the time of his departure, Shackleton had not yet decided whether to attempt the crossing immediately upon reaching Antarctica or late in 1915, but his instructions ordered Mackintosh to have the southern-most depot standing ready as soon as possible. During the *Aurora*'s month-long voyage south, Mackintosh had reluctantly surrendered his aim of stringing the depots all the way to the Beardmore before the on-set of winter in March. Mackintosh realized that the Ross Sea party could not possibly shuttle thousands of pounds of supplies and return to base in two months, so he resigned himself to setting out just the first two depots for the first season. The last cache would be about 150 miles south of the coast, well short of where Shackleton counted on finding it. Mackintosh hoped that either Shackleton's supplies would last longer than expected, or that the transcontinental party would de-lay their start until the following summer, in November 1916.

To Joyce, these calculations were moot if the dogs were incapable of pulling the loads. He objected strenuously to Mackintosh's plan to rush the dogs into sledging immediately. Joyce "could not get him to see that we were jeopardizing the dogs," and resented the interfer-ence in his bailiwick, insisting that Shackleton had placed him in charge of stores, sledges, and dogs. In response, Mackintosh produced Shackleton's letter of instructions, which clearly gave him authority over the entire expedition. Shackleton had given Mackintosh a choice: to remain on the ship and select a man to head up sledging operations, or to lead the shore party himself. Mackintosh told Joyce that he intended to oversee sledging personally, designating Stenhouse as captain of the ship in his absence and second-in-command of the expedition. If any-thing happened to Mackintosh, Stenhouse would be in overall com-mand. Joyce would have no autonomy in the field.

Joyce was stunned. Although Shackleton had written to Joyce that "the shore party for the Ross Sea would be in command of an officer, to whom you would be responsible," he had not expected Mackintosh's inflexibility. Shackleton's leadership style was far more inclusive. On the *Nimrod* expedition, Shackleton had sought advice on important issues and could be swayed by a sensible argument, regardless of a man's rank. In fact, Joyce had played a key role in Shackleton's momentous decision to establish his base at Cape Royds in 1908. Mackintosh, it seemed, would be far more hierarchical, only seeking the advice of his officers. "If I had Shacks here I would make him see my way of arguing," Joyce wrote in frustration in his diary. The dogs were not the only point of contention. He believed that mooring the *Aurora* in the Antarctic instead of sending the ship north for the winter was, as he fumed, "the silliest damn rot that could have possibly occurred."

All that was left to Joyce was his power of persuasion, and Mackintosh was adamant. Joyce backed down. He was well drilled in naval discipline, and that meant carrying out his superior officer's orders. There would be no training of the dogs, no trial runs for the men. "Mack is my Boss & I must uphold him," he wrote, adding a subversive note, "until I find that he is not fit to carry out the hard tedious work that is in front of us." Mackintosh ordered Joyce to head out with Australians Jack and Gaze as an advance team to break trail.

Joyce joined the shore party in the hectic final preparations, sorting out gear, fixing sledge runners, and bagging the carefully weighed provisions. The team leaders drew lots for the pick of the dogs. "Joyce has been considered to have selected the best," Mackintosh recorded. "But I am not as sure that he has, as most of his dogs are fighters!" Joyce counted on having the last word. "Having one eye will play merry hell with him in extreme temperatures," he observed with satisfaction.

In the morning of January 24, Joyce made a last bid to win Mackintosh over and conceded defeat, writing, "As he will not take my advice about the dogs I must let him have his way." He donned sledging gear and joined Jack and Gaze on the ice. At the sound of Joyce's first call, sheer mayhem ensued among the nine dogs, which were more eager to settle scores than tow sledges. Finally, with the team disentangled and harnessed to the sledge, Joyce's party set out, flanked by an escort to see them off. "There was no show of emotion amongst any of us," said second mate Leslie Thomson, "but we all knew that this was the beginning of the most serious part of the expedition."

The Great Barrier

On the face of it, Mackintosh's aim for the season was straightforward: to construct the first two of the six depots, one at latitude 79° south near Minna Bluff and the second at 80° south. But his plan was complicated and left little margin for error. Joyce's party would make two trips to build the Bluff Depot, a journey of about 150 miles round trip, leaving 260 pounds of supplies each time. Mackintosh's team would haul a single load of 240 pounds to the Bluff Depot, then continue on to establish the 80° South Depot, before turning for home. Mackintosh charged a third party with making at least one trip to the Bluff Depot with the motor tractor and depositing 540 pounds there.

The scheme hinged on each trip to the Bluff Depot taking sixteen days. If all went according to plan, all three teams would be back at Hut Point by mid-March, before the onset of winter. Mackintosh meticulously worked out each sledge load to the fraction of an ounce, allotting enough food and fuel for the journey and no more. If the daily mileage fell below ten, or days were lost due to blizzards, their only choice would be to break into the depot rations.

Most of the shore party ignored the intricacies and sorted the gear for sledging. "There was confusion about starting, and I for one lost grasp of the scheme as a whole," wrote Stevens. To make up for lost time, Mackintosh marshaled all available forces, enlisting twelve men instead of six. The scientists would be forced to postpone their research. Only the shortage of sledging equipment and clothing prevented him from drafting men from the ship's crew.

On January 25, the day after Joyce departed with Gaze and Jack, the

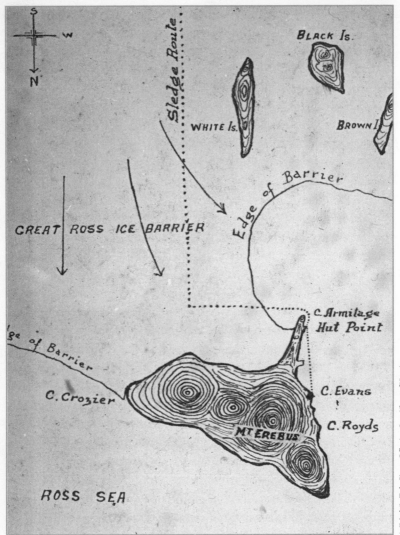

ROSS ISLAND AND THE SLEDGING ROUTE

second team, consisting of Mackintosh, Spencer-Smith, and Wild, left the ship. The third team, six men with Cope in charge, prepared to pull up the rear with the motor tractor: Ninnis, Hooke, Stevens, Richards, and Hayward. Sixteen crew remained aboard the *Aurora* to man the ship. Of the twelve sledgers, only Mackintosh and Joyce had Antarctic experience. For all of the shore party's enthusiasm and commitment, it was still unclear how they would work together—or, for that matter, how Mackintosh and Joyce would sort out their differences. Shackleton was known as a shrewd judge of character, but these men had been selected in haste, on different continents, and thrown together with no training and little chance for esprit de corps to develop. It would be, in Shackleton's words, a baptism of frost.

———

Joyce's party arrived at Hut Point after trekking six hours across the sea ice. The door of the shelter was blocked with drift, so they crawled through the window and camped for the night. By morning, snow eddied thickly around the hut. Gaze braved the storm to check on the dogs tethered outside, where he found Dasher dead and Hector's hind legs mauled, the aftermath of a fight for supremacy. It was a harsh reminder that the fiercely hierarchical animals must be staked far apart. Purchased piecemeal from multiple owners, most of the dogs had never sledged together in a team and feuded incessantly, a problem that Joyce had intended to overcome with training. Faced with this setback and a worsening blizzard, Joyce decided to regroup until the weather cleared.

His first order of business was "satisfying the inner man," fixing a meal of pemmican, sledging biscuits, and tea. "It went just A1," enthused Gaze. "The more you have of it the better you like it." Pemmican consisted of 60 percent lard mixed with crushed beef jerky. The biscuits were a kind of sailor's hardtack, a mixture of little more than flour, water, milk powder, and sodium bicarbonate. On the trail, the frozen pemmican would be heated and mashed with crushed biscuits and boiling water to form a one-pot meal called hoosh. Along with eight ounces of pemmican and a pound of biscuits, each man's daily ration consisted of five ounces of sugar, two ounces of cocoa and oatmeal, an ounce of tea and milk powder, and a pinch of salt.

They waited for two days, talking and reading and smoking. Joyce relished playing the part of the old hand with the younger men. Yarning about his adventures and advising them about "crevasses, frostbite and snow-blindness, strongly emphasizing the necessity of never travelling without first putting on their snow goggles," to avoid the excruciatingly painful disorder.

The two young Australians warmed to Joyce, impressing them with his "great capacity as a leader not only on account of his general personality & unselfishness but on account of the care & thought exercised in everything pertaining to our party." Jack and Gaze were unfazed by the crude manner that nettled others and appreciated Joyce's earthy good humor. The opinion was mutual. Forgiving the awkwardness of the two amateurs in their skiing debut, Joyce rated them as "splendid sledging mates and an excellent sample of Australians."

Irvine Owen Gaze was, like his cousin Arnold Spencer-Smith, a strapping sportsman. At six feet, four inches, they were the tallest men on the expedition, their stature disguised in the chaplain's case by a self-effacing stoop and accentuated in Gaze's by his unruly vertical hair. Spencer-Smith's mother was a Gaze. Her brother, Frederick Owen Gaze, had immigrated to Australia and settled in Melbourne with his wife, Constance. Their son, Irvine, attended Scotch College, a boys' religious boarding school near Perth, where he excelled at football, cricket, and tennis. Here the family resemblance ended. For a sportsman, Spencer-Smith was oddly accident-prone and uncoordinated, while Gaze moved with rangy grace. Spencer-Smith was the urbane Oxbridge man and Gaze, nicknamed Googs, was the proverbial colonial cousin, an outspoken lad with unpretentious tastes and unbridled enthusiasms. When his cousin told him the expedition was recruiting in Sydney, Gaze was eager to apply. "Living a real life," he called it. "Practically as close to nature as it's possible to get in these super-civilized times." His pitch for the job was short and forthright: "My general health is good & as far as I know I can stand the cold." His highest accolade was "A1," and he used it often, dealing with new experiences with easygoing good humor.

Gaze's longtime friend Keith Jack also took the physical challenges equably. Fair-haired with glacial blue eyes, Jack was twenty-nine, five years older than Gaze, and soft-spoken. The son of a successful Melbourne stationer and his wife, Jack worked for an insurance company for four years before attending Melbourne University on a chemistry scholarship. Awarded his bachelor's and master's degrees in science with first-class honors, Jack was teaching at a college and pursuing research in soil chemistry when he heard about the expedition. Jack's chemistry professor had been Orme Masson, a leading light in Australian science and a prime mover of Mawson's expedition. Masson and other mentors commended the serious young scientist to

Mackintosh. "He has excellent natural abilities, writes a good hand, and is of unimpeachable personal character in every respect," read one reference. "His parts are solid and will wear well." Yet Jack was rejected when Stevens learned of his boyhood bouts of whooping cough and a ruptured hernia. Jack was undeterred by the refusal, wiring back immediately, reply paid, "Would do any capacity assistant anything if you would find post." His persistence impressed Mackintosh, who hired Jack without further delay.

During a brief lull in the "almighty great blizzard," Joyce hiked south to survey the sledging route. Pulling the fully loaded sledges up and down the steep hills of the peninsula to reach the Barrier's edge several miles away was a daunting prospect, so he decided to journey across the sea ice, giving Cape Armitage a wide berth to avoid the fractured ice and open water near the shore. Once on the ice shelf, their course would mirror that of Shackleton and Scott before them, angling east for about twenty nautical miles before turning sharply south and heading about fifty miles toward Minna Bluff, the landmark for their first depot. Joyce had talked Mackintosh out of trying a shorter route directly between Black and White islands, which was a minefield of crevasses.

On January 27, the weather cleared and the three men loaded the twelve-foot wooden sledge with extra supplies from the hut, as instructed by Mackintosh, and hitched the eight dogs to their paired traces. Much to Joyce's annoyance, it weighed 1,079 pounds, 200 pounds more than the load he had agreed upon with Mackintosh. Contrary to expectations, the sledge coasted into a "ripping start down the slope from the Hut. All went beautifully for about a mile," marveled Gaze. "We were congratulating ourselves at getting through so easily." Then the sledge runners bogged down and the men sank up to their knees in slush. The sea ice was melting in the summer warmth. The harder the men heaved, the deeper the sledge sank. "It really broke our hearts," Gaze wrote. "For we'd pull and pull and then look round and find the dogs calmly looking on; how we cursed the swines."

After hours of toil, Joyce suggested they dump half the load, sledge forward for a while, then return to pick up the rest. Relay work meant slogging three miles to advance one mile, but it seemed the only choice. Gaze called it the most strenuous day he had ever experienced in his life, and Joyce heartily agreed. Stranded on the sea ice with the inert burden at day's end, there was no turning back. The three men pitched the tent and crawled stiffly into their sleeping bags, each with an ear out for the creaking of the shifting floes. In the morning, Joyce backtracked to warn Mackintosh's party of the troublesome ice.

Joyce learned that Mackintosh and his team had gotten off to an even less auspicious start. Mackintosh, Spencer-Smith, and Wild had departed the ship the day after Joyce's party. Within an hour the same "almighty" blizzard that had trapped the first party in the hut, engulfed the second on the sea ice. Mackintosh ordered a halt for the night, but when the storm had not abated by morning, he decided they could wait no longer. Sky and ground became indistinguishable, and the compass needle malfunctioned, making it impossible to stay on course. "We don't know where we are," Spencer-Smith wrote plaintively in his diary on January 26 as they became hopelessly lost. When the weather cleared, they saw that they had been steering in the wrong direction and were five miles out in McMurdo Sound. Arriving at the hut, they collapsed, exhausted, in their bags. A few hours later, they were awakened by Joyce's grunts as he heaved himself through the window. After hearing of their misadventure and telling Mackintosh about the unstable sea ice at Cape Armitage, Joyce headed back to join his mates.

"Joyce has just returned from Hut Point simply busting himself with laughter," Gaze wrote. "*The biggest joke on record.*" Still smarting from his clashes with Mackintosh, Joyce gloated over this evidence of the captain's inexperience. He found it incomprehensible that Mackintosh would sledge blindly in a blizzard on sea ice, violating a cardinal rule of polar travel. Joyce's companions sided with him, blaming Mackintosh for the problems with sledging. "The man who bought these dogs ought to be kicked round the town for a b— rotter, and the man who consented to take such a team down south ought to have his bumps read," declared the plain-talking Gaze. The Australians had already formed a negative opinion of Mackintosh, believing that the poor organization in Australia was his responsibility. Ridiculing Mackintosh cemented the trio's camaraderie. "We all have had a hearty laugh & every now & again one of us mentions the event & this sets us laughing again," Jack recounted. They set forth once more, determined to make headway in the slushy morass. Joyce was unconcerned that such carping would undermine Mackintosh's authority, and perhaps even the mission itself. For him, it had become a grudge match.

As Joyce's party turned in for the night, Mackintosh, Spencer-Smith, and Wild, anxious to overcome their ill-starred beginnings, turned out for breakfast. Mackintosh reasoned that colder night temperatures would crisp the snow surface and allow the sledges and skis to glide more easily. Although the sun circled around the sky in full view for twenty-four hours each day, the circuit was lopsided, sinking a few degrees lower toward the horizon at night, and it often made a discernible difference in the snow texture. During the night, Wild triumphantly noted that they "passed Joyce after a struggle."

He called the race prematurely. Mackintosh's sledge soon became mired in the icy porridge. His dogs either brawled into "a mass of rolling, struggling fur and fury" or sat down and refused to pull. He decided to relay, but the dogs crept only a half mile forward before balking, sensing the hesitancy of their novice masters. Mackintosh abandoned "gentle persuasion," as the chaplain put it, and began thrashing the dogs with a whip. "After much beating the poor beasts found stopping no use, so off they started," he noted. Wild and Spencer-Smith each donned a so-called man-harness, a canvas cummerbund with woven cotton braces over the shoulders, and pulled in tandem with the dogs. But their combined might, along with "the boot, the whip, words and blandishments," had little effect. Over the next eight hours, Mackintosh's team crawled a mere four hundred yards. By the next day, both parties were at a standstill, stalled by a fresh blizzard.

There was little they could do but lie in the tents, reading and writing in their diaries. Most of the men dutifully recorded daily events in official leather-bound journals embossed "Imperial Trans-Antarctic Expedition" in gold. These were not intended as personal memoirs, but as official records, to be collected by Shackleton for eventual use in his published accounts. The expedition members' agreement specified that all diaries, notes, scientific data, photographs, and motion pictures were the "absolute property of the commander" and barred them from writing or lecturing about the expedition without his express permission. However, the men soon let down their guards and confided more of themselves to their journals.

Wild noted the brutal extremities with a startling blandness. "Shot & skinned Captain," he recorded after the first ailing dog was put down. "Minus ate him. Later strangled himself on the way down." His curiously detached prose signaled not insensibility but imperturbability, a tolerant good nature untroubled by self-doubt or thwarted ambition. Friends and relations described him as a happy-go-lucky soul with a long fuse and few complaints with his lot in life. "Relay work no good makes one swear," he wrote during the soul-destroying efforts to push the sledge through the ice. After nineteen years in the Royal Navy, he could curse a blue streak, but ever the son of a Victorian schoolmaster, settled for an occasional decorous "b----y" or "h-ll" in his diary. His penchant for bursting into song moved Spencer-Smith to prophesy that he would "someday be Captain of [the] Salvation Army."

Spencer-Smith's own voice was one of busy optimism, jotting philosophically, "the work is cruel—but it's all in the game." His diary was a voluble record of aches and pains, philosophical ruminations, and sledging mileage. The hallmarks of his classical education at Cambridge were imprinted on every page, smatterings of Greek, Latin, and French and snippets of verse. He read avidly—popular romances, historical novels, poetry, and the Bible—though polar narratives were conspicuously absent from his portable library. There were theological tomes as well as treatises on Darwin in his bag, his Anglican faith coexisting comfortably with his intellectual curiosity. He captioned each entry with quotations from scripture to frame the day's meditations and reminders of birthdays and the anniversaries of his life in the church.

Other elliptical references suggested a rich and inscrutable interior life. There were hints of a private well of melancholy beneath his cheery fortitude, some hidden regret in his outwardly gadabout former life on the other side of the world. "A nightmare of Edinburgh last 'night.' Pereunt et imputantur!* It's curious how long the depression of such lasts." There was no trace of a crisis of faith, but the note of sadness flickered throughout his diary: "All the old questionings seem to come up for answer in this quiet place: but one is able to think more quietly than in civilisation." Of his soul-searching, he gave no further

* In Latin, "The days pass away and are set down to our account." Penned by the Roman poet Martial, it was an admonishment to live each day to the full and avoid regret.

clue. To his companions, he revealed a peculiar obsession, oddly mystical for an Anglican priest, that they shrugged off as a joke. He shared his birthday, March 17, with Captain Lawrence "Titus" Oates, the gallant cavalry officer who died on the doomed retreat from the South Pole with Scott. It was also said to be the day of Oates's death. Spencer-Smith was determined to reach the tent where the bodies of Scott and his companions had lain buried beneath the snow since 1912. According to Scott's diary, somewhere south of that camp, the dying Oates uttered his last words, "I am just going outside and may be some time," and crawled from the tent to die, presumably to unburden his companions. Something of Oates's heroic sacrifice struck a chord with Spencer-Smith, which echoed in his quotation from the Book of John, "Let us also go, that we may die with Him."

When the blizzard tailed off, Mackintosh ordered a fresh start. The new snow was as light as cornstarch, flummoxing their efforts as the sledge plowed into the drift. Bone tired, they unloaded the sledge and commenced relaying. Mackintosh despaired, confessing, "I must say I feel somewhat despondent as we are not getting on as well as I expected, nor do we find it as easy as one can gather from reading!" He was a mariner, first and foremost, and he struggled to come to grips with the work by applying the familiar regimen of naval discipline. Instead of sharing duties, he ordered Wild to care for the dogs and Spencer-Smith to handle "peggy duty," cooking all the meals. Mackintosh assumed responsibility for navigating and grappling with the obstreperous dogs on the march.

Neither in this department, nor in any other, would Mackintosh ask for Joyce's help. Not only did his training and experience preclude the notion of turning to a subordinate for guidance, but Joyce had challenged his command of the shore party and questioned his judgment. He could not appear to bow to Joyce's challenge and risk his credibility at the very start of the expedition. He remained aloof in an attempt to keep the reins of leadership firmly in hand. Mackintosh believed that outstripping Joyce each day was the best demonstration of his authority.

In practical terms, it was precisely the wrong approach. In the featureless landscape of the ice shelf, where landmarks were often obscured by haze, the dog team had no destination. Sledging dogs need a physical goal on the horizon to motivate them effectively. Faced with virgin ground and a crushing load, they rebelled. Joyce had been skiing ahead as forerunner to inspire his team, but now he cannily hung back, allowing his dogs to follow the scent of their rivals in the broken trail. Had Joyce cooperated with Mackintosh, they might have traded off the lead position to ensure that neither team became too worn out from breaking trail. But the standoff had spawned a competitive rivalry, with each man determined to prove he knew best.

Joyce's strategy paid dividends for his team. On January 30, their fourth day of sledging from Hut Point, his party arrived at the edge of the Barrier. Unlike the waves breaking on sheer cliffs at Cape Crozier, the sea seemed to have frozen in midswell at the head of McMurdo Sound, forming a gradual slope fifty feet up to the ice shelf. The overloaded sledges demanded Herculean effort in ideal conditions; even the moderate incline was hard going. Once at the top, they surveyed the Ross Ice Shelf for the first time. Stretching hundreds of miles to the south, it was a breathtaking sight, "a vast wall surrounding an immense snow plain, bediamonded by the sun," as Spencer-Smith described it when Mackintosh's party arrived. Cloud shadows played across the Barrier, like sun dappling immense leaves. To the east, the camelback profile of Ross Island reared up, the mountains wreathed in the volcanic plume of Mount Erebus. To the west, they saw the Transantarctic Mountains, stretching three thousand miles south along a great rift that divided the continent in two.

Only reluctantly did Joyce's party double back onto the shifting sea ice for the second load. The dogs liked it little better, sinking up to their bellies into slush. When Joyce stepped off his skis to assist the team, he plunged through thin ice. The fast current threatened to drag him under until Gaze hauled him out. They were forced to pitch the tent on the ice to warm Joyce. On January 31, a snowstorm pinned them in the tent. Joyce feared that the third party, following with the motor tractor, would blunder into the disintegrating ice. When it cleared, he sent Jack to caution them. Jack found the six men at the hut sleeping and woke them to report the danger.

————

Leaving Cope to decide his party's next move, Jack skied back to catch up with Joyce and Gaze, who were already under way. By midday on February 1, the party had hauled their last load up the incline to the

Barrier, taking three days to cover seven miles. The men were euphoric. The retreats over the sea ice were over. From that day onward, they would make their way south on the broad prairie of the ice shelf.

Mackintosh's team had preceded them, leaving jibes scrawled in the snow next to the trail: "Buck up. Ship will catch you up, you cripples," and, on a more encouraging note, "Pub Ahead." Gaze strained to see his cousin's camp and made out something odd in the distance. He skied over and found a dog tethered to a bamboo pole. Mackintosh, exasperated with his lead dog, had abandoned Jock with a heap of seal meat and a snow shelter, intending to pick him up on the return trip in a few weeks' time. Gaze hitched Jock in the traces and the dogs bounded forward into the chest-high snow. "The new dog Jock is working splendidly for us," wrote Jack, seeing how he invigorated the team. Mackintosh's party was still ahead and Gaze, for one, was in "no hurry to catch them up," seeing Joyce's point that the dogs ran "far better when they have a scent to follow." Soon the team was pulling a full load rather than relaying.

Farther south, they encountered a cluster of snow mounds which Joyce soon realized were the remnants of Scott's Safety Camp, stocked with cases of provisions and frozen seal meat. Nearby was a buried motor sledge, one of two defunct vehicles abandoned by Scott's men. Joyce found a new depot of seal meat with a message from Mackintosh, forbidding them to feed the meat to the dogs. The order was "b----y rot," in the words of Gaze, who "didn't take any notice of it and gave the dogs a real good feed; they deserved it too." Gaze and Jack had been disturbed to see Mackintosh "knocking hell out of [his dogs]" with a whip. Joyce only struck the dogs to break up life-threatening fights, knowing that routine whipping demoralizes more than spurs a team. Disillusioned, Gaze decided that Mackintosh "hadn't any damned idea of managing dogs; any more than he has of managing anything else."

Nonetheless, Mackintosh was the leader, and they had no choice but to follow. The sledge tracks ahead turned sharply east to detour around crevasses honeycombing the Barrier near White Island, as Shackleton had instructed. He knew from experience that geological formations created dangerous regions of disturbance in the ice shelf. The rifts were often concealed by a brittle mantle of drifted snow, too weak to bear the weight of a man. Though unseen, the colossal forces beneath the surface were audible. "No crevasses as yet, though there was one terrific 'boom' right under us at one stopping place," wrote Spencer-Smith, unnerved. Pulling up the rear, Joyce ordered his party to stay on ski to distribute their weight. They were far more likely to crash through a fragile snow bridge on foot.

The progress of Mackintosh's party was painfully slow. When it took eleven hours of grueling work to move forward just over a mile, Mackintosh decided to stop relaying and pull the full 1,200-pound load. The eight dogs were unable to move it. The dogs were now staggering under 150 pounds each, even though Joyce had advised that the maximum load for a healthy canine was eighty to one hundred pounds. The three men harnessed up alongside the animals, hindering almost as much as they helped by breaking the rhythm of the dogs' pace. Spencer-Smith found the work "cruelly hard and disappointing."

When Joyce's party caught up on February 2, Spencer-Smith watched them in amazement, "proceeding merrily" with Jock trotting smartly in the lead, hefting cargo of 1,100 pounds, and Jack and Gaze skiing alongside. "Of one thing I am sure," the chaplain decided. "The whip should be used as sparingly as possible." Mackintosh wagered a magnum of champagne that his team would reach the Bluff first.

The parties went their separate ways until 2:30 AM on February 4, when Mackintosh rousted Joyce from his sleeping bag to announce that the time had come to veer south for the Bluff, with luck, just a week's march. He borrowed Joyce's map, his own lost somewhere on the trail. For days, both parties had been on the lookout for Scott's Corner Camp to guide the swing south, but the old depot was nowhere to be seen. Even if they had Scott's maps, the depot would not have been in the same place—the ice shelf was constantly but imperceptibly moving toward the sea. Mackintosh was crestfallen. He had banked on finding Scott's old depots for navigation and to supplement the supplies. He decided to chart his own course south, certain that they had safely outflanked the crevassed region. Every mile and a half, they stopped to build a snow cairn. Like a trail of bread crumbs, the string of markers would signpost the route back to the coast.

The terrain was worsening. The ice shelf looked as if a "vast sea, slightly troubled, had frozen in an instant." Sastrugi, the Russians called the corrugated snowfields, formed by the wind sculpting the snow into undulating peaks and troughs. Up ahead, Mackintosh's party labored up to each crest and all too frequently slid back down on the hard, slippery surface. In two and a half hours, they gained just 150 yards.

Their progress was disheartening. The same distance had taken previous expeditions two or three days; Joyce's party had been toiling away for ten. At least the Bluff was finally coming into view southwest of their course, miraged by an atmospheric illusion to resemble a vast placid lake. The long rocky spur extending from Mount Discovery in the Transantarctic range was a familiar sight for Joyce. He had used it as a navigational landmark when depoting the sumptuous cache that had saved Shackleton's party in 1909. The new depot would be on the same latitude but twenty miles farther east of the Bluff to avoid the turbulent crevassing around it.

As they pushed south, the temperature was perceptibly dropping, reaching lows of minus fifteen at night after daytime highs of fifteen degrees. The sledge runners slid easily on the hard crust. The conditions were good for traveling, but miserable for the men. The sledging costume had worked tolerably well before the dramatic temperature fluctuations. The wool base layer included long underwear, "pyjama suit," heavy sweater, socks, cap, and mittens, topped with a close-woven cotton Burberry outfit, balaclava, fur mitts, and the improvised ski boots. The ensemble failed on frigid, sunny days when the water-resistant Burberry suit trapped body heat during the exertions of sledging, causing them to perspire heavily. At night, the damp garments soaked the sleeping bags. "None of us get much rest in these sleeping bags, they're rotters," complained Gaze, whose bag was also too short for his long-limbed frame. Even the simplest actions could have excruciating consequences. "It was quite painful once when I opened my mouth rather wide to shout," wrote Spencer-Smith. "Every hair on my face seemed to be tearing out at the roots!" They were all plagued by a multitude of physical woes: ice-encrusted beards, frostnipped toes, blistered feet, and weeping cuts.

The dogs suffered as well. Many of the animals were sick and losing weight after being fed biscuits left at Hut Point in 1902. Fights had left them with an assortment of bite wounds, some infected. Hector, the victim of several skirmishes, was so badly injured that Joyce reluctantly decided to leave him behind with food, intending to pick him up on the return trip.

Mackintosh's dogs fared the worst. Raw, festering patches from the harness and the whip marked their scrawny bodies. Canuck finally revolted, allowing himself to be dragged by the neck rather than pull, until the men gave in and put him on the sledge. Mackintosh blamed the problems on the animals, unaware that his regimen was running them into the ground. He sent Spencer-Smith to reclaim Jock to shore up his lagging team. Joyce felt bound to follow orders and surrendered the dog. Gaze was livid, writing that if he and Jack had their way, "that blessed dog would *not* have left our camp—he received decent treatment here, he'll get banged there." Joyce's party regularly surpassed Mackintosh's best mileage, often exceeding ten miles per day. Reduced to seven dogs, Joyce's team would be hard pressed.

On February 6, Wild visited Joyce's camp and told them that Mackintosh had talked about taking Joyce's best dogs and heading to 80° south. Joyce's party was incensed. "Mackintosh would appear to have a faculty for making plans & just as quickly unmaking them," Jack commented. "Seems foolish to break up the teams just as we have got them into working order." "Let him try," Gaze fumed. "I didn't intend again to refer to the man, but he's such a bloody ass that you can't help it." Gaze had no say in the matter. Mackintosh's team was falling apart. The sickly Canuck had run away and was presumed dead. Jock, who had been pulling so well for Joyce, succumbed after five days in the harness with Mackintosh's team. "We woke to find poor old Jock dead in the snow," Spencer-Smith wrote on February 9. "He ate his supper last night, though exhausted at the end of the march. He must have died quietly in his sleep, please God, painlessly. I put up a small cairn over his body."

On February 11, Joyce and company broke camp early. Mackintosh's party was far behind. After two hours of marching, Joyce called a halt and announced that they had reached the end of the road. Mist obscured the heights of Minna Bluff, making it impossible to be certain that they had arrived at the bearings specified by Shackleton, but Joyce felt sure they were close. The three men piled four weeks' worth of stores and built a twelve-foot snow cairn around it, surmounted by flagged bamboo poles.

As the party settled in for breakfast, Mackintosh arrived and announced his new plan to forge ahead with Joyce and Wild and lay the depot at 80° south, taking nine dogs. Joyce was adamantly opposed to the scheme. "I had him one side and tried to persuade him not to take the dogs any further south, as they were feeling the effects of the hard sledging," he recorded, his point underscored by the death of another of Mackintosh's dogs, Bobs, that very day. "However, he decided otherwise." Spencer-Smith quietly erected a makeshift headstone for Bobs. Joyce toed the line, noting curtly, "I quite see his point that the depôt must be laid at 80° as Shackleton may come across this year & expect to find food there."

Mackintosh sent the other three men, with Spencer-Smith in charge, back to Hut Point with a week's rations and four dogs. Spencer-Smith, Gaze, and Jack harnessed Duke, Gunboat, Towser, and K.K. and steered north. Joyce's former teammates were confused and "keenly disappointed," according to Jack.

> Under the capable leadership of Joyce we have had a most delightful trip & were looking forward to a quick return to Hut Point & another long journey together with sledges to lay the [80°] Depôt. Wonder what on earth has made Mackintosh change his plans this time.

"Plans are so easy to make, but quite a different matter to carrying them out," Mackintosh reflected in his diary, underestimating the blow to morale of the reshuffle. Gaze called the move "a rotten slap in the face." Jack steeled himself to "trust it is best & make the best of the new plans."

Spencer-Smith watched as Mackintosh's party receded into the distance. He would not have the chance to look for Scott's resting place until the next season. On the homeward trail, the inescapable misery of the dogs deepened his gloom. "We passed Jock's grave this morning: the blizzard had defaced the name but had uncovered one pathetic hind paw. R.I.P.," he wrote. The surviving dogs were too weak for even the lightened sledge, refusing to pull and lying down in the snow to sleep. The men unhitched Duke and K.K., who disappeared to meet the fate of Canuck, whose rigid body they found beside the tracks.

The outlook was grim for the men as well. A week after the morose parting at the Bluff Depot, they were dangerously low on food, and the fuel tins were leaking. Without kerosene it would be impossible to thaw the frozen blocks of pemmican or melt snow for water. It was growing much colder. Garments froze stiff seconds after being removed. "Wonder what 30 and 40 below will feel like—don't like to dwell on it," wrote Gaze. After Spencer-Smith's gold tooth snapped off at the root, he found consolation in Robert Browning:

Ah, but a man's reach should exceed his grasp
Or what's heaven for?

Not a moment too soon, the trio crossed paths with Cope's motor party on February 18, who shared rations and the story of their travails since leaving the ship.

———

On January 31, the third sledging party had marched across the sea ice behind the motor tractor to a chorus of cheers. Cope ran the vehicle with Ninnis as mechanic and Hooke assisting. Stevens, along with Hayward and Richards, pulled another sledge by man-hauling. The tractor aroused high expectations, promising to carry heavier loads than a dog team at a faster clip. All of the earlier vehicles tried in the Antarctic had proved ill-suited to deep snow and extreme cold. The Ross Sea party's tractor was a new experimental design, featuring a wooden body, two front strut-mounted skis, and a rear metal paddle-wheel, powered by a nine-horsepower, two-cylinder engine.

The motor tractor ran like a dream in its debut trial on the sea ice, chugging along at eight miles per hour. The addition of cargo without further testing proved its undoing. The tractor ground to a crawl with a light load in tow and bogged down in the snow with a host of mechanical problems, from wet spark plugs to a slipping clutch. A member of the crew observing the spectacle commented archly that the motor seemed most useful "as a convenience for keeping Ninnis' feet dry."

Cope decided to ditch the sledges and send Ninnis to Hut Point to find materials for repairs. With Hooke on the running board, refilling the leaking water tank with scoops of snow, the tractor broke down on arrival as vibration shook the engine apart. There were no tools or hardware in the hut. Ninnis and Hooke abandoned the vehicle and set off in a blizzard to join their party in man-hauling a 1,200-pound sledge. After sixteen grueling hours, Cope's team arrived at the hut on February 1 and collapsed in their sleeping bags, where Jack found them to warn of the unsafe ice.

The six men retreated to the ship to improvise parts. After furious tinkering back at Hut Point, Ninnis started the Rube Goldberg contrivance, which expired clamorously. He gave in. Like draft animals, Cope's party donned man-harnesses on February 7 and set out for Minna Bluff with two sledges weighing about a thousand pounds each.

Dragging a load averaging 300 to 335 pounds each, the man-haulers heaved forward against the crushing burden to little effect. They resorted to relaying the massive loads, wallowing in deep snow through triple the distance to gain a dispiriting two miles in eight hours of sledging. Tentmates Ninnis, Hayward, and Richards were unfazed and raring for heroic exploits. Ninnis and Hayward told stories about their experiences in the wilds of Canada. At twenty-one, the youngest of the trio was Richards, the affable physics teacher from the parched Australian gold rush town of Bendigo. Dubbed Richy, he had rarely even seen snow before, but excelled in soccer, boxing, running, rowing, and cricket. Fueled by banter, the Bandsmen, as they called themselves, marched forth, inventing doggerel to pass the time. The occupants of the other tent were Hooke, Stevens, and Cope. Hooke had not expected to join the sledging, but took it in stride. The weedy Stevens was in poor shape, lagging in the traces and nodding off to sleep whenever the sledge stopped. More worrisome, the formerly affable Cope had become remote and arbitrary under stress, behaving bizarrely.

Rising each morning at a progressively later hour, Cope began the day with a leisurely recitation from Dickens's *Pickwick Papers*. Once under way, Cope floundered ahead into deep virgin snow rather than follow the trail broken by the other parties and strayed into crevassed ground. He called frequent rest stops, ordering the men to pitch the tent for tea breaks, and halted earlier each day. Cope had the only watch in the party, and the men suspected he was prevaricating about the time to trick them into shorter days. The dithering and chicanery were costly; after the first week, they had only covered twenty miles, raising the possibility that they would run out of food and need to break into the depot rations.

The zealous Bandsmen champed at the bit, yet none were willing to defy Cope. Instead, they vented their discontent in their diaries and satirical songs. Ninnis called Cope's conduct "rank 'criminal' incompetency." "You cannot take a very young man out of a University town and give him charge of a party in the wilds, when he has never seen snow or travel," he complained in abject frustration. The Bandsmen were considering striking out on their own when the party encountered Spencer-Smith, Gaze, and Jack on February 18.

———

Jack was distressed to learn that Stevens and Hooke were too ill to eat, and he volunteered to stand in for one of them. After a month of toil, Jack was consigning himself to weeks on the trail when he had been home free for the season. The ailing Stevens took Jack up on his magnanimous offer. Cope shared rations and fuel to make up the other party's desperate shortfall, and Spencer-Smith, Gaze, and Stevens steered north for the hut.

The trio arrived at Hut Point on February 22 to find that they were the first to return. The sun had begun to dip below the horizon, a reminder of the approaching winter. Across McMurdo Sound, the sunset bathed the distant mountains in luminous shades of violet "like visions of Fairy-land," described Gaze. "I shall never forget this scene as long as I live." There was no sign of the *Aurora*. Broken sea ice churned in McMurdo Sound, streaked with wide leads of open water. Crossing to Cape Evans was impossible, yet the pack was clearly too dense for the ship. Gaze inventoried the provisions left by previous expeditions and found much of it spoiled by mold, leaving just four weeks of food for twelve men. "Altogether things (as usual) are going to be in a pretty mess unless Stenhouse comes along very shortly—meantime we're not worrying—what's the use," he wrote.

Mackintosh had ordered Spencer-Smith's team to return to the Bluff with an extra load if possible, but food was too short. The prospect of squatting in the drafty, uninsulated shelter for even a limited time was repellent. Their diet, at least, improved in variety. Coveted luxuries like chocolate and cigarettes surfaced. "By jove we *are* living well," Gaze crowed. "What Ho." There was little else to do but lie in their bags to keep warm and wait for a clear, calm evening to fire a rocket or light a bonfire as a signal to Stenhouse. "There's a great blizzard blowing," Gaze mused on February 25. "It will be jolly rotten for the fellows on the Barrier."

———

So it was, as they learned on March 2, when Ninnis, Richards, and Hooke arrived. Cope had sent them back, on the advice of Jack. "Our good spirit," as they called Jack, had injected new life into the party with his calm rationality. He analyzed the friction of loads and rigged a sail with tent canvas, which dramatically improved their daily mileage. Still, Jack concluded that they were losing the battle. "The load this party is hauling seems to me very absurd & far too heavy for a team of untrained, inexperienced men," he wrote. He convinced Cope to lighten the load by making a partial depot on the spot and reducing the party's number to cut the food consumption. The energies of Ninnis and Hooke, he believed, would be better spent getting the motor tractor and wireless operational before winter, so he dispatched them to Hut Point with Richards on February 24. Jack, Cope, and Hayward continued struggling south toward the Bluff Depot.

Ninnis, Hooke, and Richards arrived at the hut to find Spencer-Smith, Gaze, and Stevens in a quandary. It was plain that few of Mackintosh's original assumptions held. The round-trip journey to the Bluff Depot was planned for sixteen days; Spencer-Smith counted twenty-six. The average daily distance was about five miles, not ten as projected. Inevitably, the six men on the Barrier would run out of food. Ninnis argued for sledging out to meet them with supplies. Hesitant, Spencer-Smith searched for solace in the Bible and found his doubt reflected in the pages: "By what authority doest thou these things?" he quoted. After twilight on March 7, he decided to fire their only rocket. Darkness quenched the trailing sparks as the men watched for a reply signal from Stenhouse that never appeared.

An impatient Ninnis walked five miles north on the Hut Point Peninsula with Richards in search of a lookout point. They scrambled up a promontory called Castle Rock, rising over a thousand feet, and looked south over the Barrier. The sledging parties were nowhere in sight. To the northwest, they spied the mainmast and crow's nest of the *Aurora* spiking above the low hills of Cape Evans. Emboldened, Ninnis and Richards pushed on, ascending into the lower slopes of Mount Erebus to search for a land route to Cape Evans. As they had been warned, the coast was an impenetrable snarl of icefalls and the pair soon stumbled into crevasses. They turned back, finally convinced that Cape Evans was unreachable except across the sea ice of McMurdo Sound.

On the morning of March 11, the Barrier parties were still missing when Stevens sighted the ghostly specter of the *Aurora*, spars and rigging sheathed in ice, cruising silently toward Hut Point. Thomson and Larkman rowed ashore and reported that the *Aurora* had "had a devil of a time" riding out fierce weather in their absence. Thomson hustled the shore party back to the ship. Taking no chances, Stenhouse steamed north without delay. The six men on the Barrier were on their own.

– 6 –

Eighty Degrees South

One degree farther, Mackintosh promised, and their labors would be finished for the season. One degree of latitude meant sixty nautical miles of untrammeled snow and ice, a journey of more than a week at their current rate of travel, a pace that would be impossible to sustain in the steadily worsening conditions. Winter was approaching, bringing colder temperatures and frequent blizzards.

Although Joyce had opposed taking the dogs to 80° south on February 11, the team was their best chance of finishing the job and returning to Hut Point before the weather deteriorated. The leader, Nigger, came from Joyce's original team, a massive, domineering animal prized as "champion of the pack." The strongest puller, Shacks, was a favorite for his compliant nature and "large half-open brown eyes, like an old man looking underneath his glasses." Pat and Scotty came from Joyce's team as well, the latter a vicious fighter. Tug and Briton were practiced fugitives, regularly chewing free of their harnesses. The rest of the dogs were smaller mongrels. Alternately indolent and aggressive, Pinkey appeared to be part greyhound. Pompey, who resembled a border collie, was a small, skittish animal unused to the harness. After repeated beatings, Mackintosh reported he was "pulling for all he is worth." Pompey's partner, Major, was an obstinate runt, "a curious character" with "fiery red eyes," who bared his teeth defiantly whenever he was whipped.

The sledge had lightened considerably since the best part of the load had been dumped at the Bluff Depot, yet the dogs still suffered. The soft snow surface had hardened to a brittle, glassy crust, cutting their tender paws as they punched through and blotting the snow with blood. The hair tufts between their toes gathered clods of sharp ice that cut the soft flesh. Some of the dogs scrabbled frantically at their eyes with their paws and whined piteously as ice crystals clinging to their lashes scraped with each blink. At night, they attempted to paw their way into the tents for the warmth seeping through the canvas. Some of the problems were unwittingly aggravated by the men. An attentive handler would have trimmed the paw tufts and eyelashes.

More seriously, the Hudson's Bay dogs lacked the traits that suited huskies so well to the polar regions: obliquely set eyes to minimize abrasion and glare, an insulating double-layered coat, bushy tail, well-furred ears, and tough paws. Thus protected, huskies sleep outside, curled into tight balls in the worst blizzards without the risk of hypothermia that threatens other breeds.

Weather conditions were worsening. The temperature was dropping markedly with the advancing season, hovering around zero degrees. To conserve fuel, the Primus stove could only be used long enough to prepare meals. The damp was inescapable. Their clothing never completely dried and the wet garments rendered them far more vulnerable to hypothermia and frostbite. In the cold night air, the steam from cooking froze on the tent walls. An incautious jostle showered frost down on their heads. Perspiration and breath froze, forming an icy rime that thawed from body heat and soaked their woolens and fur sleeping bags. They tried vainly to dry the wet socks and gloves by tucking them next to their skin. Mornings became a grievous trial as they struggled to don their stiff clothes and boots. "We call it 'our hour of discontent' for getting into these frozen boards takes anything from a quarter to half an hour of hard struggle and pain," wrote Mackintosh.

Wild seemed somehow impervious to the miserable conditions. "Wild slept like a top," marveled Mackintosh. "He is a remarkable little fellow, always merry and bright, as soon as he lays down he starts snoring." A fraction over five feet tall, Wild was powerful and sinewy for his height, the result of decades of hard labor on naval ships. His bantam physique earned him the nickname Tubby and his attitude won Mackintosh's approbation. "Wild, ever jolly, cheerful, an optimist, keen and ever ready to take up anything, very humorous with a large vocabulary of naval expressions, fairly tough, and plucky as any Britisher."

Since their disagreement over the dogs, Mackintosh had been wary of Joyce. "Joyce, a different character—quite alright while humoured, when he is willing, and would do anything for any one, but he has no stability; alright while all goes well, not very hard, feels the cold very easily; but he always sticks it out." There was more than a grudge in Mackintosh's assessment. As another of the men put it, Joyce was "somewhat bombastic," and his pride needed careful grooming to keep him on an even keel. He played the hardened veteran to the hilt, but the truth was that he was not in ideal condition for the arduous work of sledging. In 1901, Joyce was known as "an all around athlete," though he was stocky, just over five feet, four inches tall and weighing in at 160 pounds. Fourteen years later he was seriously overweight, tipping the scales at 197 pounds.

Nonetheless, Joyce was tough, and he had a reputation for stoic endurance. His practical knowledge and experience gave him confidence that he could extricate himself unscathed from the most desperate situations. He seemed, in fact, to relish the ordeal, imagining himself the rough-hewn hero of the kind lionized by his favorite poet, Robert Service. Called "the Kipling of Canada," Service was inspired, like Jack London, by the roustabout life of the Yukon wilderness. The straightforward ethic appealed to Joyce, who identified with Service's heroes and quoted him as gospel:

> But the Code of a Man says: "Fight all you can,"
> And self-dissolution is barred.
> In hunger and woe, it's easy to blow—
> It's the hell-served-for-breakfast that's hard.

Unlike Joyce, Mackintosh was disheartened, a fact he studiously concealed from the other men. "What on earth am I doing here?" he wrote on his third wedding anniversary. "That's what I ask myself and such thoughts wish me back at Home to the dear ones, waiting so patiently." He had been elated when Shackleton summoned him to command the *Aurora*, without foreseeing how keenly he would miss his family once he sailed. In November, news of the birth of his second daughter had come in the midst of his travails. Now, after months of uphill battles, he was nagged by doubt and uncertainty. "What a weird situation when you come to think of it—what on earth are we so keen about?" he wondered in his diary after tossing, sleepless, in his bag at night. Eight thousand miles from home, hobbled and harried at every turn, his thoughts had the ring of regret. "In waking, you build castles, go through your whole life—past, present and future."

———

The party crossed 80° south on February 20 and celebrated with a tot of brandy, drinking to "sweethearts and wives," the traditional sailors' toast, followed inevitably by "may they never meet." Men and dogs were spent. "Excitement reigned supreme (I don't think)," Wild recorded laconically. The final twenty miles had taken twenty-six hours as they tackled chiseled ridges of sastrugi four feet high. They dumped their final load at 80°02' S, 169°25' E and christened the last cache of the season the Rocky Mountain Depot after the local terrain. Planned to contain 220 pounds, it held 135 pounds of food and fuel. They had kept a sharp lookout for Scott's One Ton Depot for days and found no trace of the trove of provisions. Neither was there a sign of Scott's last camp, where the bodies were entombed beneath the drift.

The following day, Mackintosh set up a theodolite, a sort of calibrated telescope, to record the depot's position. At their farthest south depot, Mount Hope, they would deposit maps with notes and bearings so Shackleton could locate the depots. Using the theodolite in freezing temperatures was maddeningly difficult. Mackintosh's whiskers froze to the cold metal and the lightest breath clouded the sight with a web of frost. In the end, the accuracy was doubtful; the clamp of the scope refused to hold level, the metal parts contracting in the cold at different rates so they no longer fitted together properly. In the meantime, Joyce and Wild built snow mounds every half mile to guide Shackleton to the depot, surmounting every other cairn with a flagged bamboo pole. The line of cairns would extend five miles to the east and west, crosswise to the route. If Shackleton strayed off course, even in a storm, he would likely to stumble across the transverse chain of signposts stretching ten miles across his path.

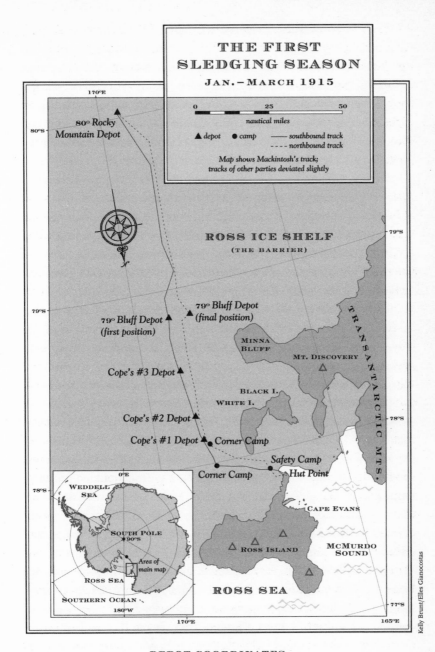

THE FIRST SLEDGING SEASON
JAN.–MARCH 1915

0	25	50

nautical miles

▲ *depot* ● *camp* —— *southbound track*
- - - - *northbound track*

*Map shows Mackintosh's track;
tracks of other parties deviated slightly*

170°E

80°S

80° Rocky
Mountain Depot ▲

ROSS ICE SHELF
(THE BARRIER)

79°S

79°S

79° Bluff Depot ▲
(first position)

79° Bluff Depot ▲
(final position)

MINNA
BLUFF

MT. DISCOVERY △

TRANSANTARCTIC MTS.

Cope's #3 Depot ▲

BLACK I.
WHITE I.

78°S

Cope's #2 Depot ▲

Cope's #1 Depot ▲ ● Corner Camp

Safety Camp ●
Corner Camp ● Hut Point

CAPE EVANS

0°E

WEDDELL
SEA

SOUTH POLE
✱90°S

Area of
main map

ROSS SEA

SOUTHERN OCEAN

180°W

78°S

△ ROSS ISLAND △

ROSS ISLAND

McMURDO
SOUND

ROSS SEA

77°S

170°E

165°E

Kelly Brunt/Elles Gianocostas

DEPOT COORDINATES

79° Bluff Depot (final position): 78°52' S 169°05' E
80° Rocky Mountain Depot: 80°02' S 169°25' E

Five miles out from the depot, the party labored over the eastern-most cairn, a grand construction fifteen feet high with notes on the bearings and course for the next depot in a tin tied to a pole. The main depot itself stood ten feet high, topped by twenty-five feet of bamboo pole festooned with banners. "Shackleton's party cannot fail to find this," Joyce wrote with evident pride. Before they could extend the cordon to the west, a blizzard closed in. There was no alternative but to wait out the storm. The dogs coiled into tight balls as the blowing snow buried them. The three men lay in their bags, shivering and unable to move around the tent enough to generate warmth. Joyce and Wild braved the storm to tend the nine dogs only to find that the familiar mounds had disappeared. The snow was accumulating fast and compacting hard, threatening to seal the breathing holes and suffocate them. After two hours of excavation with frostbitten hands, Joyce and Wild finally liberated the team.

Joyce stewed silently over Mackintosh's decision to take the dogs south. His worst fears were realized when the storm subsided. "To my sorrow discovered two of them had succumbed," he wrote on February 23. Briton had been flagging for some time, but the death of Tug, a mainstay of the team, was a shock. The survivors were weak, and devouring anything in reach: canvas, leather, and even the brass of their harnesses. A few days before, Mackintosh had cut their food by a third to make the supply last longer.

The animals were wasting away. Even the full daily ration of Spratt's or Thorpe's brand dog biscuits, weighing a pound and a half in total, was grossly inadequate. Spratt's was the first commercial dog food, patented in 1862 by American James Spratt on the same basic recipe as his food for hogs, cats, and poultry. His concoction of boiled live-stock carcasses and charcoal mixed with 75 percent flour and vegetables baked into dry cakes replaced table scraps or horseflesh as the food of choice for bourgeois pets in Victorian England. While it was arguably a healthier choice for parlor canines, the biscuits were grossly inadequate for sledge dogs. The daily biscuit ration was deficient in both quantity and quality, providing a fraction of the necessary calories and insufficient protein and fat. The dogs were also dehydrated in the dry, frigid climate, which increased their susceptibility to kidney damage. To conserve fuel, snow was only melted for human consumption.

Ordinarily, huskies obtain sufficient water by eating a high-fat diet for sledging, notably fresh seal. Eating only dry biscuits, the dogs had to rely on snow for hydration, risking hypothermia in the non-Northern breeds.

Too little and too late, the men slipped scraps of their own rations to the starving animals. The stamina of the seven survivors was difficult to judge. The dogs howled dolefully as they were hitched for the final task at the 80° south depot, laying the western cairns. "My heart aches for them," Joyce wrote. "I don't know how I've refrained from giving Mack a bit of my mind, will have to keep that in until we get back. He will have enough to think about." In the snowy spindrift, visibility was dropping. After only a mile, it became "thick as a hedge," as Joyce put it, and impossible to push on. Mackintosh had no choice but to order a halt until the weather cleared. They stalled in the tent until February 24, when Mackintosh abandoned the cairn-laying and ordered a start north in the mist. "What a place, still, what else can we expect? It's all happened before and we knew it would be so before we came so we must grin and bear it," Mackintosh lamented.

At least now, the party was homeward bound. Hut Point lay 150 miles north. "Our trouble is that we are going back with ten days' provisions, so it means shoving on for all we are worth," Mackintosh wrote. At their usual pace of five to eight miles per day, it would take at least three weeks to reach the hut—assuming conditions were always clear for travel. The record was not promising on that score. Of the thirty days since Mackintosh had left the ship, his team had spent six tent-bound. He had tried to conserve food by cutting back to one meal a day when they were immobilized. Reluctantly, he decided to take back ten days' worth of food from the depot in case of emergency. After sledging barely four miles, the blizzard forced them to camp again, this time for three days. "A rotten miserable time," Mackintosh called it.

"I could not refrain from remarking that in this degree of latitude no human being has known the weather to be calm," Joyce observed, re-calling the foul conditions that had beset previous expeditions in this region. The reason lay to the southwest, where glacial valleys in the Transantarctic Mountains funneled cold, dry air from the high interior plateau down the Byrd Glacier to blast Minna Bluff. The raging bliz-zards spawned by polar depressions had plagued Scott and Shackleton.

The Ross Sea party cut a path directly through this unsettled region at a peak time of year for the storms.

By the time the weather eased on February 27, Mackintosh had reached the painful conclusion that it would be necessary to return to the Rocky Mountain Depot again and plunder still more provisions. Joyce was deeply discouraged, writing, "It seems hard after depôting stores to rob it again . . . Mack is feeling the strain." The next day, the three men watched the sun dip below the horizon for the first time since the Ross Sea party's arrival in Antarctica, tinting the sky the color of burnished copper. The days of perpetual daylight had ended. In less than two months, the heart of the Antarctic continent would be shrouded in perpetual darkness. With the failing light came the autumn cold, sharpened by a stiff breeze. The dogs were oddly silent as they slogged through deep drifts and hardly touched their food, even when Joyce doled out double portions. To save their strength, the men rigged a sail on the sledge with bamboo poles and the tent canvas, but the burden was still too great. "Pat stopped behind so I expect he is done for," Wild recorded impassively. Joyce released him from his harness. In the next few hours, two other dogs collapsed and they, too, were left behind. In desperation, Mackintosh gave orders to camp early for the night to give the animals a chance to regain their strength. To his relief, two dogs caught up to the camp, but Pat never reappeared.

The night was sheer misery, minus twenty degrees. Joyce was suffering badly from snowblindness after leading the sledge in the glare day after day. Wild could not bring himself to write up his journal, noting simply "too cold." Back on the march in the morning, they donned reindeer-fur boots called finneskoe to keep warm. The soft boots insulated far better than the canvas ones but were a poor fit with ski bindings, causing the men to flounder awkwardly. Before long Shacks began staggering in the traces, as if punch-drunk, destined for "no better reward than to be left behind, now that he has given us of his best," in Mackintosh's words. He feared that the entire team was failing. "We cannot expect much more from them. It is hard, but this is a cruel part of the world," he wrote on March 2. The events that unfolded were harrowing nonetheless:

We went off fairly well for half an hour, then Nigger commenced to wobble about, his legs evidently giving under him. We let him out of his harness and let him travel along with us but he has given us all he can, and now poor brute can only lay down. After Nigger my friend Pompey collapsed . . . After Pompey—the bachelor and quite one of the best dogs—Major fell down, and followed in the tracks of the others, then the last but one Scotty. They are all lying down in our tracks; one thing they have a painless death, for they lie curled up in the snow and fall into sleep from which they never wake again.

A lone dog survived, Pinkey, an indifferent puller. The three men labored to take up the slack, grappling with the unwieldy sledge as the wind billowed into the sail. Caught off balance by a fierce gust, Mackintosh fell and the sledge lurched over his body. He conceded the day and called a halt. Longing to see the dogs trotting along behind, he looked back and saw dark forms lying motionless along their path.

"We shall have to call this the Dead Dog trail," pronounced Ernest Wild in his typical bald fashion. His black humor concealed a grave fear: he had lost virtually all feeling in both feet. Peeling away his sodden woolen socks, he saw that his toes were soot-colored, a sure sign of advanced frostbite. Joyce's efforts to revive the circulation failed. Only time would tell if living tissue survived under the hardening carapace on Wild's toes, or if the damage would progress to gangrene.

On March 4, the temperature fell to minus twenty-eight degrees at sundown, the lowest yet recorded on the journey. Wild's feet worsened, and by March 6, he was riding on the sledge with Pinkey, both too debilitated to walk any longer. Before the day was out, Pinkey died. As another blizzard gathered force outside, the mood in the tent was desolate. "On Polar Journeys the dogs are almost human, one never feels lonely when they are around," Joyce wrote. "I am more than sad about it, this could have been avoided if common sense had been shown."

At long last, Joyce sighted the Bluff Depot through the binoculars on March 10. They hastened to the depot and broke into the food stores. The campsite seemed eerily unchanged since they left, bound for 80° south, on February 11. Mackintosh was alarmed to find that none of the other parties had returned with more supplies for the depot. It was difficult to avoid the conclusion that some mishap had prevented them from carrying out orders. He had little time to dwell on his worries when he realized that Joyce had built the depot in the wrong place. Mist had obscured landmarks when he had checked the bearings, and it was roughly four miles east of its intended location. The three men would have to dismantle the depot and move it to the bearings Shackleton had specified. Long after nightfall, they were still trundling loads to the new position: 78°52' S, 169°05' E. After fourteen hours on the march, they finished the task just as a blizzard began, stalling them yet again.

Hut Point was about seventy-five miles north. Mackintosh was unwilling to borrow more than a week's worth of provisions from the Bluff Depot to supplement their dwindling supplies. The raided depots seemed a cruel mockery of their labors. By March 13, about ten days of food remained and they had only plodded about five miles closer to the hut. Mackintosh continued his policy of one meal per day during blizzard days laid up in the tent to conserve food, so they would go hungry until they were on the move again. "Having been without a meal since yesterday evening one's insides soon begin to feel hollow," he wrote, resolute in the face of his own misery.

Their predicament was a painful repetition of Shackleton's wretched starvation march in 1909. Although Shackleton had intended to use the new sledging diet devised by nutrition expert Wilfred Beveridge, the provisions packed into the hold of the *Aurora* varied little from the inadequate diet used on earlier expeditions. In the chaotic days before departure in 1914, Shackleton only shipped enough of the special Beveridge rations for his own depots. The Ross Sea party reverted to the traditional polar field diet staples of pemmican, biscuits, chocolate, and oatmeal. On full rations, Mackintosh and his men consumed about 4,300 calories per day, some 30–50 percent less than required for heavy manual labor in extreme cold. They were dehydrated as well.

Temperatures plummeted as the Antarctic winter drew inexorably closer, ranging in the minus thirties. The seasonal transition was not gradual. As sunlight diminished in March, the continent began a precipitous descent into intense cold, the so-called coreless winter, which would last from April to September. The men were becoming frostbitten as they lay in their sleeping bags at night. Ten pounds at the outset, each sleeping bag weighed about thirty pounds, the fur inside "a mass of ice, congealed from my breath," as Mackintosh recorded. By day, the wet clothing froze into heavy armor, and a thicket of ice clung to their beards and mustaches.

"We are 3 old crocks," Joyce wrote. Their faces were swollen by frost-bite, the fluid in the blisters freezing hard. Wild's face, hands, and feet were so blighted that his companions rubbed his body nightly to revive some deadened part. As well as painful frostbite around the socket of his missing eye, Mackintosh suffered from a toothache, which Joyce attempted to medicate with the denatured alcohol used to prime the stove. Due to its low freezing point, the liquid had reached the temperature of forty to fifty below zero. Mackintosh screamed in sheer agony as the lining of his mouth sloughed away to raw tissue. He tried to fortify the morale of his men, but privately he brooded. "I ask myself—is it worth the candle?" he wrote on March 18, swearing that he would never again return to the Antarctic.

The timing was desperately close. Mackintosh had instructed Stenhouse to return with the *Aurora* to Hut Point to pick up the sledg-ing parties on March 20. By March 19, the party had only reached Corner Camp, thirty-one miles from the hut. They could only hope that Stenhouse had delayed sailing. Food, fuel, and matches were run-ning low. Four days later, the single meal consisted of biscuit crumbs and cocoa powder, and by March 24, they finished the crumbs. Two hours later, Joyce spotted the Safety Camp flag. Inside the depot was a note from Spencer-Smith reporting that the other parties were safe. With any luck, Mackintosh's party would join them on board the ship that night.

Hours later, they reached the Barrier edge to find McMurdo Sound glazed with young ice, too thin to bear weight. They roved around the edge of the Ross Ice Shelf, looking for a route onto the Hut Point Peninsula. The ice cliffs were impossible to negotiate, so they reluc-tantly camped. It was a "woe begotten night," as Joyce called it. In the morning, they ditched the sledge and Joyce guided them to a place he remembered called Pram Point, which sloped up to the heights above Hut Point. Atop the steep volcanic hills, they could see the hut below, well over a mile away. Exhausted and aching with hunger, they threw caution to the wind and glissaded down the slope on the seats of their pants. As they trudged the last few hundred yards toward the hut, they realized that the *Aurora* was nowhere in sight.

Hut Point

They were stranded at Hut Point, but they were not alone. Cope, Hayward, and Jack had reached the coast on March 14, only to find a letter from Stenhouse informing Mackintosh that the ship had anchored there three days before, just long enough to land some stores and pick up Spencer-Smith, Gaze, Ninnis, Hooke, Richards, and Stevens. The *Aurora* had "not had a rosy time," according to Stenhouse. The advancing season had brought an onslaught of storms. Buffeted by hurricane-force winds, the ship dragged anchor time and again, forcing Stenhouse to rove McMurdo Sound in search of safe anchorage. With the hazards of the coast often obscured by blizzard, the *Aurora* was rarely out of peril as Stenhouse dodged the loose pack ice and bergs crowding the sound. Already low after charging the pack in January, the coal bunkers were severely depleted from constantly maneuvering out of harm's way. The remaining supply was just sufficient to keep the engines going for about three weeks. Stenhouse worried that it would not be enough for the return voyage to Australia. He decided to make one last run at Hut Point on March 11, then moor the ship at Cape Evans and shut down her engines for the winter. He knew then that it would be impossible to return on March 20 as he had originally agreed with Mackintosh.

Mackintosh accepted Stenhouse's decision equably. In his written orders, he had instructed his chief officer to "look after the ship, above everything else." The first priority was finding a suitable place for winter moorings; steaming back to Hut Point to pick up the sledging parties was secondary. Though he had not anticipated that the ship would be moored before his return from sledging, Mackintosh had advised Stenhouse to "be prepared for all eventualities" and trusted his judgment. In any event, it had always been possible that the sledgers would return late, and, missing the ship, be forced to wait until McMurdo Sound froze solidly enough to cross on foot to Cape Evans.

Still, a long sojourn in the hut was a bleak prospect for men in their condition. The first sight of Mackintosh's party shocked Hayward. "I cannot describe their ghastly appearance, the Skipper looks dazed," he wrote in his diary. Wild's condition seemed the worst, his feet "raw like steak" and his right ear tinged green and oozing viscous fluid. The frostbite damage had almost certainly progressed to gangrene. Joyce's hands, nose, and feet were beyond feeling, and his fingers were bloated and misshapen. Mackintosh's face was disfigured into a swollen mass of mottled, livid flesh. The socket of his missing eye was badly stricken. Cope tended their injuries, although his clinical practice had thus far been limited to performing a postmortem on a dog with a copy of *Modern Surgery* at hand. He amputated one of Wild's toes and part of his ear.

The first night of their reunion was appalling. The group had only three sleeping bags between them, so sodden and worn that Hayward called them "indescribable unless Dante's Inferno would meet the case." The six men shared the bags, sleeping and pacing by turns as the temperature fell to seventy below zero. "We are still alive this morning, so must be thankful," wrote Hayward.

To Joyce, the hut seemed palatial "after the great gruelling," but in truth it was only a marginal improvement over a tent on the Barrier. Situated on the tip of a narrow low headland, the exposed shelter took the full brunt of southerly winds streaming from the interior of the continent and storms brewing over McMurdo Sound. Built in 1902 by Scott's *Discovery* expedition, the storehouse was never outfitted with insulation, furniture, or bunks. Patterned on the traditional Australian settler's bungalow, the shedlike structure's pyramidal roof and surrounding veranda were designed to promote cooling and block out the sun, both of which it did admirably well. The open-raftered space, thirty-four feet square, was impossible to heat above freezing. A dank chill penetrated the gapped planks of the floor. To make a snow-free corner of the drafty interior habitable, the men piled crates around a space roughly eight by fifteen feet and hived it off with cloth. Inside this shanty within the hut, the six men spent their hours huddled around the jerry-rigged blubber stove.

With next to no coal and just twenty gallons of kerosene for heat and cooking, fuel was the most urgent concern. Food was more plentiful, as Stenhouse had dropped off enough to last two months. Supplies left by Scott's expeditions would last another two months. Though the bulk of the food was over a decade old, some of it was still edible. A seal colony near the hut promised a regular supply of fresh meat and blubber. Leaving the invalids at the hut, the strongest men—Mackintosh, Jack, and Hayward—began hunting soon after their arrival. The passive Weddell seals put up little resistance. Fast and agile underwater, they were clumsy on land. The hunter would stun the slow-moving animal with a blow to the nose, then swiftly draw a blade across the throat. It was a gruesome task. "It really is murder killing these innocent harmless brutes who roll their eyes and start with fright when they see you," Mackintosh lamented. "At first I detested the job, especially when the seals looked beseechingly at one with their large eyes, but after starving in the tent I am afraid the tender instincts, if any, in us vanish."

However odious the butchery, seal meat was the best medicine for their wasted bodies. After living on the limited sledging diet since January, all of the men were malnourished. Joyce's insistence on serving the meat bloody rare ensured that the nutrients were not cooked away in the pan. The flavor was another matter. "Believe the taste is an acquired one," wrote Jack. "Sincerely do I hope so." The motley assortment of leftover provisions added variety: flour, sheep's tongues, corned beef, porridge, bottled fruit, canned salmon, sardines, dried onions, and beans. As they chipped away at the ice around the hut, odd luxuries including a leg of mutton and a Christmas pudding surfaced. Tobacco, to universal disappointment, was in short supply. Mackintosh mandated only two meals each day to conserve the stockpile, with Joyce's specialty, scones fried in blubber, served in between.

With this regimen, the invalids slowly convalesced, though in Joyce's words, "It was a couple of weeks before our faces straightened out again." During their waking hours, the shuttered gloom of the hut limited activities inside. Mackintosh took long walks, often with Hayward. Jack monitored the weather instruments and salvaged scrap wood and wire to build dartboards and games. All of the men were enthusiastic card players, gambling with lumps of sugar. Wild constantly debated for sport, provoking fierce arguments on inconsequential topics. With his favorite verbal sparring partners, Joyce and Hayward, he vigorously defended the proposition that Antarctic weather was worse than Canadian, and that sailors were indeed tougher specimens than cowboys.

Of far greater moment was the debate over Shackleton's whereabouts. "If the T. C. [transcontinental] Party should not arrive by the 1st April at the foot of the Beardmore Glacier," Shackleton had written to Mackintosh, "it is most probable that they are not crossing that season." From their own experience, Mackintosh and Joyce knew that this was overly optimistic. Scott's polar party had braved winter conditions on the Barrier into late March and perished. If Shackleton had arrived at the Beardmore as late as April 1, as he had told Mackintosh, he and his men would likely have died on their march to the coast. If they had arrived earlier, they would have found no supplies until they reached the Rocky Mountain Depot. The Ross Sea party was not willing to accept either prospect. The alternative—that Shackleton had postponed the crossing until October or November of 1915—they considered with immense relief. Then there was still another season, another chance to get the depots in position. "I have been thinking of Shack's trip across the Antarctic Continent which will be a really marvelous feat of endurance & pluck," wrote Hayward, adding, "I sincerely hope that his dogs will pan out better than ours have."

The realization was unavoidable. Their own first season had been a debacle. Sixteen dogs had died, leaving only eight survivors on the ship for the next season's work. The men bore lasting scars from their privations and sacrifices, which, in the end, had achieved very little: 105 pounds of provisions and fuel at the Rocky Mountain Depot for Shackleton and 158 pounds at the Bluff Depot. Under Cope's leadership, the motor party had turned in a disappointing performance as well. Charged with depositing 540 pounds of supplies at the Bluff Depot, they failed even to reach it. They had struggled about twenty-four miles south and dumped a total of 450 pounds of stores in three small depots along the way. As one member of the motor party calculated, they consumed 550 pounds of food and fuel in the process.

The outcome was a grim vindication for Joyce. He saw the dogs as key to the success of the enterprise and harbored a powerful sense of resentment over Mackintosh's dismissal of his judgment. He pointedly reminded Mackintosh of his own 1909 depot journey, a two-day dogsled sprint in which he had covered over forty miles in a single day. Their recent horrific journey had taken twelve days to cover the same distance. Despite his vows to give Mackintosh a bit of his mind, however, Joyce was careful not to push too far, and Mackintosh studiously ignored the digs. He had perceived the essential conflict in Joyce's nature between the maverick and the clock-watching sailor. Joyce longed to be his own man, yet was reluctant to shoulder the burdens of leadership. He was a particular brand of petty officer, reflexively sniping behind his superior's back and then knocking off to enjoy a pint with the seamen while the "Old Man" bore the responsibility.

Predictably, Joyce laid the blame for their current plight on Mackintosh, which was the source of "very heated arguments," as he reported in his diary. "Stenhouse knowing the conditions ought to have anchored here until the sledging parties were picked up, there being good anchorage in the bay, or landed coal, stores, lanterns, clothes, & stoves." Ignoring the fact that he had no idea what ice and weather conditions had been when Stenhouse made his decisions, Joyce concluded that the party had been stranded because Mackintosh had not worked out a plan of action with the first officer before handing over command. Joyce badgered Mackintosh about the specific orders he had given Stenhouse, and Mackintosh, called to account by a subordinate, refused to tell him. Joyce took his reticence to mean that he had given no orders and declared "Mack is in the wrong for not leaving instructions."

Joyce's agitation fanned the discontent of the other men, particularly Gaze and Jack, who were annoyed by Mackintosh's "shabby treatment" of their party in the field. "Many of the skipper's actions were discussed & not at all to his advantage," as Jack put it bluntly. The handling of the dogs and the organization of sledging were the chief complaints. Cope's erratic leadership was also a lightning rod for discontent, especially for Jack, whose energies had been so squandered by mismanagement. The biologist seemed singularly unsuited for the task of leadership of a field party in the Antarctic. Cope had volunteered for the expedition on a whim after being forced to leave university, motivated by lack of direction rather than driving ambition. In a situation that would test the mettle of the most dedicated, he was out of his depth.

Some faulted Mackintosh for delegating authority to Cope in the first place. There had been early signs that Cope was ill-prepared to handle the responsibility. On the voyage south, Mackintosh himself had remarked that "[Cope] thinks he is here for a picnic: so had to assure him this was not the case." Although he admitted to Stevens "he had no confidence in Cope," Mackintosh believed it was only proper to put the Oxbridge graduate in charge of a party. In the merchant naval hierarchy, the ship's surgeon was an officer, and Shackleton himself had pronounced Cope "an excellent man." But Mackintosh's logic was unfathomable to the civilians of the shore party and only served to alienate them.

The men kept their grievances to themselves. None were prepared to risk an open challenge of the commander. Nevertheless, as Jack confided to his diary, "one cannot help asking oneself if Mackintosh is a very capable leader & if Shackleton's selection was a very wise one." Attitudes toward Mackintosh had undergone an undeniable shift. They no longer hesitated to question his everyday decisions. "Arguments are rife," Mackintosh wrote in his diary. "We have so much to settle and decide when we get back to the ship. I am sure if we should remember all that we have argued upon and not decided it will take us many days finding out." Faced with opposition, Mackintosh chose to accommodate the dissenters in order to keep the peace in the cramped quarters rather than force the issue.

———

The autumn light was fading steadily in April. The small supply of candles was soon finished, leaving only the smoky blubber lamps. Even these "evil smelling" lanterns became a luxury they could ill afford. As temperatures fell well below freezing, the seals deserted the colony near the hut for the warmer sea below the ice. The stockpile of blubber was nearly gone. "We are rather up against it," Hayward reported. The men scraped the fat drippings from under the stove and chopped up supply cases for firewood.

The uncertain supply of precious fuel meant that it was out of the question to melt snow for wash water. In any case, the effort would have met with uncertain results. The blubber stove smoked badly and coated everything inside the hut with a clinging sheen of greasy soot. They ate with filthy hands, though Hayward tried to improvise utensils with twisted wire. "Here we now are leading a life of a primitive people," wrote Mackintosh in disgust. The other men tolerated the squalor, but he found it disturbing. His revulsion was symptomatic of a deeper discontent. "I am thinking of this time next year," he wrote in his diary. "Home, Sweet Home, all that life's worth living and hoping for!" The desire for adventure and acclaim that inspired him to join the expedition faded as he yearned for his family and questioned the "queer places and positions man will place himself in for choice." The discomforts became insufferable. "It's too terrible, everything we touch is blubber," he wrote, his diary becoming a litany of distress over their living conditions. "What a crowd of utter tramps we look: long matted hair, uncropped straggling beards, grease all over ourselves, clothes— dirtiness personified."

Mackintosh became obsessed with reaching the ship, although Joyce guessed that McMurdo Sound would not freeze solidly for months. "Little chance of getting to C. Evans for some time yet," Mackintosh wrote in dismay on March 31 as he watched a vista of whitecaps on the open water where ice had been a few days before. The sea ice was known to be unstable from December until May. Not long after they arrived at the hut, the ice was two inches thick, strong enough to bear the weight of a man, but it was soon demolished by the wind-driven swell and swept out to sea.

Mackintosh knew how changeable the ice could be from hard experience. In January 1909, he and a companion had set out from the *Nimrod* to traverse frozen McMurdo Sound and carry mail to Cape Royds. The sea ice broke up, trapping the two men on a floe far from shore. As swift currents drove them farther out to sea, they leaped from floe to floe, across widening leads, and only just reached land. Stranded miles up the coast they defied all common wisdom and traversed the crevassed mountain slopes to make for the hut. "But oh!

What one pays for inexperience," Mackintosh wrote ruefully. Climbing unroped and snowblind at times, they made their way toward Cape Royds, dropping into crevasses more than once. By a stroke of good fortune, their ten-day journey ended safely when they were found by their comrades. His luck astounded them. "Mackintosh was always the man to take the hundredth chance," as John King Davis observed. "And on this occasion he had got away with it."

Nor was unstable ice the only danger. "Late last evening I watched the killer whales in the Bay here, sporting about & breaking up the sea-ice, 8 inches thick & these whales simply made great lanes through it with the utmost ease," Hayward wrote, unnerved. Orcas routinely cruised McMurdo Sound and had been known to jostle the floes to dislodge an unsuspecting man. Though the orcas would swiftly lose interest upon learning that the quarry was not seal, immersion in the freezing water was hazard enough.

———

On April 22, the winter sun edged below the horizon. Four months would pass before another sunrise. "It seems as though some vitalizing force had been withdrawn from our lives," Joyce wrote. By early May, the light was fading rapidly, and a lunar desolation settled over the dimly twilit Hut Point. Mackintosh suspected that their grasp of the calendar had slipped. "The people who have been keeping their logs going are all different," observed Wild, who abandoned his own diary soon after his party arrived at the hut. It seemed to Joyce "as though all the days were rolled into one long night."

They read voraciously to pass the time and soon exhausted their supply of books. Fortunately, Scott's expedition had left heaps of magazines, newspapers, and novels in the hut. A popular romantic thriller, *Soldiers of Fortune,* made the rounds, and its merits were debated at length. The book's original owner, *Terra Nova* scientist Frank Debenham, had inscribed an arch comment on the flyleaf: "Quite the nicest novel I have ever read, the hero and heroine are perfect pets," slang for hotheads. Scrawled below, in another hand, was the line "like some people on the Expedition." Discord was not uncommon on polar expeditions and unsurprising given the intense physiological and psychological stresses. What was rare among British polar explorers was the open admission that all was not the picture of harmony. Apsley Cherry-

Garrard was the very model of the Edwardian gentleman explorer when, writing about his first horrific sledging journey in 1911, he noted that they "did not forget the Please and Thank you, which means much in such circumstances," adding, "We kept our tempers—even with God."

Scott concurred in his account of the expedition, published posthumously in 1913. "I am very much impressed with the extraordinary and general cordiality of the relations which exist amongst our people. I do not suppose that a statement of the real truth, namely, that there is no friction at all—will be credited," he wrote. "There are no strained relations in this hut, and nothing more emphatically evident than the universally amicable spirit which is shown on all occasions." But as one of Scott's men revealed, there had been "great trouble and unhappiness" on the expedition, much of it caused by Scott's mercurial temper. Neither was Shackleton's expedition exempt from strife. He extolled the "cheerful readiness" and camaraderie during the *Nimrod* expedition, yet fisticuffs broke out on more than one occasion.

"It's all in the game," as the British expression went, and playing the game meant toeing the line, keeping the proverbial stiff upper lip, and never telling tales out of school. The code did not sit so comfortably with the Australians. A commonwealth since 1901, Australia had a culture that had grown a world apart from Britain—"colonial morals," as Spencer-Smith quaintly put it in a lively discussion. Individualism, enterprise, and informality had become the hallmarks of the nation's way of life. On the depot-laying journey, the Australians Jack and Gaze were the most candid in their diaries, even though they knew Shackleton might read their critical remarks. Jack had no qualms about sharing his views with his companions, as his discussion of Mackintosh's foibles with his mates on the trail demonstrated.

It was unsurprising, then, that after nearly six weeks of confinement in the miserable hut, Jack was the first to speak plainly to Mackintosh. He had tolerated Cope's failings in the field and tried to help him improve the party's performance. At Hut Point, Cope alternated between boisterous antics and begging off sick from duties. In Jack's view, the behavior formed a pattern of malingering and rudeness that extended back to the earliest days of the voyage. When Jack complained to Mackintosh, "Cope interjected once or twice in a rather objectionable manner & thus came to loggerheads with Skipper." Cope escalated to "words warfare" and Mackintosh demoted him, stripping him of rank and demonstrating that he had limits. Cope would not lead a sledging party in the next season.

As the bitterly cold month of May wore on, their tolerance was rubbed raw. "The conditions generally here do not conduce to the betterment of one's temper," noted Jack, who erupted in an uncharacteristic spat with Wild over an idle comment. Wild began writing in his diary again as an outlet for his frustrations. "Everybody has had a go at me (except the skipper) for making too much noise, so I thought I might as well start the log again . . . I suppose the light (or absence of it) food & dirt made us all badtempered." Even Wild's sunny disposition was sorely tested. Out on a walk, he broke through the sea ice up to the shoulders. At minus thirty, his garments froze instantly as Joyce hauled him out. Back at the hut, Mackintosh refused to loan him extra long underwear or let him keep warm in their shared sleeping bag while his clothes dried. "I can't understand the people here at all," Wild complained. "They've got no business down here at all. I don't know what they came for (with one or two exceptions)."

Some of the disputes were of greater consequence. "[Mackintosh] has got some daft ideas about getting back to the ship," Wild wrote. "He has got all sorts of impracticable schemes." Just a week after Wild had fallen through the ice, Mackintosh lobbied for one or two of the men to join him on a dash to Cape Evans, moving fast and light with no gear. If a blizzard came on, he proposed they crouch down and pull their Burberry tops over their heads for protection until the storm passed. Jack objected strenuously. "I maintain it would be madness, especially as we have been told by him never to leave our tent or sleeping bags at any distance. In spite of this he wants to start on a walk . . . carrying a few sticks of chocolate—truly a strange contradiction. Can't fathom him at all. He will risk his own life, ours, & those of the party of six coming overland. Sheer daftness!!" A journey of thirteen miles hardly qualified as a dash. With a sudden shift of wind, the ice could be on the move, as Mackintosh's own past experience had shown. In a blizzard with no visibility, it would be disastrous. In the end, Mackintosh was dissuaded, but the episode had undermined faith in his judgment.

As Jack noted, "Mackintosh probably thinks me too cautious but he is rash to a degree, & fails to learn from any experience."

At long last, the ice appeared safe at the end of May. When Mackintosh announced that he intended to make the journey to Cape Evans, there was no mention of traveling unequipped, and this time, there were no dissidents. On June 2, all six men ventured out onto the sea ice under a shrouded moon, towing a sledge packed with gear and food. About nine miles out, they came to the ice peninsula of Erebus Glacier Tongue. Farther north, the Dellbridge Islands loomed, like an elliptical ring of great standing stones. Just beyond the islands, a mile or two away from Cape Evans, they spotted a flag staked into the ice. Attached was a note from Stenhouse, directing them to a rope ladder to ascend the coastal cliffs if the sea ice was broken away from the cape. Then, after nearly twelve hours' plodding on the ice, Wild heard barking. They steered blindly toward the sound on the dim hummocky coast and reached the hut, where they were greeted by the chained dogs. Stevens, Spencer-Smith, Gaze, and Richards were astonished to see them. The ravaged men looked "like Anarchists" to Spencer-Smith. "The hut to us seemed like a palace," said Joyce, donning his tinted goggles in the dazzle of the acetylene lighting. Wild asked for a smoke and a drink. Then Stevens told them that the *Aurora* was gone.

— 8 —

"An Ideal Place in a Blizzard"

JANUARY 25—MAY 9, 1915

"Your first and paramount duty will be to look after the safety of the ship, this comes before everything else," Mackintosh admonished Stenhouse on the eve of his departure for sledging on January 25. To that end, he tasked Stenhouse with seeking secure winter moorings for the *Aurora* before sailing to Hut Point on March 20 to pick up the sledging parties. Even if Shackleton's party arrived by then, it would be too late in the season to sail north. The ship would have to be moored stoutly enough to withstand the furious winter gales of McMurdo Sound.

It was a formidable mandate, but Mackintosh's confidence in Stenhouse as master of the ship and second-in-command of the entire expedition was absolute. "He is an excellent fellow," he wrote in his diary. "'Tis such a comfort to have him." Stenhouse was a worldly-wise and accomplished merchant officer with nearly half his lifetime at sea, serving much of his career in tall ships that dwarfed the *Aurora*. He had not commanded a square-rigger or sailed in Antarctic waters before, but that could be said of every polar navigator at some point in his career.

It was this promise of a sterner challenge to his seamanship that had prompted Stenhouse to join the expedition. At twenty-seven, he was searching for a higher purpose. He had earned his master's ticket and advancement was on the horizon, yet he wondered whether a career commanding steamships would prove fulfilling. Improbably, this Scotsman found inspiration from an American, Theodore Roosevelt, who had lectured in London in June 1914. Stenhouse was galvanized by Roosevelt's philosophy of "the strenuous life." According to his

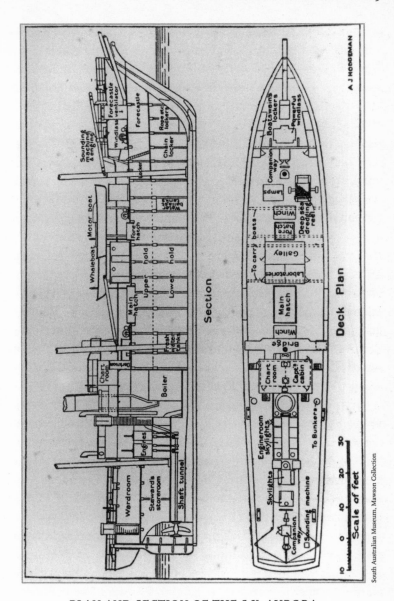

PLAN AND SECTION OF THE S.Y. *AURORA*

code "the life of toil and effort" was the only worthwhile aspiration for a man, rewarding its followers with "that highest form of success which comes, not to the man who desires mere easy peace, but to the man who does not shrink from danger, from hardship, or from bitter toil, and who out of these wins the splendid ultimate triumph." Stenhouse resolved to avoid at all costs the life of "ignoble ease" scorned by the former Rough Rider, and in September 1914, he joined the Ross Sea party.

Yet, Stenhouse's taste for adventure did not extend to reckless thrill-seeking. Honor and duty came first and foremost. His upbringing imbued him with a gallantry that few failed to notice. To the lower deck he was a *pukka sahib,* colonial slang that signified the consummate upper-class gentleman. Stenhouse and Mackintosh both wore the moniker comfortably, schooled as they had been in the cultivated manners of the Raj during their time in India. Yet they differed in their approach to command. Mackintosh had the common touch, earning popularity with the crew while still enforcing the hierarchy. Carpenter C. C. Mauger called him "a good man & skipper, strict and straight and generous to a fault." Stenhouse, on the other hand, had a certain reserve that some misread as arrogance. Still, none questioned his sense of fair play. If his high-mindedness led him to hold others to exacting standards, Stenhouse was no less demanding of himself when following the orders of his own superiors.

Those orders ultimately came from Shackleton. His directive on the location of the winter moorings had been most emphatic: "You must not winter her south of Glacier Tongue." Shackleton was determined that the *Aurora* not share the fate of Scott's *Discovery,* which had been moored at Hut Point in 1902 and became frozen fast in ice ten feet thick until 1904. It took the combined forces of two rescue ships sent from England and a load of explosives to free the trapped vessel.

Later expeditions avoided the danger by disembarking expeditioners during the Antarctic summer, then returning to port in Australia or New Zealand for the winter, only sailing back to pick up the explorers in more navigable summer waters. This strategy was a luxury Shackleton could not afford. He had no funding to fuel a second round trip of over four thousand miles and another refit of the ice-battered ship. On the other hand, Shackleton lacked the resources for a rescue mission if disaster befell either his ships or the shore parties. Unlike Scott, he could not count on the British government to dispatch a vessel to bail out his private expedition free of charge. Still, Shackleton felt that he had to gamble and instruct Mackintosh to winter the ship in McMurdo Sound. He believed that the *Aurora* could avoid the dangers of the pack by staying clear of Hut Point. In his experience, there was less danger of impenetrable ice choking McMurdo Sound north of Erebus Glacier Tongue.

There was a fundamental flaw in Shackleton's plan: north of Glacier Tongue, there was no suitable bay protected from the elements. A ship moored on the coast would be vulnerable to pack ice movements and high winds, as well as the gusts roaring down from the heights of Mount Erebus. Dominant southerlies flowing from the interior were diverted to the east and west of Ross Island by its mountainous terrain, buffeting the McMurdo coastline and Cape Crozier. Exposed features like Cape Evans and Cape Royds bore the full brunt of the winds.

Shackleton suggested two places to seek refuge: Horseshoe Bay, an indentation in the coast near Cape Royds, or Glacier Tongue. In the lee of the glacier, he believed, the ship would be protected from the southeasterlies. Shackleton was confident that either place would shelter the ship for the winter "unless the most abnormal circumstances" arose, and the *Aurora* would be free for sailing in February or March. Mackintosh had his own views and modified the orders. Horseshoe Bay, while good for the ship, was too far from the Barrier, a twenty-mile slog across the sea ice. In his opinion, Glacier Tongue was the only viable option, so when the last sledging party left the ship on February 5, Stenhouse steamed there for reconnaissance.

The *Aurora*'s first sally was unpromising. Ice filled the cove on the north side of Glacier Tongue, holding the ship at bay about two miles west of the end of the floating glacier. Stenhouse waited for a change of wind to sweep out the ice. The *Aurora* was secured with ice anchors, barbed iron hooks driven into an ice floe. The following days brought heavy squalls that set the pack surging into motion. But the ice crowding Glacier Tongue failed to move out, and the swell broke up the floe to which the ship was anchored, setting her precariously adrift.

Day after day, Stenhouse maneuvered the ship through the ice, searching for an opening. The firemen stoked the boilers constantly to keep the engines at full steam so that he could dodge and weave at a moment's notice. Finally, he gave up and tried to anchor to a floe to conserve fuel. Time and again, the floe would disintegrate under the anchors, and the untethered vessel would drift dangerously close to the coast, forcing Stenhouse to try for another anchorage. Despite the efforts of her harried crew, the ship broke loose six times in one day. Finally, on February 14, Stenhouse charged into an opening and made a run for Glacier Tongue. Within hours, the ship was alongside the glacier and the crew had planted three anchors securely in the ice.

While Stenhouse heaved a sigh of relief, the mood in the wardroom was anxious. "We've named the Tongue 'Apprehension Point'—it seems likely to fall off anywhen," quipped chief engineer Alf Larkman uneasily. Glacier Tongue looked stable, but it was essentially a tentacle of ice floating on the sea. With no warning, a three-mile stretch of the glacier's snout had broken off in 1911. Shackleton was well aware of the dangers. He acknowledged that anchoring there would be no mean feat and advised the officers to "keep the ship from being right under the Glacier Tongue ice wall, as heavy winter snow-falls are liable to bury the ship." Even so, he believed it was the best choice for the *Aurora*.

Boatswain Scotty Paton disagreed. With Mackintosh and Joyce ashore, he was the only experienced polar seaman aboard the *Aurora* and he knew McMurdo Sound. Paton, the veteran of four Antarctic voyages, had been awarded two Polar Medals and had sailed south with both Scott and Shackleton. Paton's Peak on Beaufort Island in McMurdo Sound was named in his honor, and he had returned the compliment by naming his eldest daughter Myrtle Beaufort. As the crew finished fixing a heavy wire cable to one of the anchors, his disapproving frown was unmistakable. Yet navigation was not his bailiwick, so he hesitated to hold forth. Stenhouse finally asked his opinion. "'Well Bo'sun, what do you think of this place?'" Paton recounted in his diary. "Of course I have reasons for knowing it, being acquainted with it for several years. I told him straight out I did not like it, in fact I told him he could never winter the ship here, I condemned it. He did not like what I said, of that I am certain, because he looked straight at me and said 'Don't be pessimistic.' I know this place well, however, we shall wait and see."

As Paton knew, the area was largely uncharted. The few soundings showed it to be too deep to anchor, in the region of two hundred fathoms under the keel. To make fast, the ship would be forced to lie too close to Glacier Tongue for Paton's comfort. The next day, a stiff northwest wind proved his point, slamming the ship broadside into the glacier repeatedly. Stenhouse ordered the anchors weighed and steamed against the wind to avoid being dashed against the coast during the gale. He had little recourse. The ship could shelter north of the glacier from the frequent southeasterlies, but she would be laid open to storms from the northwest. Stenhouse decided it was the lesser of two evils. Paton thought otherwise, but kept his peace, consigning his skepticism to his diary. "Still, Mr. Stenhouse is not without hope and swears he will give it a fair trial, 'well let him,'" he wrote. "He may yet winter the Ship there. It is certainly an ideal place in a blizzard, but a Nor'Wester 'Oh my,' look out."

The gale ushered in a return to the grind of breaking adrift and "dodging trouble," as acting chief officer Leslie "Tommy" Thomson put it. With no third officer on board to relieve them, Stenhouse and Thomson stood constant vigil. Before joining the *Aurora*, Thomson had been second officer aboard the *Kakapo* for the Union Steam Ship Company of New Zealand. Although it was his first voyage into the ice, the thoughtful, easygoing Australian was an accomplished seaman. He agreed with Paton, doubting they would ever be able to moor the ship securely to Glacier Tongue. The failed attempts were consuming an alarming amount of fuel, "playing old Harry with the coal," in Paton's words. By Thomson's reckoning, they would soon cut into the reserves for the return trip to Australia. Without coal, the *Aurora* would be forced to make the voyage under sail, at the mercy of the winds, not a bright prospect with the ship's rig.

Thomson had his eye on another mooring place: Cape Evans, where the Ross Sea party would make its winter base in the hut. The available charts were nearly devoid of depth markings in the waters off Cape Evans called North Bay, so Stenhouse sent Thomson out in a whaleboat for soundings. To him, Cape Evans looked promising. Thomson urged Stenhouse to "drop both bower anchors and moor stern onto the beach putting two good kedges into the shore and set everything tight with the windlass." On a coastline dominated by sheer icefalls, Cape Evans boasted a rare tract of ice-free ground. The low hills of the headland sloped gently into gravel beaches, providing a solid shore in which to sink the anchors. The major drawback was that, as a mere bulge in the coast, Cape Evans afforded less protection from the southeast and north-west winds than had Glacier Tongue. The bay was also unprotected from mass ice movements in McMurdo Sound. A sudden migration of the pack could drag the ship out to sea or drive her aground. But in Thomson's view, grounding the ship was infinitely preferable to smashing her against a glacier or being crushed beneath thousands of tons of calved ice. Thomson could not understand why Stenhouse resisted Cape Evans. "Beggars can't be choosers," he wrote.

Contrary to Paton's doomsaying and Thomson's doubts, Stenhouse succeeded in making fast to Glacier Tongue with three anchors on February 21. The crew was granted shore leave and they sledded down the glacier with abandon, finding amusement as their companions abruptly disappeared into crevasses and reappeared unscathed. Stenhouse and Thomson watched them blow off steam, savoring "every prospect of having a quiet evening." The next day, Stenhouse wasted no time in securing the ship with another ice anchor and the heaviest wire cable. Thomson was still wary. Explosive booms sounded at intervals as giant bergs calved off the glacier. "Things looked like being more comfortable here than sculling about the Sound but it is not the place that looks like a permanent quarters," he wrote skeptically. Paton held his tongue, noting archly, "It certainly looks an ideal place."

In the end, the last word belonged to the boatswain. Paton's dreaded nor'wester struck on February 24. At two AM, Thomson woke to the terrible sound of the hull grinding and splintering. The wind had hurled the ship broadside as a surge of pack ice pinned her against Glacier Tongue. The *Aurora* was trapped, steep swells slamming the hull against the glacier. Her sides, though reinforced, were only two feet thick, far thinner than the reinforced prow. Over and over, she was battered against the wall of ice until Paton feared "her ribs would crack." Stenhouse ordered all hands to lower spars and hatches over the sides as shock-absorbing fenders. Then, the wind wheeled around to the southeast, and with the retreating pack, the *Aurora* was freed from the snare. It was but a brief respite. The wind suddenly flew around to the northwest once again, forcing the ship back alongside the glacier with the pack closing in. The shelflike stern of the ship, the counter, overshot the glacier. Thomson watched as "she rose and fell with the swell every time sitting down upon the Glacier with her counter and grinding and crunching her stern post and rudder against [it]. And this time things looked like drawing to a speedy end."

Pack ice encircled the *Aurora* and cut off her escape. Stenhouse could not risk raising steam for fear that the revolving propeller would be snapped off as the counter seesawed down and slammed onto the glacier. The officers were powerless to save the ship: it was a question of saving the men. Stenhouse and Thomson prepared to give orders to heave stores overboard onto Glacier Tongue and abandon ship. Then, improbably, the pack brought the ship's deliverance. The dense blanket of ice damped the swells, so that the ship was no longer forced into the glacier. "Luck which seemed to have forsaken us for a time came again to our rescue," Thomson wrote, amazed. Twelve hours after the storm began, Stenhouse steered the ship north as a battery of floes assailed the hull. "It is easy to see now that this is an impossible place to winter, and I don't think another attempt will be made," Paton commented dryly.

He guessed rightly. Stenhouse immediately steamed north, and by six o'clock, the ship was anchored off Cape Evans in twenty-three fathoms. He had resolved to winter the ship at the cape. The delay came at a high cost, as Thomson was aware. "This is the place I have been advising the [commanding officer] to come to now for nearly a month and all the worry of steaming, drifting and making fast to the sea ice might have been saved, not mentioning incidentally an amount of coal," he wrote. "I do not now see how we can have enough to steam home and we will have to depend on our sails to a big extent especially with fair winds." The shortage of coal also meant the ship was very light in the water, at the mercy of the frequent squalls. The crew stowed sacks full of coal ashes below as ballast, a desperate measure that could not make up for the weight of the coal. All the same, Thomson was optimistic, deciding that they had "not had very bad luck" and hoping that their good fortune might hold until the return to Australia.

The homeward journey was a year, possibly two, in the future. In the meantime, there was much to be done before the ship's run to Hut Point on March 20 to pick up the sledgers. The lateness of the season meant that the crew would have to batten down the ship and shut down her engines for the winter immediately on their return to Cape Evans. Thus, Stenhouse had to ensure that the groundwork for the moorings was laid in just over three weeks. Adopting Thomson's scheme, he decided to position the *Aurora* stern to the beach, bows out to sea, with two bower anchors dropped to the sea bottom and two planted in the beach. Hard anchoring, it was called, and the aim was to immobilize the ship and keep her from being thrashed about by the elements. Once North Bay froze during the winter, Stenhouse and Thomson counted on fast ice—the solid sheet anchored solidly to the coast—to secure the *Aurora* and protect the ship from any jostling pack ice in the sound.

The first task was sinking the two anchors firmly in the beach. The whaleboats were too light to carry the one-ton iron anchors, so carpenter Mauger built a raft out of hatches and fourteen oil drums to ferry them to shore. Guiding the raft was miserable for the sailors, who waded through the surf in sea boots sloshing with achingly cold water. Near the hut, they chipped at the frozen gravel until their shovels clanged against rock. Once buried in the deep pits, the anchors were fixed in place with stone rubble, cement, and seawater. Thomson had never seen a storm mighty enough to dislodge such moorings. "Very satisfied with these anchors and feeling sure that they would hold until they broke."

Scotty Paton surveyed their handiwork as he paced the deck on night watch. Relieved as he was to turn tail on Glacier Tongue, he did not share Thomson's sanguine view of Cape Evans. Paton knew the hazards lurking there, recalling vividly when the *Terra Nova* grounded near the cape. Mackintosh, too, had run the ship aground off Cape Evans in January, where the bottom sloped abruptly from thirty-two fathoms to less than five. Paton also knew that North Bay was a bad holding ground due to the rock bottom and that the sea ice tended to migrate out to sea en masse during winter gales.

Once again, Stenhouse found it hard to avoid Paton's discontented grimace, and asked what he thought of Cape Evans for the winter moorings. Paton answered plainly: "I told him I was not in love with it, that the only place in the [sound] that I thought fit for a ship to winter was Back-Door Bay." Eight miles north, Backdoor Bay was a small notch in the coast at Cape Royds close to Horseshoe Bay, one of Shackleton's recommendations. Neither was much more than a cove, but the land encircling the tiny bays still afforded some protection from weather and ice movements in the sound. Stenhouse dismissed the suggestion immediately, saying, "Mackintosh would not hear of such a thing." Thenceforth, Paton confined his doubts to his diary.

Yet, Paton could not entirely avoid being the bearer of ill tidings. At 4:30 AM on March 10, the boatswain's cry of "She's away!" roused Thomson and Stenhouse from a twilight sleep. The ship had dragged anchor and was adrift, wallowing in a nasty swell. At the helm, Stenhouse steamed steadily against the wind, straight toward the Barne Glacier, a sheer wall of ice just north of Cape Evans, in an effort to keep the ship from being hurled across the sound. The gale-force winds whirled the spindrift and snow into mist and the coastline was blotted from view. Thomson worried about blindly running foul of an iceberg. Then, as the wind fell off momentarily, the mist cleared and Stenhouse saw the Barne looming dead ahead like the cliffs of Dover. He steamed clear just in time, but the ship was not out of danger. The wind escalated to hurricane force and gradually the *Aurora* began to heel over. To Thomson's horror, the masts swayed as the deck tilted toward the vertical, "nearly upon her beam ends." Once the spars submerged, she would be very near her end. The seamen clung to the iced ratlines and struggled barehanded with the flapping sails as unlashed crates tumbled down the upending deck. By some miracle, the howling wind moderated and she righted. By midnight, the ship was finally safely at anchor.

Stenhouse hung fire on the bridge. He knew he could no longer wait without jeopardizing the ship. The *Aurora* could not withstand another storm without being moored securely. He considered making the run to Hut Point immediately to pick up the sledging parties, then secure the ship on his return. It was ten days earlier than Mackintosh had instructed, but there was a chance that the parties had returned. Opportunity beckoned; the blow had been fierce enough to clear McMurdo Sound of pack ice all the way to Hut Point. If he waited, it was liable to close just as quickly. In the early hours of March 11, Stenhouse ordered Thomson to steam south.

As the *Aurora* approached Hut Point just before noon, a figure lurched into view. Soon five others appeared. When Thomson and Larkman rowed ashore, they found Spencer-Smith, Gaze, Richards, Hooke, Ninnis, and Stevens waiting. Six men were still out on the Barrier: Mackintosh, Joyce, Wild, Jack, Cope, and Hayward. Thomson hurried the crew to unload two months' worth of stores for the missing men and embark the sledgers before the weather changed.

After their ordeal, the six men were relieved to be in warm bunks aboard the *Aurora* again. Then at six in the morning, the company was roused by Paton's cry of "She's away Sorr!" The *Aurora* was ranging heavily and dragging anchors as the wind freshened into a gale from the south. The storm ratcheted upward to a more furious pitch than they had seen before. The mounded black swells rolling toward the ship and breaking over the deck were steeper than anything Thomson had imagined possible. Sudden gusts "threatened to tear the masts and spars out of the ship." The ship was blown some thirty nautical miles north, to the entrance of McMurdo Sound, before it was over. Thomson was on deck for over twenty hours, standing beside the helmsman on the bridge with only a flapping canvas screen to shield them from the furious weather. Stenhouse relieved his haggard first mate and steamed for Cape Evans. Defying the bullying wind and ice, the ship could only make half her usual speed. By evening the *Aurora* listed drunkenly off Cape Evans. The queasy members of the shore party ventured to the wardroom for a meal, where they departed from custom and toasted the ship's company and "absent shipmates on the Barrier."

The predicament of the stranded shore party was ever present in Stenhouse's mind. He knew the *Aurora* could not return to Hut Point again that season; it was far too risky to expose the ship to the winter ice and weather. They would have to reach Cape Evans on foot, but McMurdo Sound was a churning morass of floes and open water. It might be months before the ice was sufficiently stable for a safe crossing. With no means of communication, he could only hope that Mackintosh received the letters he had deposited at Hut Point, explaining the ship's troubles and outlining his plans. He had suggested that Mackintosh light a bonfire to signal the party's return. Then, he superseded that scheme with another, in which he said he would fire rockets at 9:00 each night when the ice around Cape Evans appeared solid enough for the trek.

Stenhouse agonized over another possibility: taking the *Aurora* to New Zealand for the winter. Shackleton was clearly opposed to this course of action for financial reasons, but Stenhouse began to doubt whether the ship could take yet another beating. Returning to port seemed to fulfill Mackintosh's commandment to protect the ship "above all else," but the shore party would be left to their own devices. Stenhouse knew they were unprepared for a long stay at Hut Point. Due to the conditions, Thomson had been unable to unload the intended year's worth of stores for six men at Hut Point. And arriving soon could be Shackleton and his party, anxious to sail north after the transcontinental crossing.

On March 14, Stenhouse made his decision. The *Aurora* backed into position for winter moorings at Cape Evans. Running from the shore anchors to the stern were five wires, both four-inch steel hawsers and smaller cables. Later, a heavy chain was made fast to the stern and hauled ashore. Like guy ropes on a tent, the wires could be tightened individually until the optimal tension stayed the ship. The two bower anchors obtained their purchase less from the flukes digging into the sea bottom than the weight of the seventy fathoms of cable securing each one. Thomson declared "there were enough hawsers and anchors out to hold a battleship." Once the ship was firmly embedded in fast ice attached to the shore, he and Stenhouse believed the moorings would be unassailable.

On March 20, Stenhouse told the engineers that he was finished with steam on the main engine for the season. As the boilers were blown down, Larkman echoed Stenhouse's judgment that they were "not justified in leaving safe moorings" to attempt to pick up the missing men. The risk to the ship was too great and there was still much to do to ready her for winter—clearing the cargo on deck, making the sails snug, constructing protective canvas awnings, installing the wireless, and overhauling the electric lighting motors. Stevens, Spencer-Smith, Gaze, and Richards went ashore to set up the scientific station in the Cape Evans hut. Though the ship was just offshore, the scientific staff was expected to be largely self-sufficient. Ice conditions could make access unpredictable. Eighteen hands remained aboard the *Aurora*. With the moorings requiring constant attention, there were few chances for the crew to leave the ship.

The *Aurora* was woefully undermanned. The burden weighed heaviest on Stenhouse and Thomson. "My watch on deck runs into about 15 or 16 hours a day in good, bad, or worse weather," Stenhouse had noted early in the voyage. During the recent crises, he had pulled stints of fifty consecutive hours. In the engine room, Larkman and Donnelly had routinely worked for twelve hours or more at a stretch since January. Carpenter Mauger had his hands full repairing damage to the ship. In addition to their deck duties, Stenhouse detailed Ninnis to assist Hooke with the auxiliary generators for lighting and wireless.

Under the watchful eye of Scotty Paton, just five sailors kept the ship in trim. "There is no Sunday in the Antarctic," Paton griped as another seven-day week ground to a close. The ringleader of the sailors was William "Ginger" Kavanagh, a good-natured able seaman "with scores of impossible yarns and usual nautical knowledge of language." Kavanagh, Sydney Atkin, and "Shorty" Warren were the toughest and hardest working of the lot, seemingly insensible to the cold as they worked aloft. Nineteen-year-old Arthur "Jack" Downing was an American expatriate who called Cornwall home. Charles Glidden was an Australian ordinary seaman looking to work his way up and boost his wages from £4 per month to £6.

Stenhouse rated them a good company overall, but some, he complained, were "'pore ignorant seamen' & nothing else" who could not be expected "to enthuse over anything or even worry about anything except their pay day." The adventure was more than they had bargained for, with the constant drudgery of shoveling tons of snow off the decks and spending days sawing the mooring lines out of the ice. Most often, the entire crew was standing by twenty-four hours a day on so-called Calashee watches, liable to be called at any time. Grumbling increased with the frequency of seal meat on the menu, as Stenhouse instituted rationing to stretch the provisions. Even the tots of rum doled out as rewards, regarded by every sailor as his due, became a rarity.

"This strain continually is beginning to tell on everybody," Thomson observed, finding it increasingly difficult to keep discipline. One sailor obstinately refused duty because of the cold, insisting "he could not do himself justice as a sailorman in such weather." The firemen were the hard-core malcontents. "From such as these one can expect nothing," Stenhouse reflected bitterly. "The scum of the earth & like dogs that bite the hands that feed them. To such people as these, the welfare of the Southern Party is of no concern." Stoking was unskilled labor and often attracted dockside toughs down on their luck like Grady, who had no sea experience. Tattooed William Mugridge was an ex–Royal Navy stoker who had been discharged for chronic misconduct. Harry Shaw had been a stoker before, but made no bones about hating the job. The former police constable tormented his shipmates and shirked his work. Mackintosh regretted hiring Shaw after putting to sea, calling him "rather a nuisance." Larkman found Shaw to be "ignorant & insolent" and suspended him more than once.

Managing the troublemakers and boosting morale was a tall order for Stenhouse and Thomson when the ship demanded their full attention. Neither officer had the time to monitor the subtle undercurrents of mood and head off trouble. It was some time before Thomson linked the cook's morning illnesses to the dwindling rum supply. They were both new to the roles that had been thrust upon them. It was one thing to execute the orders of the Old Man, or to discuss the O.M.'s decisions in hindsight. Shouldering the onerous burden of command and assuming full responsibility for the ship and all of the souls aboard was an entirely different matter. Stenhouse may have passed his master's examination with flying colors, but his mettle had yet to be tested as a leader. Preoccupied with the welfare of the ship, he seemed remote, and some of the crew chafed at being treated "like coolies." Nevertheless, there was little he could do in a time of storm and stress. Once the ship was rooted securely in ice, there would be time to attend to morale.

To Stenhouse's chagrin, the ice was not the panacea that he had expected. As the bay congealed around the ship in late March, the pressure began in earnest. The internal stresses generated as the ice collided with the unyielding coast were apparent, as slabs were thrust upward like colossal dominoes. As the ice clamped the ship in its crushing grip, stretching the cables taut, the massive timbers of the vessel flexed and groaned. Thomson came to dread "the ungodly noise" of the pressure and "respect and almost fear its power." Stenhouse ordered another wire cable ashore, for a total of seven wires pinning the ship. Still, by the following day, the ship was dragging anchor, shoved relentlessly toward the shore by the pressure. Paton was pessimistic. "I am afraid that no moorings, tie up how you may, could resist [the ice]," he wrote in his diary on March 31.

After barely four months in the Antarctic, the *Aurora* resembled a wrecked hulk, the stern thrust aground, listing sharply to starboard, the rigging and tattered sails rimed thickly with frost. The breaking seas had encased the crates and gear in ice. As the crew climbed up the steeply pitched deck, there was an uneasy sense that the *Aurora* was at the mercy of incomprehensible forces. Spells of violent tremors shook the ship and created heavy swells in the sound. High above Cape Evans, Mount Erebus was erupting, belching forth dense black smoke. A red glare glowed on the underbelly of the clouds, reflected from the crater. "Wintering a ship here is nothing short of a huge dream or nightmare," wrote Paton, expressing the anxious mood of his shipmates.

At least, Stenhouse reflected, the spreading ice meant there was a chance of the men at Hut Point making their way to Cape Evans. In early April, he walked four miles south to Glacier Tongue. The ice was fast as far as he could see down the coast. He sent men out to mark the final leg of the journey to Cape Evans with flags, and on the seventeenth, he ordered the firing of a rocket from Wind Vane Hill, high on the cape, at 9:00 PM. On the next clear evenings, April 17 and 24, he ordered rockets fired as well as the burning of a blue Bengal light to signal "respond immediately." Though Stenhouse scrutinized the horizon every night, he saw no answering signal. The shore party was now more than a month overdue, and he "seriously believed some misfortune had occurred." One or more of the men may have been incapacitated and unable to make the journey, or worse, they had all perished on the Barrier. They might have attempted the journey to Cape Evans only to become caught in a breakup of the ice during a gale. There was little Stenhouse could do. He resolved to send a search party to Hut Point and Corner Camp, but bad weather and changing ice conditions delayed their departure indefinitely.

The winter darkness was encroaching. In the officers' cabins flanking the wardroom, the temperature fell below zero. The ink in Ninnis's pen froze as he hunched over his diary just four feet from the wardroom stove. Stenhouse tried to assign duties in the forecastle to keep the seamen inside as much as possible, but the constant adjustments to the moorings and the rigging of the wireless aerial forced them out into the cold. Stenhouse ordered the moorings tightened daily, taking the slack out until the wires were "taut as any fiddlestring," in Paton's words. "All the cables are now singing and twanging," wrote Ninnis as the wind intensified, "and we wonder if the night will see us drifting out in the pack ice." The humming sometimes rose to the high-pitched vibrato warning that the wires were at breaking strain. A seaman knew to clear out when it started; a flailing cable could slice a man in two as neatly as a guillotine. One by one, the wires snapped, like so much bailing twine, and by mid-April, most had given way.

After weeks of agonizing work the crew finally completed the wireless masts on May 6. It only remained to mount the aerial. By evening, a southeasterly gale was gathering force. Before retiring, Stenhouse checked the moorings. Paton paced the deck on his usual nightly watch. Below, the men shifted uneasily in their bunks. There was something unsettling about the hysterical fury of the storm. Ninnis lay awake, ready to bolt as he watched the deck beams buckle overhead. Just after nine o'clock, Hooke heard the wires "commence their song" from the wardroom and went to fetch Thomson. Wakened by the woozy swing of the ship straining at her moorings, he was already dressing when Hooke tapped at his door. The ice was on the move again. Before they reached the companionway, two explosive reports boomed like artillery blasts. On deck, broken wires whipsailed shrilly through the air as Paton raced toward the bulkhead, hurricane lamp in hand, shouting, "She's away wi' it!" Then, another succession of shots sounded and the *Aurora* was adrift, held fast in a massive floe as it pulled clean away from the shore. The bower anchors were still attached and threatened to topple the ship over as they dragged the sea bottom. Stenhouse was on deck, shouting for all hands to clap relieving tackles on the cables to keep the strain from tearing the windlass out of the ship. As Thomson arrived on deck, he saw the lighted windows of the hut receding into the whirling snow and darkness.

Stenhouse ordered steam on the main engines, knowing full well it was impossible. Not only had Larkman blown the boiler down for the winter, but the engines were partially dismantled for repairs and the water intake valve had frozen solid. Larkman and his crew began painstakingly thawing the pipes with lamps. Two days passed before they could raise steam. By then, the *Aurora* was twenty miles north of Cape Evans, caught in the grip of the ice. Stenhouse knew there was no chance of battling back into McMurdo Sound. "Fast in the pack and drifting to God knows where," he wrote on May 9. "What of the poor beggars at Cape Evans and the Southern Party? . . . A dismal prospect for them."

Marooned

MAY 7–AUGUST 31, 1915

At three o'clock in the morning on May 7, Richards ventured out into the blizzard. Every four hours, around the clock, one of the scientists headed up Wind Vane Hill, about two hundred yards from the hut, to check the meteorological instruments. Richards skirted the sculpted snow dunes, some twenty to thirty feet high, until he found a route up the hill. In the pewter light of the waning moon, he looked seaward for the reassuring sight of the ship's topgallant masts spiking above the drifts. The *Aurora* was gone. "There was no sign of the ship and what little could be seen was open water," he wrote afterward. In disbelief, he hurried down to the shore, where he found the massive anchors uprooted. The chain cable had been wrenched apart and the steel hawsers frayed, "all snapped like cotton."

When the storm subsided, the four marooned men flashed messages in Morse code seaward with a lantern. There was no reply. Soon their signaling efforts were put to an end by a fresh storm out of the southeast, blowing for three full days with unprecedented ferocity. Wind speeds registered at ninety miles per hour, then exceeded the anemometer gauge. When the blizzard moderated, they tried to signal the ship by wireless without success. Knowing that the engines had been shut down, they realized that the powerless ship had probably been carried so far north by the pack that there was no chance of her return until the next summer. That was the optimistic view. Some, like Richards, believed that "she had been lost altogether."

By June, the Cape Evans residents had resigned themselves to being stranded in the Antarctic for a year, possibly two, until the Ross Sea party's return to civilization was overdue and a rescue ship sent south. But the news came as a shock to the six members of the Hut Point company, who arrived at Cape Evans on June 2. The ship represented safe haven, a beachhead of civilization and security in a hostile land. Mackintosh alone seemed unshaken, assuring his men that the *Aurora* was built to withstand the ice. By summer, he insisted, the ship would be back at Cape Evans to pick them up. In his diary, however, he admitted that the ship's disappearance was "a knock out blow" and that he believed rescue would be much longer in coming. "We must put the best face on it: others, I console myself, have been in a worse position. But two years here is a sorry problem under the conditions that have been forced upon us."

However passionately he longed to be done with the expedition and sailing for home, Mackintosh found no fault with Stenhouse. Shifting blame was alien to his chivalrous nature. He realized, too, that Stenhouse was hamstrung by his and Shackleton's orders:

> If one goes to a place where it's safe, you run the risk of being frozen in the next season, for instance like the *Discovery*; and that I was not permitted to do. Then on the other hand, the remaining course is to take the best pick out of available shelters round about these slopes—which were never too sure—as regards wintering a ship. Yet I must admit—this has quite been a surprise I did not calculate for in spite of everything.

Stenhouse's next move was anyone's best guess—if indeed the *Aurora* had survived. Richards gave Mackintosh a heartening clue. When conditions worsened, Stenhouse had apparently considered sailing north to New Zealand for the winter. "I pray & hope he has done this, for if he has his work will be a masterpiece of skill," Mackintosh wrote in his diary.

Two years in the Antarctic was an ominous prospect. As a scientist on the first expedition to endure a winter on the continent wrote, "I can imagine no greater punishment than to be 'left alone to live forgotten and die forlorn' on that desolate shore." Though lurid, his imaginings were realistic. The psychological stresses could be as punishing as the physical deprivation. A telegraphist on Mawson's expedition had lost his sanity in Antarctica and spent years in asylums on his return home. Men unprepared for the long Antarctic night fared the worst. In 1898, Belgian explorer Adrien de Gerlache steered his ship into the pack below the Antarctic Circle. For thirteen months, the men of the *Belgica* were unwilling captives of the ice. Scurvy swiftly claimed a first victim as the survivors descended into despondency. The ship's surgeon, Frederick Cook, likened the atmosphere to that of a madhouse. Cook had the presence of mind and force of personality to seize the reins and arrest their downward spiral. In the end, it was all they could muster to save themselves.

For the Ross Sea party, saving themselves was merely a means to an end, as all were acutely aware. "We are ten men who have to relieve Shackleton at the Beardmore Glacier 400 miles distant without any equipment to speak of," wrote Hayward. The party was left with only the clothing on their backs. A fortunate few had the odd spare personal items, but most of the sledging clothes, including Burberry outerwear and finneskoe, were still on the ship. The cast-off clothing from Scott's expedition was all but gone, most of it taken aboard the *Aurora* to make up shortages of warm clothing for the ship's crew. Sledges, Primus stoves, tents, cookers, and the rest of the essential gear for sledging were still stowed on the ship. They had, however, unloaded a set of *Encyclopedia Britannica,* which proved to be the "only text-book on the Antarctic" on hand as a guide to survival.

The bulk of their provisions were still aboard the *Aurora* as well. Mackintosh's intention had been to use the *Aurora* as a floating warehouse, moving supplies off the ship as needed. Only the scientists were intended to live ashore for their research. In April, the scientific staff quartered in the hut had submitted a request for more supplies to Stenhouse and Thomson. Stenhouse ignored the appeal, since Mackintosh had instructed the officers not to stockpile rations at Cape Evans. The shore party's case was not helped by the epicurean tone of their requests—including whiskey, toffee, chocolates, tobacco, and a gramophone—which left the ship's officers disinclined to seriously consider a plea for essentials like matches.

Likewise, Stenhouse had rejected a request for extra coal. Ten tons, along with other fuel, had already been landed on the beach. Most of it had been carried away by an unusually high tide. With the dire shortage on the ship, Stenhouse could not afford to spare any coal and even considered reclaiming some from the shore party to fire the *Aurora*'s engines. Seal blubber would be their principal source of fuel.

Shackleton's specially designed sledging rations for the depots had already been landed and were stored in the hut. The rations included newfangled patent items, including Streimer's Nut Food and Trumilk, which proved all too tantalizing. Yet Mackintosh allowed them only a small sampling of the forbidden fruit, which they unanimously pronounced "excellent both in bulk and taste." Sausages made of lard and powdered beef could hardly be described as a delectable treat, but, as Richards put it, the Ross Sea party's own scavenged pemmican was "somewhat less fresh than could have been desired."

Fortunately for the marooned party, Scott's well-funded expedition had left a great deal behind when the *Terra Nova* departed. The hut itself would be their foothold on survival. Though strictly utilitarian, Scott's main base was very comfortable in comparison with the crude shed at Hut Point. Fifty by twenty-five feet, the spacious interior housed bunks for twenty-five men, a photographer's darkroom, science laboratory, and galley. The timber structure was equipped with insulated walls, sprung mattresses, a cast-iron range, a coal stove, and acetylene gas lighting. A latrine shed stood nearby. Stables and a porch enclosed two sides of the hut to protect the interior from the wind. The hut was stacked with provision cases, the contents an odd mix: enough flour and biscuits to feed ten men for two years, pemmican and sledging biscuits for a year, and a bounty of jam and bottled fruits. Meat, oatmeal, tea, coffee, sugar, potatoes, and butter were in extremely short supply.

To make up the shortages, the party would have to slaughter seals every week. Hayward and Gaze immediately set to the daily ritual of hunting the animals and cleaning the carcasses. Mackintosh excused Spencer-Smith, who anguished over the "grisly and greasy job." Instead, the chaplain busied himself with photography and cooking for the group. The other men, too, sought solace in routine, the tensions of life at Hut Point forgotten for the moment. Joyce and Wild sewed trousers fashioned from ripped canvas tents for the entire company. "So here we are only a bit better off than we were at Ht. Pt.," Wild remarked with his usual sangfroid, pleased to have a smoke and a drink at long last. "However I suppose we will get over it alright." Richards immersed himself in the meteorological observations. The youngest member of the party was a "hard conscientious worker," as Mackintosh observed, "doing twice as much as ordinary people." Richards excelled in the impromptu navigation lessons taught by Mackintosh. Physicist Jack, too, welcomed the first real chance to delve into scientific work since leaving his job as a science lecturer in Australia. All in all, Mackintosh found the Australian trio of Richards, Gaze, and Jack a "splendid asset," remarking that "once they start on anything they do not leave it until it is completed." Chief scientist Stevens and biologist Cope had not yet settled down to scientific study and frequently joined Spencer-Smith in the galley. The three had been friends since the voyage south on the *Aurora*, when they were assigned to the same watch. A lapsed Presbyterian, Stevens sparred amiably with the chaplain over religious doctrine.

Mackintosh allowed himself a measure of optimism, writing in early June, "All is working smoothly here and everyone has taken the situation very philosophically." For the first time in 139 days, he soaped the grime from his emaciated limbs, emerging from the steel tub "a new man in body & spirit." For the moment, it was enough. "We are not going to commence any work at present, for the sledging operations, until we know more definitely the fate of the *Aurora*. She is constantly on my mind & I daren't think any disaster has occurred."

On June 21, Mackintosh rallied the group for a Midwinter's Day party to relieve the gloom and lift spirits. With homesick intensity, the men vied for prizes in relay races and pin the tail on the donkey. Afterward, they settled in for an elaborate meal of tinned beef tongue and plum pudding, crowned with a tatter of holly left by Scott's men. Gaze, nicknamed "the local millionaire" for his hoard of tobacco, produced his tin and, to universal astonishment, a bottle of Scotch. "I could not have wished a more pleasant day in these parts," observed Mackintosh with satisfaction as the morose group gradually came to life.

Mackintosh himself was preoccupied throughout the festivities. In spite of the present handicaps, there was no question in his mind of abandoning the depot-laying for Shackleton. As expectations of the ship's return faded, Mackintosh pored over the charts and mileage records of the first season and tried to reconcile the figures with the season's quotas. The net result of their labors was 718 pounds of supplies in position. Considering that a total of 4,500 pounds were required for the depots and their own sustenance, the lion's share of the work had thus been postponed until the second season. It would be a wildly ambitious undertaking, even had the proper equipment and supplies been available. There was much discussion of "the sledging problem"— "a pretty big problem, too," as Spencer-Smith saw it—but as yet, no one had figured out a solution.

Mackintosh worked feverishly to craft a plan to resolve their liabilities. "Working out sledging programme—more disappointments," he recorded on Midwinter's Day. His imprudence in the first season had deprived the men of their chief asset: canine pulling power. Twenty of the original twenty-six dogs had died. Only two dogs had survived sledging, and four remained from the *Aurora*. Mackintosh dismissed the four males and two females as useless, predicting "none of these will be any good." He had no desire to wage a battle of wills with the obstreperous animals again. The dogs would remain at the hut as the men man-hauled the supplies.

The nagging dilemma for Mackintosh was how to sustain an average pace of ten to fifteen miles per day. Shackleton had based his plans on traveling at least fifteen miles every day, but in the previous season, Mackintosh's teams had averaged just over five. The window of opportunity for depot laying was October through February. "Over March is too late to be out," Mackintosh had decided after the first season. Thus, they had five months to traverse over a thousand miles, crisscrossing the Barrier on multiple trips. If the party proceeded too slowly, they would consume too much food and fuel and cut into the depot stores. As Mackintosh immersed himself daily in pages of runic calculations, he was overwhelmed by the magnitude of the task, noting in agitated shorthand,

Such setbacks—& surprises where life & death are mingled so closely I have not experienced before. And it's hard to be existing with a sword of Damocles suspended over one as S[hackleton]'s life—I compare to it—for the responsibility lies on my shoulders. For myself—it would be different—though God knows, I have much to lose!

Mackintosh's anxiety was not limited to the logistical conundrum. However he configured the sledging, he would be demanding enormous sacrifices from the men. At a time when it was vital that he inspire and motivate the party, he was plagued by doubt. Mackintosh was new to command and did not wear the mantle of leadership lightly. Though he had earned his master's certificate in 1907, he had only attained the rank of third officer when he left the merchant marine to join the *Nimrod* as second mate. By dint of Shackleton's confidence, he leapfrogged to the rank of expedition commander and master of the *Aurora*. He was forced to find his way as a leader in the crucible of crisis. Shackleton had likewise vaulted into command on the *Nimrod* expedition, and, feet to the fire, had proven himself, weathering the challenges and inspiring cohesion and positive morale through canny leadership. His was an intensely personal style, cultivating relationships with men of all ranks and making them feel that their needs were met and their voices heard. Some of his adroitness in the role sprang naturally from his personality. As Mackintosh's friend John King Davis put it, "The leadership of men must rely fundamentally upon force of character. *Everything* is in that attribute. Derived from it and hardly less important, is manner: the manner of instinctive command." It was a testament to Shackleton's robust persona that the stranded party was motivated by allegiance to a man that most hardly knew. Mackintosh and Joyce had once stood by his side; five others had met him only briefly. The three Australians—Gaze, Jack, and Richards—had never laid eyes on "the Boss." Though they all felt bound by their pledge to support Shackleton, it would require more than an absent leader to fortify their resolve. With the demands of the new sledging season looming, Mackintosh depended on personal loyalty to motivate them.

Unfortunately, the party's trust in him was at a low ebb. For all of his bravery and dedication, some of the men viewed him as uninspiring. Richards felt he was "plucky as they come," but with "no personality, no ability, and no qualities of leadership whatever." The debacle of the first season seemed to result directly from Mackintosh's decisions. Then there was the directive not to land stores at Cape Evans and his unrealistic scheme to cross the sea ice to Cape Evans too early in the season. Though he had aborted the plan, the hint of impetuosity eroded their confidence in his judgment. An undercurrent of reproach from Joyce was evident, and a notion arose among the men, as yet unexpressed, that Mackintosh should have delegated sledging operations to Joyce and stayed on the *Aurora*. The idea grew from Joyce's false insinuations that Mackintosh was overstepping Shackleton's orders and usurping Joyce's authority in the field.

As the men debated his routine decisions, sometimes openly, and Mackintosh shrank from demanding compliance, his authority steadily weakened. "Under the circumstances I don't like imposing tasks," he wrote at Hut Point. By the time they reached Cape Evans, persuasion, not obedience, was the order of the day.

> Everyone doing as they pleased, it is good to see the ready spirit in which every one lends a hand, for instance clearing up the table & clearing up dishes after the meals—I have made no hard rules but in the day after my arrival here—when all the fellows were together, I gave an outline of the situation as it is at present, and emphasized the necessity they should practice in the use of fuel light & stores in order that these commodities should last out—if necessity forces itself to that extreme—for two years. I also explained that the comfort of our little populace depended on each of us, to give & take, & to keep cheerful & bright—But I have no fears that this will not be so— Yet I realize one of the hardest matters is the social problem & this happy medium between strict discipline & leniency.

His seemingly reasonable words masked a reluctance to impose discipline. "The social problem," as Mackintosh called it, was at the root of his hesitancy. Seven of the men were civilians who had joined the expedition as volunteers for nominal wages. Only Joyce and Wild were salaried. As Mackintosh saw it, he "should have been in a better position with an executive officer" to support him, as he felt compelled to treat the volunteers "with delicacy." Wild and Joyce were former Royal Navy men, reflexively carrying out the orders of their superiors. Mackintosh felt distinctly uncomfortable dealing directly with the rest, who were unschooled in traditional naval discipline. "I miss the services of an officer, although these are a sterling lot of chaps it requires an intermediary," he wrote. In the merchant ship's hierarchy, the captain relied on his subordinate officers to assign duties, dole out rewards, and mete out punishment. In the absence of Stenhouse and Thomson, Mackintosh deputized a reluctant Stevens, whom he ordered to "act as his second, as Stenhouse had done on the ship," as Stevens recounted, taken aback. When he balked, Mackintosh told him "it was a pity [he] wasn't a sailor, and that [he] did not understand."

Mackintosh not only longed for the support of a strong first officer, he was ill at ease with the demands thrust upon him as commander. "The loneliness and austerity inseparable from command," as John King Davis expressed it. At times, he seemed to wish himself back in the slipstream of a resolute superior officer. "I miss having another officer—What a weight off my chest when we meet 'Shacks'—And will we? Please God we do," Mackintosh wrote. In the meantime, he was alone, and Shackleton's fate depended on Mackintosh's ability to enlist the support of his men.

The day of reckoning arrived on June 26. Mackintosh summoned the group for a meeting. Even in the face of all the crippling setbacks, he announced that he intended to go forward with the second season of depot-laying for Shackleton. After reviewing the supplies, poring over the maps, and working out the logistics, he assured them it was entirely possible to lay the depots. Nine of them would join the sledging team, with Richards left behind to maintain the meteorological observations at Cape Evans. With any luck, they could repair the motor tractor and transport much heavier loads than in the first season. He hoped to begin ferrying stores to Hut Point in September. Then, the party would get under way on the Barrier in October and return to base by March. Mackintosh expected to meet Shackleton on the homeward trail and march with his triumphant party to the coast. He did not mince words, however, about the chances of success. He admitted it would require a record-breaking feat of polar travel to accomplish their mission.

In a tacit plea for cooperation, he presented his plan and invited comments. There were many. Some felt October was too early in the spring to be sledging. The memory of roaming the Barrier in the raw, aching cold of the previous season was still fresh. Several men also objected to Mackintosh's decision to leave Richards behind. In their view, all hands would be essential to carry out Mackintosh's daunting plan.

Mackintosh put the question to a vote. All nine men pledged their unreserved support, "mutually & unanimously," deciding to sort out the fine points after a trial run on the ice in September. Richards expressed the group's consensus: "All were agreed that the one object that must be attained, no matter what else was sacrificed, was to place food depôts for the six men of Shackleton's party." They were all certain that the transcontinental party would be "utterly dependent on our food depôts for their survival." It was for them a point of honor. That their own survival might hang in the balance was not a factor in the decision, as Spencer-Smith put it: "The job will D.V. [God willing] be done, tho' there'll be 10 very much played out men at the end of it. It's all in the game." Hayward spoke for the group in vowing to "do their damnedest" for the "stupendous undertaking."

Mackintosh had the necessary support, though time was short. Just two months remained to prepare for the initial push to Hut Point. Mackintosh put Joyce in charge of equipment, who enlisted Wild in a search for discarded sledging gear outside the hut. A systematic rummage turned up three tents, one intact and two badly ripped, and battered sledges. They excavated frozen mounds of Scott's damaged goods. Jack was overjoyed to receive a sopping wet blanket with pickax holes. Joyce and Wild proved to be masters of improvisation, reconstructing damaged goods into serviceable gear. "Wild has made an excellent shoe out of an old horse rug he found here, and this is being copied by other men," Mackintosh wrote. They patched holes and soaked blubbery garments in gasoline to remove the grease. From the canvas tent scraps, they stitched more trousers, pullover jackets, and outer boots for every man as a substitute for the missing water-resistant Burberry outerwear. A mangy reindeer fur sleeping bag yielded multiple pairs of warm inner boots and mitts. Other bags were meticulously repaired. They crafted leather strips into sandals to slip over the soft boots for traction on the ice. Woven cotton strips of lamp wick were stitched into sledging harnesses. Richards and Jack weighed Shackleton's rations precisely and packed them into the hundreds of calico bags that Joyce and Wild had sewn.

The Primus stoves were a more serious problem. The lightweight burners, which vaporized kerosene to produce a powerful, fuel-efficient flame, were the cornerstones of the party's survival. Using the parts of several broken stoves, Joyce and Wild carefully rebuilt three functioning units and hoped they would hold. Some of their improvisations were more dubious. For a group of diehard smokers, the scarcity of tobacco was aggravating in the extreme. Quitting at a time of extreme duress was a doubtful proposition. As Joyce noted, "One can forgive and forget many indiscretions over this soothing weed." At Hut Point, he satisfied his own addiction by smoking dried mixed vegetables, "but was speedily requested to cease." Wild concocted a more satisfying blend of tea, coffee, sawdust, and sennegrass. Dubbed "Hut Point mixture," it revived the ritual of smoking but did little to satisfy nicotine cravings. With the addition of dried herbs at Cape Evans, Wild's special blend enjoyed a newfound popularity.

Some members of the science staff proved to be ingenious inventors as well. Richards and Jack delved into scientific work even though most of the laboratory equipment was still aboard the *Aurora*. Only the weather instruments had been landed—thermometers for measuring maximum and minimum temperature, anemometers, a barograph, and a thermograph. Scott's outstanding scientific staff had left much of value behind, including labware, instruments, and tools. Jack painstakingly ground his own precision mirror for an Aitken dust counter, allowing him to measure the density of atmospheric particles for the study of electrical processes and the formation of precipitation. Meanwhile, Richards built an electric microthermometer, which gauged temperature to one-thousandth of a degree. Together, they devised experiments in oceanography and glaciology. Where measurement was impossible for lack of instruments, they used the most elementary of scientific techniques, observation. They described the activity of Mount Erebus, noting the plume direction as an indicator of air currents high in the atmosphere. As the aurora australis streamed across the heavens in phosphorescent ribbons of light, Jack described the pulsations minutely, watching for patterns that might give clues to the mysterious phenomenon. For Jack and Richards, just starting their careers, the Antarctic was a wonderland, that rare corner of the earth that had not already been swarmed over by generations of scientists and studied, cataloged, explained, and described in volumes on the shelves of the Royal Society.

Cope, on the other hand, was still laid low, both physically and mentally, and had yet to begin his biological work. Stevens was stalled by his own bitterness at the secondary place research had taken to depot laying. He felt wasted after the exhausting drudgery on the Barrier and resented being thrown into the role of "general factotum in unpleasant work" by Mackintosh. In late June, the intense Scotsman turned on Mackintosh, complaining that he had been "hoodwinked" by Shackleton. Explorers were "all frauds," he fumed, enlisting scientists "under false pretences, not doing all they are expected to: but doing work of more a manual character." Mackintosh accused him of lacking initiative, since their days were mostly free in winter. As Mackintosh pointed out, the other scientists had suffered disappointments as well but made the best of it. Richards had joined the expedition with his hopes pinned on an ambitious series of experiments in terrestrial magnetism, rendered impossible without the proper equipment, and Cope would have no opportunity to study emperor penguins at Cape Crozier. To keep the peace, Mackintosh tried to mollify Stevens while defending Shackleton. Privately, he conceded in his diary, "Perhaps he's right—but no one on such an expedition can do exactly what they have been taken on for."

Stevens's attitude widened the gulf between the younger scientists and their chief, and his inconstant allegiances estranged them. Stevens could be combative with Mackintosh one day, defending the rights of his junior scientists, then currying favor the next. Jack, for one, mistrusted Stevens: "I cannot fathom him, appears to be playing a double game—a deep customer truly." Stevens retreated to the galley, where he sought refuge with his fellow odd man out, Cope, and the kindly Spencer-Smith.

In the absence of charismatic leadership from Mackintosh, the men found purpose and solidarity in like-minded companions. Former schoolmates Jack and Gaze were inseparable friends and Richards befriended fellow physicist Jack. Hayward teamed up with Gaze for seal-hunting duties, but spent off hours walking with Mackintosh. Hayward's opinion of the Skipper, as he called him, was undimmed by the recent crises, his respect for authority deeply ingrained by his upbringing. As the only members of the party who were in love with women they left behind, they also shared a common yearning for home. The lovelorn Hayward read romance novels to remind himself of his fiancée. "Down here, where everything breathes of the unknown and appears so vast and limitless, how nice it is to be reminded in such a nice way of other & more pleasant scenes," he wrote to her after reading *Lorna Doone*, inserting carbon paper under each page of his diary so every entry became part of a continuing letter. He elaborated his dreams for their future together: "House Cricklewood, Shower bath essential, Holiday cottage in Broads district with yacht." The pages were illuminated with boyish sketches of bowler-hatted financiers and dapper gents, monocled and top-hatted for a night on the town. But then there were other doodlings, polar explorers and Canadian Mounties on horseback, and a note to contact the Hudson's Bay Company in Canada on his return. The lure of the wilderness was still strong.

Mackintosh, for the most part, shied away from close relationships. The officer's ethos was too deeply rooted; he felt uncomfortable living cheek by jowl with the men and fraternizing with his subordinates day in and day out. In this respect he was not unlike Scott, who had arranged the living quarters inside the hut according to the rigid Royal Navy hierarchy, separating the bunks of officers and men and ordering segregated meals. Mackintosh's most personal relationship was with Spencer-Smith, with whom he celebrated Holy Communion in the tiny darkroom, where the chaplain had carefully arranged his vessels and brass candlesticks.

Though the party was not particularly observant, there was universal agreement about their chaplain. As one man said, "Spencer-Smith is the finest person that I have ever met." Tolerant and kind-hearted, he was unfailingly hard-working, if somewhat clumsy. Affectionately called the Padre or Smithy, he regarded every aspect of the experience, good and bad, with an optimistic enthusiasm that endeared him to all. Wild was also universally popular. "A cheerful willing soul," Mackintosh called him. "Nothing ever worries or upsets him, and he is ever singing or making some joke or performing some amusing prank." The stolid Wild was unruffled by strife and unfailingly patient, an attribute no doubt honed as a middle child of thirteen Wild siblings. His joys were simple: finding a cigarette butt with shreds of tobacco intact, or another pair of threadbare socks to add to the party's clothing collection. He spent most of his time with Joyce as an apprentice in the polar arts. Joyce, in turn, happily assumed the role of mentor and protector of the younger brother of his old comrade in arms, Frank.

While the bonds were sustaining for the men, they proved divisive for the group. As the weather worsened and trapped them inside the hut, the men withdrew into cliques. The isolation and conditions amplified irritations. On a typical August day, the stove roared at full blast, yet a few feet away, the nails in the walls were rimed with frost, and the temperature inside the drafty hut peaked at twenty-three degrees. Joyce was an ever-present thorn in Mackintosh's side, lording his experience over him at every opportunity. Relations within the scientific staff deteriorated to the point where Richards refused to share his data with Stevens. Gaze and Hayward resented both Stevens and Cope, perceiving them as layabouts who failed to pull their own weight.

Mackintosh was reluctant to referee conflicts and discipline the men. "For peace's sake I am not saying anything," as he put it. He had drifted apart from most of his men and gravitated to Stevens, querulous and unpopular but highest in rank below him. In early August, the two hatched a plan to hike to the Cape Royds hut to inventory supplies. It was not an urgent errand, but cabin fever made it seem so. On August 13, Mackintosh announced that he and Stevens were going out and would be back in a few hours, in order to dodge any objections from Joyce. The direct route, six miles across the sea ice, was an obstacle course of open leads and imposing ridges some thirty feet high. So the two men climbed the foothills of Mount Erebus behind the hut, intending to keep to the land and follow the coastline.

The pair was heading toward the same ground Mackintosh had traversed in 1909, when he blundered into the crevassed foothills with a companion. Stevens quickly learned what Mackintosh already knew about glaciated terrain, as one leg plunged through a snow bridge into thin air. They carried no rope; there would be no means of dragging a fallen man back from oblivion. The pair climbed higher up the steep slopes, trying to outflank a swarm of crevasses. Instead, they were moving into nastier ground. Somehow, they eventually descended safely to Cape Royds, where a sudden storm pinned them in the hut. They found food but no fuel, taking turns trying to sleep under thin blankets and pacing to keep warm. When the weather cleared two days later, Mackintosh was anxious to return to Cape Evans and decided to take the most direct route across the sea ice. Once out on North Bay, the thin ice flexed and crackled beneath each foot tread. The two men backtracked and followed the shoreline to Cape Evans.

In the end, Mackintosh and Stevens came through unscathed, though they had little to show for the perilous adventure. Without a rope, it had been impossible to haul back any supplies. Instead, the two men could only fill their pockets with tobacco and soap. Stevens later admitted that the journey was "a very stupid thing," calling it "utterly inexcusable." There had been no compelling reason to go immediately. Just a week later, the ice was solid enough for a sledge party to march safely across to Cape Royds for a load of essentials.

After days of worry, Joyce could hardly contain his outrage and dropped barbed remarks about the incident. Mackintosh had not only endangered two lives, but he had risked the entire depot-laying operation, placing Shackleton's party in jeopardy as well. Joyce's overbearing safety lectures, while high-handed, had a point: one thoughtless act could cost all of their lives. Just a few days before, Spencer-Smith had shown just that. Smelling gas, he inspected the acetylene lighting pipes in his darkroom for leaks with a candle. He found one when a jet of flame shot out and engulfed a wall. Thanks to Stevens's quick action, they doused the blaze before it reached the combustible motion picture film and chemicals. "It might easily have been worse," Mackintosh acknowledged, "and with a burnt out hut here, it would indeed be the last straw." Henceforth, Joyce had little compunction about chastising reckless behavior by any of the company. "I did not go out—Joyce proved troublesome—it is a mistake ever allowing him to come," Mackintosh complained, chafing at Joyce's scolding presence.

———

Day by day, the horizon brightened. Behind the western mountains, the sun hovered, bruising the sky with livid hues of purple and green, then warming to a hard orange glow. "We saw the Sun the first time for over four months. It was a grand sight for us," Wild wrote on August 26. For Mackintosh, it was a balm for the soul. "The same old sun but how welcome—it brings one nearer Home: enters the body mind & spirit." The frustrations of winter ebbed away, and the men celebrated the return of the sun with a feast. Spencer-Smith designed an elaborate French menu to mark the occasion, and they topped off the meal with liquor and cigars from Cape Royds.

In a burst of optimism, Mackintosh proposed an early start to sledging on September 1. The unpredictable storms of late winter made it too dangerous to be on the Barrier, but he believed they could safely get a head start by moving supplies from Cape Evans to Hut Point. Even if the weather took a turn for the worse, they would never be more than seven miles from shelter.

At Mackintosh's request, Cope examined the party to assess their fitness for sledging. Using his rudimentary diagnostic skills and a medical kit provided by Burroughs Wellcome, he pronounced Mackintosh, Jack, Wild, Gaze, Hayward, and himself able-bodied. Spencer-Smith appeared fit in all respects, except for what he described as an "intermittent heart." The chaplain reassured Cope that the condition had already been diagnosed by a London specialist. He had hoped that the physical exercise would improve it. Cope saw no reason for undue concern and advised Spencer-Smith to return "at the earliest possible moment" if his heart symptoms recurred. Stevens, Cope stated, was "the least physically fit." Tall and rake-thin, he was by far the most debilitated by sledging.

Bowing to resistance, Mackintosh declared that the physicals were strictly voluntary. Joyce and Richards promptly refused to be examined. Joyce had been leery of physicians ever since Dr. Eric Marshall had refused to pass him as fit in 1908 for Shackleton's attempt on the South Pole. Marshall had detected "resistance in liver area," an indication of cirrhosis, and high blood pressure, which he took to be a sign of early heart disease, both possibly related to Joyce's history of overindulgence. Joyce had not mended his ways in the intervening years.

Richards seemed to have little to fear from a physical. He was a keen athlete in his prime, but he had suffered a hernia playing football not long before the expedition. He struggled to hide the bouts of excruciating pain from his companions during the first season. Richards was afraid that if Mackintosh found out, he would have no chance of sledging. In the end, Cope recommended that Stevens stay behind at Cape Evans. Mackintosh was undoubtedly relieved. Despite their friendship, Stevens's attitude had not improved—"the best is not made of opportunities & much is put down as excuses not to do certain duties," as Mackintosh put it—making him a potential liability on the trail. For his part, Stevens no longer cared about the promise of geological work at Mount Hope that had brought him to Antarctica.

As the day of departure drew near, Mackintosh began a letter to Stenhouse. "If this letter reaches you, I am sure you will have a pretty story to tell, also if you come through, I must here congratulate you," he began.

> *For wherever you have been or whatever you have done your safety and those under you, as well as the ship will be a remarkable accomplishment. . . . And my congratulations to the crew who have stuck by you— To the other officers, I hope soon personally to shake by the hands, & express to every one my gratitude, which I sincerely feel I owe—I fully realize the tough time you all must have endured, and to come through all in safety reflects the greatest credit on your good self.*

Mackintosh placed Stenhouse in command of the entire expedition while he was on the Barrier, and requested that he ready the ship for Shackleton's arrival. He instructed Stenhouse to send a party to the Bluff Depot to leave a message when the ship arrived. Mackintosh expected to return as early as March 10, but instructed Stenhouse to steam to Hut Point to pick them up on March 20. With any luck, Mackintosh hoped, they could "head straight out of McMurdo Sound for Home."

By August 31, the letters were finished and the sledges packed. Mackintosh dreamed of Shackleton, arriving triumphantly from the long journey across the continent and assuring him that all was well. On September 1, Mackintosh decreed, sledging would begin.

— 10 —

Return to the Barrier

September was cruelly cold, and the sledging went forward in hurried dashes in the lulls between blizzards. "Not a very promising start," Mackintosh noted grimly in his diary as they prepared to ferry another load of supplies from Cape Evans to Hut Point, the staging post for sledging operations. The makeshift gear offered little protection against the severe spring temperatures, and the socket of Mackintosh's missing eye was badly frostbitten. The handmade canvas trousers "froze like boards," as Joyce put it, in temperatures dipping to fifty below zero. Joyce felt the subzero weather keenly. He rubbed at the numb patches constantly, knowing that a previously frostbitten finger or toe was unlikely to survive a second freezing and could succumb to gangrene.

The younger men were slow to grasp the need for this kind of vigilance. Inside of a week, Gaze's heels were so badly frostbitten in his canvas boots that he could hardly walk. He was sidelined at Cape Evans indefinitely and had to be replaced by a reluctant Stevens, a poor surrogate for Gaze in both strength and enthusiasm. Mackintosh held out hope that the motor tractor would compensate for the deficiencies in personnel and set Richards and Gaze to work on the disabled vehicle. Back at Cape Evans, they rifled through volumes of the *Encyclopedia Britannica* for some clue to the workings of the engine. To rebuild the clutch, they painstakingly bored dozens of holes in the leather with a red-hot nail. But in the end, the motor tractor was beyond repair. Mackintosh would rely solely on manpower to lay the depots.

His plan was to supplement the two depots laid in the first season at 78°52' S, 169°05' E, the Bluff Depot, and 80°02' S, 169°25' E, the

Rocky Mountain Depot, and build four new caches at approximately 81°, 82°, 83°, and 83°30' south latitude, the last being Mount Hope. The distinctive peak near the end of the Beardmore Glacier would be on Shackleton's route as he followed the glacier from the Polar Plateau to the Ross Ice Shelf. After a journey of 1,100 miles across the continent, Shackleton's party would reach Mount Hope in desperate need of food and fuel, their sledges nearly empty, and would have no time to waste searching for the first cache.

Because the Ross Sea party was self-sufficient, dragging their own as well as Shackleton's supplies, marching 360 miles straight to Mount Hope would have frittered the stores away to nothing. Mackintosh proposed to employ a pyramid system, first stockpiling at least four thousand pounds of supplies at the Bluff Depot as an advance base, then making shorter forays south from there to establish the southernmost depots. His strategy would require three sledges, pulled by three men each, to make four trips to the Bluff Depot. From there, all three sledges would make two trips to latitude 80°. After advancing together to 81°, one sledge would return to Hut Point. The other two sledges would push on to 82°, with one turning for home after depositing one load. Mackintosh's team would journey alone to lay the final depot at Mount Hope, then retreat to Hut Point. There was scant margin for delays, despite the fact that the early winter blizzards of February and March had shut down sledging one day in every five during the previous season. Even if all went according to plan, the long homeward march from Mount Hope would strand them on the Barrier during the frigid early winter days of March.

The only means of completing work on the Barrier sooner, Mackintosh realized, were to travel faster or carry heavier loads. Joyce pinned his hopes on the surviving dogs to speed the depot laying. During the *Nimrod* expedition, his team had galloped up to twenty-five miles per day with a heavy sledge, and forty-five miles per day with a light load. But Mackintosh rejected Joyce's idea. He believed that the awkward combination of man and dog in the harness had hobbled the gait of both and slowed the pace in their first season. He was resolutely opposed to taking the remaining dogs out on the Barrier. He favored heavier loads, increasing the weight on the sledges so that they pulled an average of 174 to 187 pounds per man. In Joyce's estimation, 150 pounds was manageable.

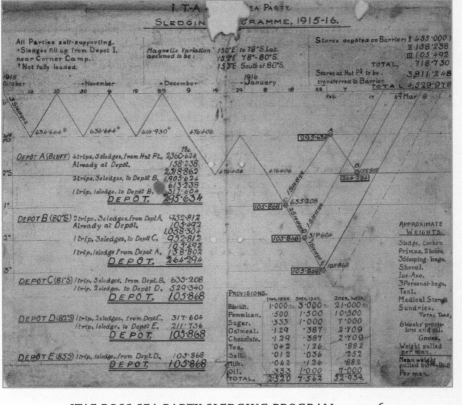

ITAE ROSS SEA PARTY SLEDGING PROGRAM, 1915–16

Man-hauling had long been favored by British polar explorers. Scott once wrote, "No journey ever made with dogs can approach the height of that fine conception which is realised when a party of men go forth to face hardships, dangers, and difficulties with their own unaided efforts, and by days and weeks of hard physical labour succeed in solving some problem of the great unknown. Surely in this case the conquest is more nobly and splendidly won." The high-flown rhetoric cloaked the fact that man-hauling was a last resort rather than the method of choice among Antarctic travelers—even for Scott, who also tried dogs, ponies, and motor vehicles—inspiring far more dread, on balance, than edification. As Cope's party had learned so painfully in the first season, it was slow, backbreaking work, more suited to a penitential pilgrimage.

And yet by the end of September, the party had indeed hauled 3,800 pounds of stores to Hut Point, on schedule and according to Mackintosh's plan. Joyce could muster little argument with success. Still, Mackintosh was beleaguered by private misgivings as he prepared for the big push to Mount Hope. Joyce might snap at Mackintosh's heels, but he was not personally answerable for the lives of the men or the outcome of the expedition. As the commander, Mackintosh alone stood accountable and felt the full burden of living up to the faith Shackleton vested in him. Poorly clad and badly equipped, his men faced "an unusual element of risk", as the ordinarily buoyant Spencer-Smith wrote to his parents.

On the last day of September, Mackintosh wearily composed a letter to Shackleton:

> *At present, while the programme has been drawn up, I should not like to predict how far it can be carried out, the fitness of the men, and how they will stand the heavy hauling we shall have to do being unknown factors. It is my wish to reach Mt. Hope at least, as prearranged; but circumstances, with which I shall herein acquaint you, have so forced themselves upon us, that our means are considerably reduced.*

Mackintosh began at the beginning: the chaotic refit and provisioning of the *Aurora* in Australia, the late arrival in McMurdo Sound, the abandonment of the penguin research at Cape Crozier, the failure of the motor, the disastrous loss of the dogs, the disappearance of the *Aurora*, the haphazard preparation for sledging. "Nevertheless, I don't complain," he wrote, making light of the staggering task before them. Of the men, he had only praise, declaring "every one has tried his best." The failures he took upon himself without reservation:

> *I am sorry I have not a more happy report to give you, sir. I stand fully responsible for all that has happened, and for any mistakes I am prepared to answer. I am leaving this letter with a sketch of our sledging plans, and shall add at a later date an account of where the depôts are situated, etc. With all my heart I trust your gallant self and party are safe, and that we shall all meet. Then any trouble or difficulty that has been experienced will be as nothing. For me it will be a joyous moment.*

The next day, the men rose at dawn to begin their second season on the Barrier. The sky above the snowy summit of Observation Hill was cloudless and vivid blue, but to the south, an opaque haze obscured Minna Bluff. It portended a blizzard blowing in from the south, the first in a wave of foul weather that kept the party hut-bound for over a week.

They were slipping behind before they had even begun. When the party finally set out on October 9, Mackintosh was desperate to make up for lost time. Once again, he increased the load of the sledges, packing each one to 600 pounds, a full 150 pounds heavier than Joyce had urged, and ordered them hitched single file with all nine men pulling together. Joyce refrained from comment and donned the man-harness with the others.

The sledges hardly budged, miring deeper in the snow. The men struggled forward, but it was punishing toil, worsened by the dazzling heat of the sun. By day's end, they had covered a half mile. In the morning, all nine men once again stepped into the traces and heaved with all their might. After an hour of marching in agonizing lockstep, they had crept just a hundred yards. Joyce took Mackintosh aside and persuaded him to give up on the sledge train, but they fared little better when the three sledges were separated. In five grueling hours, they gained a mile and a half. "I don't think in all my experience down here I have had harder pulling," Joyce wrote. Ordinarily in the lead, he spent the day pulling up the rear with a defective sledge. His mates were appalled by the stupendous exertions required for so little profit. "Devilish hard pulling," Jack called it, wondering how long they could possibly continue. For all the titanic effort, the total for the day was five and a half miles.

Clearly, the prospects for the season were bleak. Even Hayward believed it would be "impossible to carry out the season's programme." The conditions only deepened their discouragement. After the shimmering heat of the day, the night temperature dropped to a frigid minus thirty-six, freezing the sweat-soaked clothing into a brittle sheath. Throughout the night, they shivered in their bags, trying to avoid touching bare skin to the icy garments.

The physical miseries only made the men more querulous. As the slog resumed early on the eleventh, even the phlegmatic Wild was irritable. "We pulled the heaviest sledge & the others couldn't keep up. It's the foreigners that do it, they give everybody a bad heart." Joyce's forbearance was clearly worn threadbare by the "physical farce," as he called it. His defective sledge mulishly refused to budge. Lagging far behind, he shouted to Mackintosh for a halt and insisted that they stop and redistribute the weight on each sledge. After weighing the loads, he was appalled to discover that they were pulling 2,000 pounds in total, or 222 pounds per man—22 pounds more than he had reluctantly agreed with Mackintosh, and 72 pounds more than he felt was manageable. The fitful truce between the two was at an end. Joyce turned on Mackintosh, furious that he would sacrifice the men. "I think the man must be mad to think it is possible to pull such a load in the conditions in which we are placed. Only one thing he thinks of that is the men at the other end but he won't take good solid advice how to make good out of a bad thing."

The stress on the equipment was excessive as well; the ramshackle sledges could only take so much abuse. He pressed Mackintosh to reduce the loads to 150 pounds per man. "Hearts are willing but strength will not avail," he warned. By his reckoning, they could make up for the reduction with five trips to the Bluff Depot instead of four. Mackintosh rejected the plan. "I suppose he will learn to his regret that he has not taken good advice," Joyce wrote. A snowstorm forestalled the argument. Joyce retreated to his tent, "weary, worn and sad," as he put it, shivering in minus thirty degrees.

Disheartened but undeterred, Joyce braved the storm the next morning and crawled into Mackintosh's tent "to tell the Skipper off." By the time he emerged, any promise of détente had evaporated. Mackintosh announced he was splitting the group into two parties. He would push south with Spencer-Smith and Wild, and Joyce would take charge of the other five men: Hayward, Stevens, Cope, Richards, and Jack. With that, he dumped 60 pounds from his sledge, paring his party's load down to 180 pounds per man, and departed. "We've left the others behind with poor old Ern in charge," wrote Wild, admitting he was "d——d glad" to leave the tensions behind.

Joyce was left with two sledges and 1,392 pounds of stores. Each man was faced with pulling 232 pounds, a staggering burden that could only be shifted by relaying. Nonetheless, all of the men in Joyce's charge were apparently relieved to join him. "Skipper has little idea of system or order whatsoever," observed Jack, later adding, "It's 100 to 1 that he never makes it on the course he is steering. Think little of him either as navigator or leader." Leaving Joyce's team with the heaviest hauling fostered resentment, and the men complained that Mackintosh was deserting them.

Joyce's first order of business was to lighten the sledges. Without provisions, each sledge weighed 244 pounds with essential gear, including the tent, sleeping bags, Primus stove, ice axes, shovels, bamboo poles, rope, medical kit, and spare clothing. The ration of food and fuel for three men for one week in the field weighed fifty-three pounds. He told the men that he had decided to jettison all spare clothing and gear, reduce their daily rations, and depot two weeks' worth of Shackleton's provisions on the spot. The cutbacks would mean an extra trip later, but it brought the average weight per man down to 170 pounds. Joyce also informed the men that he would deviate from the previous season's course, which detoured well away from White Island to avoid crevasses, deciding that they had added unnecessary mileage to the journey.

Morale soared as the party found sledging "not the heart-breaking strain as hitherto." The change of course was a different matter. By October 16, they were threading through a maze of crevasses. Joyce edged tentatively forward, tethered to his companions by a twenty-foot Alpine rope. A pristine surface could conceal a deep chasm. He ordered the rest of the party to rope up as well, anchoring to each other or to the sledges. When he dropped into a crevasse, the man to whom Joyce was belayed swiftly braced himself with an ice ax driven into the ground, then pulled Joyce out. The technique worked fairly well, unless the crevasse was so long that belayer and belayed went down together. To avoid this hazard, Joyce observed the direction of the visible crevasses and skewed the course to run across at right angles, since they occurred in swarms running parallel.

The immensity of the fissures staggered even Joyce. "We were naming them different streets as we crossed over them," he wrote after the men safely dodged a gaping crevasse the breadth of London's Regent Street. The men went sprawling as their fur finneskoe skated over the slick surface, skittering to the rims of the carved thoroughfares. The constant daylight glare flattened light and shadow, rendering the crevasses indistinguishable. Richards and Jack removed their goggles to better discern the terrain and soon felt the sensation of burning embers under their eyelids that meant snowblindness.

Joyce's sight was unaffected, and on October 21, he spotted the black pennants fluttering on the Bluff Depot. There were signs that Mackintosh's party had come and gone, not surprising given their head start and lighter load. Wrestling heavier sledges through crevassed country with biting headwinds blowing the drift into their faces, Joyce's party struggled to cover the requisite ten miles per day and make up for the poor start. Briskly, they unpacked the supplies for the depot, 373 pounds, and prepared for the journey back to Hut Point.

Before leaving, Joyce slipped a letter for the *Aurora*'s crew into the depot. He knew that Stenhouse would send a party out to the Bluff Depot for messages when the ship returned. "My dear Paton and Bhoys," he greeted the Scottish boatswain chummily, with the nickname for his favorite Glasgow football club. "Just a line hoping you and all the Bhoys are in the very pink." He informed his mate of the new order of separate parties. With facetious swagger, he sketched their travails and expectations:

> *I think we shall just be able by a stroke of good fortune to carry out our programme that is lay a Depôt at 83°30' I suppose it ~~is~~ will be really the biggest thing ever been done. Here we are sledging, the last bath and shift of clothes I had was Jan 20th 1915 on the ship in the Galley, the last pipe of tobacco in April, ah well!*

The crossout was an uncharacteristic touch of modesty. More brazen than usual, however, was his subsequent assertion that his party would be southward bound after the next load to the Bluff, only their second. This would be news to Mackintosh, who expected Joyce's team to haul four loads in total. But Mackintosh intended only one party to lay the ultimate depot at Mount Hope, and Joyce, it seemed, was considering a dash ahead of Mackintosh to meet Shackleton himself.

There was still much ground to traverse before then, and Joyce meant to do it as quickly as possible. He hewed to the same course as the outbound journey, and despite sighting forty-seven crevasses on October 24 and blundering into many of them, the party traveled fifteen miles. By October 26, Joyce sighted Corner Camp and angled sharply west. Beyond was Safety Camp at the edge of the Barrier. The cold was stinging, twenty-six below zero, and, as Joyce observed, cruel motivation to keep moving. With good surface conditions, they would reach Hut Point by the next day.

Instead, the snow surface had frozen into piecrust, as they called the detested stuff, requiring a forceful stamp to break through with each step. The effort was exhausting. Up ahead, someone spied an upturned sledge to the northwest. Digging down into eight feet of drift, Jack realized that it was Scott's motor depot, which was soon confirmed by a note:

Dear Sir,
We leave this morning with the dogs for Hut Point. We have laid no Depôt on the way in being off the course all the way. I have not been able to leave a note before.

The letter was signed by Apsley Cherry-Garrard, who had been sent out in February 1912 to assist Scott's party on their retreat from the South Pole. Unschooled in navigation and dog driving, the earnest, bespectacled twenty-six-year-old was unprepared for the dilemma that confronted him. When Scott failed to appear at the rendezvous point, the so-called One Ton Depot, Cherry-Garrard was at a loss. Unbeknownst to him, Scott's party was some sixty miles south, desperate for aid. Instead of pushing south, Cherry-Garrard and his companion turned back, leaving the note on the homeward trail. By late March, Scott and his men lay tent-bound in a blizzard about eleven miles short of the well-stocked depot, starved and frostbitten. They died before the month was out.

The mute plea of the would-be rescuers served as a stark reminder that no search parties would seek out the Ross Sea party if they failed to appear at the hut in March, and that Shackleton could be racing, like Scott before him, for the salvation of the depot ahead. The first he would reach would be the yet unlaid cache at Mount Hope.

Joyce was unfazed by the plaintive memento mori. "Rather pathetic," he commented. He reserved his enthusiasm for another find, four cases of dog biscuits, and set to work digging the booty out of the ice. Suddenly, the ground beneath their feet collapsed by a foot. Scared witless, Richards and Cope bolted from the pit as a rumbling tremor radiated around them. The entire Barrier seemed to be dropping. Nerves frayed, Joyce pushed the team onward for a short distance before turning in for the night.

October 27 brought fair weather, and by late afternoon, the party had reached Safety Camp. A half hour later, a northerly blizzard closed down visibility. Determined to make Hut Point, Joyce pressed on. Suddenly Hayward shouted, "He's down, he's down, look out he's gone, he's disappeared." Stevens had fallen into a crevasse, his body wedged tightly in the ice. In a momentary clearing, Joyce saw that the entire party had almost glided over a steep section of the Barrier edge, a forty-foot drop. By the time the scientist was hoisted to the surface, badly shaken, a fierce blizzard had enveloped McMurdo Sound. It was late evening before they dragged the sledge up to Hut Point.

When the party arrived, they found Mackintosh, Spencer-Smith, and Wild recovering from their first trip. They had arrived at the hut three days before. Recuperated, Gaze had come from Cape Evans with the dogs. There were hearty greetings all around, but something had changed in Mackintosh's demeanor. He handed a note to Stevens:

> Private *I hope you did not associate yourself with those who made those uncalled for, disgraceful remarks on the Barrier—not even what one would expect of dock labourers or coalminers. It's disgusted me so much that unless I get an apology—every one of that party who took part—will be finished with any privileges they have hitherto enjoyed—I should like a note from you, confidentially, if I do not see you—also please keep any account you have of that trip.*

The furtive murmurs of disgruntlement had flashed into public carping. The derision that followed Mackintosh's split with Joyce on the Barrier had somehow reached his ears, perhaps through Stevens himself. Of all the men in his charge, only two enjoyed his full confidence: Spencer-Smith, whom he called "invaluable in his advice," and Wild. In the event of some calamity befalling him on the Barrier,

Mackintosh informed Stevens that the chaplain would take command of the sledging parties. He relied upon the malcontent Stevens only out of necessity, not trust. The chief scientist would be the only member of the expedition to remain at Cape Evans during sledging, monitoring the meteorological instruments for the season. In his letter to Stevens, Mackintosh instructed that any members returning to the hut during the season—presumably due to a breach with Mackintosh— "should have nothing to do with any work of the Expedition."

Mackintosh then addressed a letter to Stenhouse. Embattled, he seemed relieved to address a sympathetic confidante, a man who was perhaps no longer alive. The sledging program, he lamented, "has not been a success, owing to the teams not being able to pull their loads in this manner." Girding himself against his alienated crew, he cautioned Stenhouse to "have nothing to do whatever with anything [returning members] tell you unless they have a note from me." He seemed to be preparing himself for mutiny and a race to the ship by dissidents pleading their case with Stenhouse. Surely reflecting on his own woeful predicament, Mackintosh advised that Stenhouse keep strict discipline on the ship and avoid becoming "familiar more than you can help" with any of his subordinates.

In the cramped quarters of the hut, the tension was inescapable. Gaze, newly arrived after his month alone at Cape Evans, was disturbed by the hostility furring the atmosphere. "Seems to be something wrong somewhere—there's no pulling together," he observed. Now and again, conflict sparked into the open. Mackintosh, reversing his pessimism, decided to take the dogs on his next trip. Joyce argued the point until Mackintosh surrendered the team to him. But Mackintosh would not be dissuaded from his high expectations and still insisted on loads of 186 pounds per man. He communicated his orders in a formal letter to Joyce, who slept a few feet away on a wooden pallet. In closing, he added, "Trusting this will be plain to you & that you will ever bear in mind that the lives of Sir Ernest Shackleton & his party are to a great extent in our hands."

Mackintosh packed and left with Spencer-Smith and Wild on October 29. The next day, Stevens headed back to Cape Evans for the season. Joyce, for one, was relieved to see the ineffectual Stevens go. "I can't express words on this man, but will do so in front of Shackleton," he vowed. The dogs, however, he welcomed with open arms. The wiry Bitchie was a good puller and held her own in the team, but the four males had made a poor impression during the first season. Towser, a tawny mongrel with traces of collie, was rated by Mackintosh as "quite hopeless for this task. Has absolutely no brains or energy." Sledging was far easier for Gunboat and Oscar, stout brutes thought to be part Newfoundland, both well over a hundred pounds and lethargic. Richards called Oscar "an unlovely specimen, a bit shambly, with a broad leonine head and a low criminal-type forehead." Conrad was untried, left behind the first season when a fight left him lame. Unlike the rest, he was descended from Greenland stock, with a strong physique and dense coat. He had been purchased in Hobart from a seller claiming the dog had been born of Amundsen's team of Inuit dogs. The second female, Nell, was heavily pregnant and left behind.

After blizzards delayed their start, Joyce's party got under way on November 5. Although the floury new snow was hard going for the five dogs, who sank up to their bellies, they wallowed gamely forth at a steady pace. Con proved to be a natural lead dog, highly intelligent and capable of unifying the motley team. Joyce's faith in the dogs was justified. The party hauled a load to the Bluff, arriving back at Hut Point at a surprising clip, sixteen days round trip rather than the three weeks Mackintosh had expected. The sledge fairly flew over crevasses. "We were going so fast that the dogs that went through was jerked out," Joyce exulted. It was an astonishing performance. The next day, they made a record distance of nearly eighteen miles. The dogs were doing the work of four men. Joyce's party had deposited 624 pounds at the Bluff Depot, a full 300 pounds more than Mackintosh had ordered. Mackintosh's party had apparently preceded them, dropping off a load of 188 pounds and heading out again.

Joyce's confidence in the expedition soared as they sped along at a "ripping pace." The animals were fighting fit after their first journey. His determination was infectious. "I for one don't see why we shouldn't be able to do this," Gaze wrote in his diary.

The high morale belied the party's physical condition. Joyce was so snowblind that he could hardly open his eyes. He had been leading for fifty consecutive days, and staring straight into the "white ocean," as he called it, had seriously impaired his vision. The party had no spare goggles, so Joyce improvised, tying a leather strap with a slit cut in the center over his eyes in the Inuit style. This, too, had been broken and discarded, and his eyes were soon seared by the ultraviolet light reflected off the snow. The zinc sulphate ointment did little to soothe the pain, so he flushed his eyes with cocaine dissolved in water to anesthetize them. By the time the party reached crevassed country again, he was completely blind and blundering about on frostbitten feet with heel blisters "as big as potatoes." He had worn out two pairs of the homemade fur boots already and had only two pairs left for the next five months, with a thousand miles yet to march. Hayward was badly snowblind as well, and Cope was very nearly crippled by frostbite. Joyce decreed a prolonged stay at Hut Point after the load had been depoted, counting on prodigious quantities of seal meat and round-the-clock sleep to restore body and soul.

On November 20, as Joyce and company rested at the hut, Mackintosh's party depoted their third load of stores and turned north once again. But Mackintosh was not headed for Hut Point. At Spencer-Smith's suggestion, he had modified his plan. Desperate to speed up the depot laying, he decided to return only as far as Safety Camp after each trip, where Joyce had stockpiled some stores. His scheme eliminated the time-consuming nuisance of negotiating the sea ice to Hut Point. He also decided that it would be unnecessary to sledge as far as the Bluff on each trip, and dumped the last load at one of Cope's first-season depots. Still, the loads were fiendishly difficult for the three men to haul. Even though relations were harmonious, Spencer-Smith found the work demoralizing. "Very sick of life," he confided to his diary. Throughout November, he was nagged by a sore Achilles tendon that seemed stubbornly slow to heal as they shuttled across the Barrier.

"Puzzled and a little anxious about the other party," wrote Spencer-Smith on November 24 at Safety Camp, where they were laid up in a snowstorm. Mackintosh had trusted that he would cross paths with Joyce on the Barrier. Failing that, he left notes at the depots for crucial communication. But as the weeks wore on, he and his companions grew worried. They had not seen Joyce's party for almost a month. If either party met with misfortune, they were on their own. The next day, they broke camp and Mackintosh left written instructions for Joyce.

Hours later, Joyce's party arrived at Safety Camp, fresh from four days' rest at Hut Point and headed out on the Ross Ice Shelf with their third load, bound for the Bluff Depot. The scuffled snow and fresh tracks made it obvious that they had just missed the other party. Joyce soon found Mackintosh's note. "We now must push on, time is getting desperately short, and we have a long way to go," Mackintosh entreated Joyce. "I can only conceive that something serious must have delayed you." Once again, Mackintosh revised his battle plan and issued new orders. Reading the details, Joyce was infuriated. "Letters from Skipper, usual whining tones," he commented sourly. With the numerous shifts in strategy, Mackintosh's plan had acquired a Byzantine complexity, and heightened the probability of costly errors.

Most galling to Joyce was Mackintosh's command to stop using the dogs, presumably to save the weight of carrying dog food. After their recent outstanding performance, Joyce was more convinced than ever that the dogs were "the mainstay of the work" and tended them obsessively. With reckless disregard for the welfare of the animals, Mackintosh suggested that Joyce tie them up at Hut Point before heading south again, trusting that they would eventually break free and find their way back to Stevens at Cape Evans.

The constant improvisation had eroded the men's confidence. Richards regarded Mackintosh as "impetuous and not very tactful," and Gaze decried "the absurdity of his scheme." Still, they felt duty-bound to follow it. Now a hasty line in his letter struck at the heart of morale. Mackintosh questioned whether Joyce had left a sufficient quantity of stores at the Bluff Depot. After the staunch efforts and remarkable results to date, Joyce's party took it badly. Gaze poured his pent-up frustrations into his diary. "He'll have to be pretty careful or else he'll get his head punched," he raged. "There was a direct insult offered to Joyce in the note he left and I for one wouldn't take it from the little swine."

In fact, Joyce had resolved to make a stand. Defiantly, he harnessed the dogs and headed south with the third load for the Bluff. After months of wearing negotiation with Mackintosh, he was prepared to go his own way. His resolution did not come easily. As much as he argued, criticized, and cajoled, he was a navy man to the bone. Mackintosh was still "the Skipper" or "the Boss" in his diary, and "Sir" in his letters. The hierarchical order had been drilled into him so thoroughly that flagrant insubordination went against his grain.

Three days later, on November 28, the paths of the two parties converged. Huddled in his tent with Mackintosh, Joyce proposed a new approach. "Richards & I gave him a really good working plan to go on, but as usual he thinks he knows best, but will find out before long he is in the wrong—& on one occasion he tried to ride the usual high horse but I wasn't having any," Joyce wrote afterward. "I told him straight, he would be getting into trouble one of these days through his foolishness." Mackintosh proposed that they adopt a new scheme he was already employing—ignoring the calendar and working as if each day was twenty-one hours long, thereby working an eighth day in each week. Joyce rejected the idea. When Mackintosh accused him of sabotaging the plans, Joyce countered with his results. Including the current loads, Joyce's party had hauled 1,726 pounds to the Bluff while Mackintosh's party had depoted 553. Stunned, Mackintosh backed down and apologized. Then, he capitulated and agreed to Joyce's new program. They arranged to meet again at the Rocky Mountain Depot around December 23.

Though triumphant, Joyce was concerned about the outlook. Mackintosh's hectic regimen had taken an obvious toll on his party. It had been a month since Mackintosh's men had their last sojourn at Hut Point, and he did not intend to return there until the depot laying was finished in March. Spencer-Smith looked spent, and Wild admitted that they had been having a difficult time. When Joyce asked him about the eight-day workweek, Wild denied that they were working on that schedule and said Mackintosh was "off his head." Joyce was stymied, unsure what to make of Mackintosh's "silly lies," but he brushed his concerns aside and instead focused on the welfare of his own party. He was a firm believer in routine and rest. His team was fresh from their recent break at Hut Point, and he planned to head back for another spell at the hut in early December.

"I never in my experience come across such a idiot in charge of men!!" Joyce railed in his diary. They went their separate ways: Mackintosh's team bound for the Rocky Mountain Depot, the first step in the big push south, and Joyce's party headed north to pick up their fourth and final load from Hut Point. As the summer sun performed an endless circuit in the sky above, never rising or setting, temperatures climbed steadily above freezing. All hands stripped off their winter clothes and marched in underwear and boots, the absurd spectacle distracting them from the tedium of covering the same ground for the third consecutive month. Joyce tried a shortcut, only to find they were straying into more crevasse-pocked terrain than ever before. Far more unsettling, though, was the failure of a Primus stove on December 4. Cobbled together with tin can scraps, the remaining functional stoves seemed liable to fail at any time, given that they were used twice as long each day to prepare food and water for six men.

Joyce urged the dogs on to Hut Point at an unprecedented pace, topping seventeen miles for three consecutive days, and reached the hut on December 7. This time, Joyce kept the party there for a week, urging the men to eat as much seal meat as possible. The next journey would take them away from the hut and fresh food for at least three months, and Joyce worried about scurvy. He saw to it that the dogs were also gorged with seal meat, as it was plain to him from the death toll in the past season that a diet solely of dry biscuits was deficient. The working team was down to four males: Conrad, Oscar, Gunner, and Towser. Bitchie, like Nell, was pregnant and close to whelping, so she would be left behind. Joyce ordered the party to stockpile bags of fried seal meat for the journey.

Joyce's entourage arrived at the Bluff Depot on December 28. A note from Mackintosh was waiting with another change in plans, though he adopted a conciliatory tone: "I was very pleased on arrival here to find the two good loads your party have brought forward—I must admit, they were beyond my expectations!" Mackintosh's apprehensions had been calmed by Joyce's impressive efforts. All of the essential stores had been moved to the Bluff, ready for the next stage. Mackintosh could now move resolutely toward Shackleton.

To speed the depot laying, he asked Joyce's party to haul the necessary stores to the Rocky Mountain Depot in a single trip, instead of two. This would mean hefting two thousand pounds on the sledges— "damned heavy," in Gaze's words. Joyce readily accepted the plan, confident that the dogs were well conditioned. Although Joyce's party had averaged nearly eleven miles per day in December, twelve days had been lost due to weather delays or rest breaks.

The next morning, they harnessed up and started south for the Rocky Mountain Depot. Mackintosh had made one round trip out to the depot already, but his tracks were filled. Joyce regarded the miles between the 79° and 80° depots with almost superstitious unease. His own experience, and the stories of earlier explorers, convinced him that the weather was bound to be stormy and unpredictable.

The turbulent weather had left its mark on the landscape. As the party advanced south, the terrain was "too awful for words," according to Gaze. The sledges bogged down in the glutinous snow. That evening, they spied Mackintosh's tent about two miles ahead. For the next two days, Joyce's party followed in his tracks and camped warily apart. On New Year's Eve, however, they caught up, and Joyce ventured to Mackintosh's tent, deciding "it would be a fine thing to be at peace with all the world" for the new year. Joyce hoped to enlist Mackintosh in cooperating with a new plan.

He crawled into the tent, where Wild offered him a pipe of tea leaves to smoke. Mackintosh seemed defeated, confessing that he had only a "slight hope" of sledging as far as the Beardmore Glacier. He offered little objection as they hammered out a compromise plan favoring Joyce's ideas. It was decided that Mackintosh's party of three men would make the entire journey to Mount Hope. Joyce's party would move stores as far as 81°, where three of the six men would return to Hut Point. Joyce and two men would then push on to 82° and hand the dogs over to Mackintosh. At that point, Mackintosh left it up to Joyce whether his party would continue to Mount Hope or return to base. More manpower meant hauling more supplies, but it also meant consuming twice as much and exposing everyone to greater risk. The entire expedition could be derailed if a man was stranded so far from Hut Point with a serious injury. Mackintosh entrusted this fateful decision to Joyce. "Made all arrangements for the final sprint which after a talk came to a mutual understanding (at last). I think it is high time he woke up," Joyce wrote with satisfaction. "I suppose I am now starting on the biggest job of my life that is to get to 83° S & if poss to relieve S[hackleton]."

Gaze celebrated the new year with Spencer-Smith and Jack. As usual, he produced an unlikely array of delectables: gentleman's relish, pâté, and a stack of books. After the cheerless Christmas on the march, their reunion was the most welcome gift of all, and they kept up an excited stream of talk for hours about evolutionary theory, religion, and, inevitably, home. A year before, on the eve of 1915, the *Aurora* had sailed into the Southern Ocean, beyond the reach of wireless. "We have severed our last link with civilization," Jack had written. A year later, as he trudged in front of the sledge each day, keeping silent to conserve his strength for the hauling, his mind was filled with thoughts of family. "What is everyone doing, what has been the result of the war & how is it progressing if not already ended? These are questions that occur to me continuously on the march & at all times." Hayward's thoughts were of his fiancée as he wrote his daily letter to her in his diary, though the entries had dwindled to a terse line or two since the start of sledging. Even Joyce grew pensive. "I often think of the brother & wife & nephew and wondering how the world is treating them," he wrote of his younger brother Joe, who had enlisted at the outbreak of the war. Wild thought of his brother Frank who, if all was well, was marching with Shackleton toward the Pole.

On the first day of 1916, the parties broke camp. Joyce harnessed the dogs to the overburdened sledge. Shortly after the start, Towser began wheezing and choking. Joyce plunged his fingers down the dog's airway, which seemed to relieve his breathing, though bluish gums did not bode well. "I sincerely hope they will not give up," worried Joyce. But it was not the animals that gave in.

11

Mount Hope

"Our primus has turned dog on us," wrote Gaze on January 6. A vital metal part inside the jerry-rigged stove had burned out. Mackintosh's party was already using the only available spare burner, their stove having also given out. Joyce repaired the stove again, wondering how long it would last. He knew it would be risky to march farther south with six men depending on one working stove.

Confirming their compromise, Mackintosh's last orders directed Joyce's party to continue to 81° and send Cope, Jack, and Gaze back; only Joyce, Richards, and Hayward would push on to 82°. Now Joyce considered sending the trio back at 80°, even though their manpower was sorely needed to heft the necessary loads. He put the question to Richards. Although the former schoolteacher had only just turned twenty-two, Joyce had come to value his judgment. It was a curious alliance, the oldest veteran and the youngest novice of the party, but they worked well together. Joyce admired the young scientist's logical perspective on problems and willingness to give his all. Richards saw past the older man's often crass manner to appreciate his genuine expertise.

Ever since Joyce had gained the upper hand with Mackintosh, he had become more arrogant with the others. In December, he had ordered Gaze and Jack to leave some gear behind to save weight. When Gaze protested, Joyce decided he was "getting too big for his shoes [and] had to be taken down a peg." Gaze took exception to his "high and mighty attitude," inciting Richards to side with Joyce. In the end, the quarrel petered out when both sides realized the triviality of bickering in the face of the immense task ahead. But the balance of power had shifted irretrievably. With Richards at his side, Joyce grew bolder about imposing his will on the group and worried little about diplomacy. As one man put it, "There was nothing in his training, experience and outlook which specially fitted him to occupy a difficult position with discretion, and his temperament was not such as made it easier for him or for others." As a Royal Navy petty officer, he compelled un-questioning obedience from roughneck sailors. College boys were a different matter.

The party's mission hung on the corroded metal ring of the Primus stove. After hashing out the risks, Joyce and Richards concluded that only three men should advance south with one problematic stove. Joyce ordered Jack, Gaze, and Cope back to Hut Point. Cope, who had been struggling to keep up physically, took the news with equanimity and per-haps relief. Jack and Gaze were thunderstruck. Both had been deter-mined to see the job through to the end and were furious with Joyce. In the daily grind of sledging, they had all suffered from frayed nerves, and careless words were routinely forgiven and forgotten. Only it appeared that Joyce had been taking names. "I am very pleased to get rid of Gaze & Jack as they have not been playing the game," he wrote in his diary. Feeling his control threatened, he was unmoved by their strenuous ap-peals. Neither would he heed their pleas that he double back to assist Mackintosh's lagging party. Joyce declared his intention to move south as fast as possible with Richards and Hayward at his side.

At the best of times, Joyce could be overbearing. Jack and Gaze's tol-erance had worn out. The mutual admiration and camaraderie that Joyce and the two Australians had shared since their first outing the previous season dissolved. "He's a rotter that man. I wouldn't trust him as far as I could kick him," Gaze seethed. "Joyce is not playing the game by any manner of means—nothing is done openly—all on the quiet." Gaze suspected that Joyce was "out to play DIRT on the Skipper," as he put it, racing ahead in order to try and meet Shackleton at Mount Hope first and leaving Mackintosh "to do the Hack work."

On reaching the Rocky Mountain Depot, Jack and Gaze unloaded their sledge and reluctantly packed up for the 150-mile journey to Hut Point with Cope. Joyce could not spare a compass to navigate their way back, so they had no choice but to follow their outbound tracks north before drift covered all traces of the route. If a blizzard struck, they would have no means of keeping a true course and would be immobilized until it passed. Spurred on by the dismal prospect of being lost in a trackless waste with a faulty stove, the homebound party made haste away from camp.

An hour later, Jack sighted Mackintosh's party on the horizon. When they met, Gaze shared his suspicions with Spencer-Smith. As always, the chaplain kept any murmur of dissension out of his diary. "They gave us news of Joyce's plans and also 1 lb. of onions," he recorded. He found the intrigue and deteriorating relations disturbing. On the flyleaf of his diary, he had inscribed, "It's all in the game. Play on!" His understanding of the credo, though, was at variance with both Joyce and Gaze. He added an enigmatic line to his diary entry in Latin: "Timeo Danaos et dona ferentes," "I fear the Greeks, even when they bear gifts," followed by a coded reference to Joyce's party. The cousins said their farewells and parted, one northbound, the other headed deeper south.

"Quibus autem cognitis, ne fraude summam laudem alii acciperent (!) nos longo itinere ad eorum castra progressi sumus," wrote Spencer-Smith later that day: "Learning of this, however, to prevent others winning the first prize by trickery (!) we advanced a long way toward their camp." Gutted by a hard run of over fifteen miles, they camped for eleven hours while Joyce's party raced ahead. Rebounding from their health problems, the dogs appeared to be hitting their stride. Joyce had been lavishing extra care on them, allowing them to run unharnessed, bringing them into the tent at night, and feeding them hot meals. "It is wonderful the amount of work they are doing," Joyce wrote. As the terrain worsened, they no longer achieved the exceptional records of seventeen miles, but still averaged a steady eleven miles per day. Joyce refused to slow the pace for Mackintosh's straggling team.

Two full days passed before the parties met on January 8. Catching up with Joyce's dog-powered sledge was gut-wrenching toil. After pitching the tent, Mackintosh presented a new plan to Joyce. "I think I shall have to disobey him again as I am sure if we are left to go as we are going we can easily lay this depôt. If it is not laid it will be to his bungling," he wrote afterward. He took fifty pounds from Mackintosh's sledge to ease their burden and left camp, speeding onward with the aid of the dogs, who bounded through the chest-high snow. Late in the day, Mackintosh and his men straggled up to Joyce's camp, short of breath and obviously exhausted.

Every ounce of available manpower would be essential to lay the depots. Yet, in their condition, Mackintosh's party could become a liability. Joyce told Mackintosh to hitch his sledge in tow behind theirs. Wearily, the three men fastened their man-harnesses to trudge alongside the dogs. The following morning, Mackintosh took Joyce aside. "Skipper asked me to take over the parties which I will do until the depôts are laid," he wrote. Defeated by his own physical limitations, Mackintosh had finally withdrawn from the battle of wills. He was limping badly, saying he had sprained his knee, and also suffered from painful hemorrhoids, a common problem of polar sledging. "Parties working harmonious together," Joyce wrote with satisfaction, though he was annoyed by the slower pace. "It is a pity though they did not let us carry on as we were going."

Beyond latitude 80° south, the familiar silhouette of Minna Bluff disappeared behind them, and they ventured into a strange featureless landscape. The Transantarctic Mountains had also slipped below the western horizon. Without landmarks, there was no means of fixing their location except dead reckoning, which meant carefully plotting the direction and daily mileage on the map. It was an inexact art, particularly since the sledgemeter, rolling behind the sledge, malfunctioned when snow or dog feces clogged the wheel and stopped the distance measurement. For insurance, they built cairns more frequently. Sighting down the line of black flags on the snow mounds, the navigator could plainly see if the party wavered from the true course, but the process was a laborious one and slowed the pace.

In 1908, Shackleton had passed through these latitudes and found the experience unnerving. "The Barrier is a dead, smooth, white plain, weird beyond description, and having no land in sight, we feel such tiny specks in the immensity around us," he wrote. Each day, Joyce faced the void first, taking the lead at the end of a twenty-five yard rope. Mackintosh, Spencer-Smith, and Wild were tied to the sledge. With the compass in hand, Mackintosh watched the rope and shouted corrections, left or right, to keep Joyce on the correct heading. Joyce fixed his eyes on a distinctive cloud to approximate a straight path. The dogs were hitched behind the men. Tethered by eight-foot ropes to either side of the sledge were Richards and Hayward. After shouting themselves hoarse over the howling wind, they abandoned attempts at communication, other than commands to the dogs or Joyce, to conserve their energies. When the wind died, "the silence was profound," as Richards described it, the only sounds being "the soft crunch of feet in the snow and the faint swish of the sledge-runners."

During the "endless hours" of each day's march, Richards withdrew, compulsively performing complex mathematical calculations in his head, "an automatic reaction to the monotony that was forced on us, and an anodyne to the weariness of the body." Spencer-Smith recited poetry and scripture under his breath. He read voraciously when the party stopped for a spell or camped for the night, crowding his mind with ideas to blot out the drudgery. He seldom complained, but his deterioration was evident to the others. "Skipper & Smith very crocked," observed Hayward, seeing their traces slacken and drag on the ground. Mackintosh, Spencer-Smith, and Wild had not rested at Hut Point since late October. Determined to keep up the pace, Joyce pushed the party forward with few concessions to the weaker members.

Ironically, Joyce was soon utterly dependent on Mackintosh. His snowblindness had returned and his eyes swelled shut. Increasing doses of the cocaine wash failed to blunt the pain, so Joyce inserted the cocaine slivers directly under his eyelids. Mackintosh took over the navigation as Joyce blundered behind. On January 12, Mackintosh steered the group to 81°00' S, 169°50' E, where they laid the depot as planned with 106 pounds of food and fuel, enough to last Shackleton's party of six for a week. For good measure, they left an additional 23 pounds of extra biscuits and two weeks' surplus of fuel, as well as extra provisions for their own return trip.

After a day in the lead, Mackintosh willingly surrendered the reins to Joyce. The constant vigilance was an intense strain on his lone sighted eye, aggravated by having left his monocle at the hut. Joyce watched their decline with growing concern and noted their condition in his diary each day. "Skipper not going it very strong," he wrote on January 15. Two days later, they had worsened: "2 members of our force are resting on their arms." The party would soon reach latitude 82° south. If the original plan held, it would be the end of the line for Joyce's party, leaving Mackintosh's party to proceed alone to Mount Hope. Mackintosh had left it up to Joyce whether to turn for home at 82°. When Joyce discussed the issue with his tentmates, Richards said what all were loath to face: Mackintosh's party was in no condition to reach Mount Hope, some ninety miles south, and return 360 miles to Hut Point on their own. If they went on alone, as Richards put it bluntly, "it would have meant suicide," a catastrophe for Shackleton as well themselves. Joyce agreed. An unspoken possibility remained: sending Mackintosh's party back to Hut Point while Joyce's party carried on, but it would have taken mutiny. Another complicating factor was the fragility of the stoves. Either could expire at any moment. A party stranded so far south on the Barrier with a failed stove would die in a matter of days. They would have to stay together to survive.

Richards's influence came to the fore in the decision. He had already assumed the role of navigator, taking bearings and marking their progress each day on the chart. He doubted that Joyce had the drive to organize and motivate the entourage onward. After Mackintosh's abdication, he asserted himself more strongly but subtly, realizing that a direct challenge was unwise. Joyce, he perceived, was "a malleable character," and he became adept at handling Joyce "without appearing to handle him." Quietly, Richards became the "moving spirit" in the drive to reach Mount Hope.

———

On the morning of January 18, Richards took bearings to work out their position. The lofty peaks of Mount Longstaff and Mount Markham had, at last, appeared on the horizon. Studying the map, Richards figured they were just north of 82°. After lunch, they built the depot at 82°08' S, 169°45' E with 106 pounds of food and fuel plus extra biscuits and kerosene. Joyce informed Mackintosh that he had decided to proceed all the way to Mount Hope with his party. Mackintosh acquiesced, then asked Richards to come into his tent. To Richards's astonishment, Mackintosh produced a sheet of foolscap and insisted that he read and sign the document.

"I R. W. Richards at present serving as a member of the above expedition under the Command of Sir Ernest Shackleton CVO do in consideration of the salary by me of £ __ per annum, undertake to obey the lawful commands of the above named Sir Ernest Shackleton CVO or those appointed by him," read the spidery penciled text. It was a paraphrased version of the agreement that all members of Shackleton's expedition had been required to sign, hastily handwritten by Mackintosh in the tent. Most of the crew had signed the official agreement before the *Aurora* sailed for the Antarctic, but in the frantic rush to leave port, Mackintosh had neglected to obtain Richards's signature. The document also specified that all diaries, photographs, notes, and scientific data were owned by Shackleton.

Richards recognized that this was no straightforward exercise in administrative housekeeping. As he interpreted Mackintosh's attitude, "'Well this bugger Richards, I suppose he might blow the gaff when he gets back, we had better seal him up.'" Mackintosh had acquiesced to the challenge to his leadership, but did not want the full inglorious story of the expedition's troubles told. Richards, however, saw no shame and felt no vindication in Mackintosh's failures. He signed the document. In seven days, he meant to reach Mount Hope.

Beyond latitude 82° south, land once again ballasted the sky to the horizon. To the west, the Transantarctic Mountains arced south. The party's path would converge with the range about thirty miles beyond latitude 83° south. Shackleton had been the first to trace a route through the barricade of mountains into the heart of the continent in 1908. His "Highway to the South," as he called it, was the great Beardmore Glacier, the pathway to the Polar Plateau and the South Pole. Flowing from the higher altitude of the interior, the Beardmore coursed before him through the range like an immense river flowing downstream to the Ross Ice Shelf. Shackleton discovered a hidden passage in the turbulence below, where the glacier churned the ice shelf like a waterfall, in the shadow of a peak he called Mount Hope. Now the Ross Sea party would have to find it.

"Difficult to pick out Mt. Hope as our chart is an old one and not marked," wrote Joyce. Their map of the interior was the 1901 "South Polar Chart," revised in 1910 to include the new features recorded on Shackleton's march south. "The whole country seems to be made up of range after range of mountains, one behind the other," as Shackleton described it. Of the scores of peaks in the horizon, only the highest and most distinctive summits were named on the map. Mount Hope was somewhere hidden among them. At the center of the chart was carto-graphic oblivion, devoid of markings save for Shackleton's track end-ing ninety-seven miles short of the South Pole. The Ross Sea party was seeing the vista ahead with Shackleton's eyes, guided by his words.

By all accounts, the Beardmore would be impossible to miss. "The glacier which we now saw must be the largest in the world; it is 30 miles in width & we can see over 100 miles of its length, beyond that must be the Great Plateau," wrote an awestruck Frank Wild, who was at Shackleton's side on the polar journey. The Ross Sea party struggled to relate the map to the vast landscape around them. It transcended all human notions of scale and proportion. The peculiar Antarctic atmo-sphere distorted the perception of distance and foreshortened geo-graphical features, crowding the horizon with hundreds of unnamed mountains and glaciers. If they chose wrongly, a mistake would mean a costly diversion when the ration amounts had been cut so very fine.

Food was not the only limited commodity. The men's stamina was rapidly eroding. The day after leaving the 82° Depot, Spencer-Smith admitted that his limbs ached more each morning. Mackintosh, too, was wincing as he hobbled along on his lame knee. Joyce feared that both men were too debilitated to march all the way back to Hut Point. If their health continued to deteriorate, they would have to be dragged on sledges. Regardless of their condition, Joyce was adamant: the party could not turn back before laying the last depot. On January 21, he had sighted a mountain that answered to Shackleton's description of Mount Hope.

Up ahead, in a low line of snowy peaks, he spied a red summit, about 3,000 feet high. Joyce guessed it was thirty miles away, two or three days' march at their current pace. The cumbersome ritual of stopping every few hundred yards to build cairns had acquired a me-chanical rhythm, and the mileage had been climbing steadily for a week. In spite of a lowering sky and a stinging wind, Joyce ordered an early start. By noon, they had dashed ahead six miles, crossing 83° south. It was far too strenuous for Spencer-Smith. "Nearly fainted at 11 a.m. and had to tell at lunch how weak I am," he wrote, still feverish. "Heart rather ricked, I fear, and knees bad—swollen like a great bruise above and below knee, especially the right." After the two parties joined forces, his arrhythmia had worsened, a fact he carefully concealed from the others. The stiffness and pain in his knees he attributed to frostbite, since his trousers had worn through and exposed patches of bare skin. He marched doggedly onward, keeping pace with the other men, yet letting his traces slacken. That night, the party camped at 83°05' S 171°05' E. They decided to lay no depot there, instead depositing as much as they could carry at Mount Hope.

On the march the next morning, Spencer-Smith staggered and collapsed. He confessed that he could go no farther. Joyce and Richards decided to leave him there in a tent to recover his strength while the party moved south, to which the chaplain readily agreed. They urged Mackintosh to stay and care for Spencer-Smith, but he would not yield to persuasion. He insisted that he was duty-bound to ensure that every depot was laid. Until then, he had accepted Joyce's inroads on his authority in the name of achieving their goal, but staying behind seemed a breach of his solemn oath to Shackleton. "A weak character," Richards called him, but he had not recognized the "will of iron" that Shackleton had seen. Joyce and Richards realized that, "short of restraining him," they could not prevent Mackintosh from marching south to meet the Boss.

Hurriedly, the five men erected a tent. "I should be all right bar loneliness and disappointment (probably merited)" Spencer-Smith wrote forlornly. Mackintosh and Wild settled him into his sleeping bag and arranged provisions within reach. They meant to return in under a week, homeward bound. In little more than an hour, "they rattled off at a tremendous pace; with the dogs scrapping en route," as the melancholy Spencer-Smith watched the caravan recede into the distance.

After advancing eleven miles for the day, the party camped. With five bodies crammed into a three-man tent, it was a restive night. Joyce overslept and emerged from the tent disoriented. Impenetrable fog cloaked the landscape, blotting out the cherished sight of the mountains. By afternoon, a blizzard shut down visibility entirely. Joyce reluctantly called a halt for the day. The storm roared through the night and the following day. "Nothing doing. Can't see," recorded Wild.

Wild's thoughts were elsewhere. He had not expected that Shackleton, accompanied by his brother Frank, would be delayed so long. It was still possible that the two parties would meet in a few days' time. There was another possibility: The delay may have meant that Shackleton's party had met with disaster. Too doggedly optimistic to contemplate the worst, Wild wrote a letter to his brother.

On January 25, the snow and mist cleared to reveal a stunning tableau. The mountains glowed ocher in the morning sun. The earthy color of the low summit was a balm after months of gazing upon the antiseptic landscape. After weeks of doubt, they stood transfixed by the sight. Just as Shackleton had described, a pass lay to the west of the peak, looking for all the world like his description of the so-called Golden Gateway. Joyce guessed it was some twenty miles away.

The men hitched the team and altered course slightly to the west, aiming straight for the pass. By noon, they had sprinted eight miles south. In the afternoon, the going became more arduous. They realized they were traveling up and down a series of hills. Like mounded swells in a frozen sea, the undulations were a sign of a major disturbance in the flat expanse of the ice shelf. Laboring up the ridges of ice, Mackintosh remembered Shackleton's caution about crevasses near the Gateway's entrance. As they drew closer to the mountains, colossal slabs of ice heaved skyward in fantastic shapes, separated by deep gashes in the surface, some narrow fissures, others broad chasms. It would be no easy task to find safe passage through the chaotic terrain. After advancing seventeen miles, they camped in the tumult, waking to sounds like pistol shots as the ice was riven by unseen pressure.

Only the great Beardmore Glacier could have created such enormous pressure, Joyce reasoned. The mighty glacier surged around Mount Hope like rapids eddying around a boulder. If the party was on course, the terrain would only get worse as they ascended the offshoot toward the main body of the glacier. "A fearful mess," Frank Wild had called it as he struggled upstream through the Gateway with Shackleton in 1908, plunging into crevasses so deep that he felt certain both he and Shackleton "were going straight to hell."

Rather than risk leading the entire party into a treacherous series of blind alleys, Joyce decided to make a reconnaissance journey. Roped to Mackintosh and Richards, he led the way into the labyrinth of crevasses. "All around us was such a scene as one sees in a Pantomime, but cannot imagine in real life," he later marveled. "We seemed to be in the centre of a vortex of ice, churned into caves, all of blue appearance, dark and light." They peered into an abyss, seeing no bottom, as Joyce insistently pulled the rope. Solid ground was illusory; the sloping catwalks were snow bridges. Joyce eased forward, catching his breath for a moment after each hesitant step in anticipation of freefall. The plunges followed all too often. Richards and Mackintosh hauled him back up, time after time. Searching for a path, he spotted a ramp of ice that seemed to correspond with Shackleton's description of "a long slope of about 2 miles in length & a rise of 2000 feet." As they ascended, Richards spied something odd in the snow nearby. Moving closer, they made an electrifying discovery: a pair of upended sledges, likely abandoned by Scott's last expedition en route to the Pole.

Finally, Joyce was certain. With Mackintosh and Richards in tow, he headed for the saddle west of Mount Hope. The ice underfoot was like polished quartz as they "climbed the glacier on the slope & saw the great Beardmore Glacier stretching to the south." The Beardmore glinted in the sunlight, snaking over a hundred miles long on the broad plain of the interior plateau. Here, at its terminus, the glacier tumbled down to the Ross Ice Shelf in a cataract of ice twelve miles wide. Below, bands of pressure flared for miles, like mammoth frozen ripples in a pool at the foot of a waterfall.

When Shackleton saw his "open road to the South," for the first time, he thought it might not be seen by human eyes again. Now Joyce, Mackintosh, and Richards joined the privileged circle of living men to look upon the deepest interior of the Antarctic continent. After a year of thankless work, they were no longer laborers but explorers. "A most wonderful sight," Joyce wrote in awe. Perched on a granite outcrop, Richards trained his binoculars on the southern vista. His heart quickened as he fixed on a dome-shaped blot in the distance. Focusing, he realized it was not a tent but a boulder. Shackleton's party was nowhere in sight. Mackintosh had long cherished the improbable hope of meeting Shackleton there or awaiting his arrival; now to linger even an hour was unthinkable.

It was late in the day, and the job was yet to be done. Joyce guided his roped companions down the slope and through the icefalls, surefooted now that the trail had been blazed. They arrived back at camp at

THE SECOND
SLEDGING SEASON
SEPT. 1915 – MARCH 1916

0   25   50   ▲ depot   ● camp
nautical miles

*Parties made multiple trips and deviated slightly
from the track shown between depots.*

BEARDMORE
GLACIER

Mt Hope Depot ▲
Spencer-Smith Camp ●
83° Camp ●

▲ 82° Depot

▲ 81° Depot

ROSS ICE SHELF
(THE BARRIER)

▲ 80° Rocky Mountain Depot

Blizzard Camp
Feb. 23–29 ●
79° Bluff Depot ▲
Cope's #3 Depot
▲
Mackintosh Camp
Mar. 8–16
Corner Camp
Spencer-Smith Camp
Mar. 8–9

MINNA BLUFF
MT. DISCOVERY
BLACK I.
WHITE I.
Safety Camp

CAPE
EVANS
ROSS I.
MCMURDO
SOUND

TRANSANTARCTIC MOUNTAINS

ROSS SEA

Kelly Brunt/Elles Gianocostas

DEPOT COORDINATES

79° *Bluff Depot:*   78°52' S 169°05' E   82° *Depot:*   82°08' S 169°45' E
80° *Rocky Mountain Depot:*   80°02' S 169°25' E   83° *Camp:*   83°05' S 171°05' E
81° *Depot:*   81°00' S 169°50' E   *Mount Hope Depot:*   83°31'30" S 171°00 E
                                                          (EST. WITHIN 2 MILES)

three o'clock in the afternoon, ravenously hungry after the twelve-mile journey. The five men hastily broke camp and packed the lightened sledge. Mackintosh was visibly pained by every step. Reaching Scott's old depot three hours later, the party set up the tent again. Joyce chose Hayward and Wild to accompany him to lay the depot, leaving Mackintosh in camp with Richards.

The Gateway was a moderate climb for fit men, but these were not fit men. They plodded up the slope in a pained crawl. The loaded sledge, down to about 150 pounds per man, was an onerous burden in their condition. In a narrowing of the pass, Joyce ordered a halt. They had reached the journey's end. He had chosen "a place which could not be missed by anyone coming from the S[outh]." They built a snow cairn, fifteen feet high, and marked it with one of Scott's sledges and a flag. The depot was stocked with one week's worth of provisions for six men, seventeen pounds of extra biscuits, an extra ten days' worth of fuel, a few books, and the list of bearings for the lifeline of depots spanning the ice shelf. Wild quietly slipped his letter to his brother into the depot.

"As I said when we started sledging I would do this & so with the help of 2 good pals we carried it out," Joyce declared on reaching camp at 10:30 PM. He was utterly spent, the only man to have made both trips, well over twenty miles in one day. Mackintosh was elated. The fears and anxieties of the past eighteen months had been banished, the burden of the lives of Shackleton and his men lifted from his shoulders, and he had Joyce to thank for it. "The Skipper telling us how good it was of us to bring him along. This is his first acknowledgment of the work we done," Joyce wrote, more in ill temper than truth. "Still I don't want his praise all I wanted to see was the work carried out what men are depending on."

On January 27, the Ross Sea party turned north. "We are now Homeward Bound 360 miles to go I think with the help of Good Old Providence we ought to be in by the 27th of February," wrote Joyce. Snowblind again, both eyes bound, he was unbowed. "Hanging on to the harness for guidance but pulling my whack," he scrawled in his diary in a sightless haze. The worst seemed over. Their duty to Shackleton done, they had only to look after themselves.

12

"Homeward Bound"

JANUARY 25–MARCH 18, 1916

As Joyce guided the party into the icefalls below Mount Hope, Spencer-Smith lay in his tent at latitude 83°12' south on the trail, his optimism undimmed. "A year ago to-day we set off from the ship on our first journey—all clothes new and clean; a team of 9 dogs and high hopes," he remembered on January 25. "Only the last remain and even they should be accomplished by the O.M. and Wild by now." After the party left for Mount Hope, he spent most of his time inside the tent, save for short forays to collect snow for melting into water. In a matter of days, the tent neared collapse under the drifting snow. In the hushed cocoon, he lay in his sleeping bag, dozing fitfully and waiting. "Dreamt that we met Sir Ernest and Frank Wild with one motor and one dog sledge—both clean and neat," he wrote, flushed with excitement. In his waking hours, he read novellas, composed sermons in French, and wondered where he was, performing elaborate computations to try and fix his position and that of the rest of the party, assuring himself at first that their return was imminent, and then that the delay did not mean disaster.

By necessity, the rations left for him were minimal. As the food dwindled, Spencer-Smith cut back to two meals each day. They had also left him a bottle of lime juice extract, from which he took precise doses with meals, "in case my complaint is some form of scurvy—which I doubt altogether." The prescribed dosage of a quarter ounce per day provided half a milligram of vitamin C at most, when he would have needed at least twenty times that amount to keep scurvy at bay. The extract could do nothing to check the disease, which, in fact, already held Spencer-Smith in its grip.

Ernest Joyce (seated) aboard the *Nimrod* before sailing for Antarctica in 1907. In 1914, Joyce was hired for the Ross Sea party for his sledging and dog expertise. Shackleton considered Dr. Eric Marshall (far left) for command. John King Davis (to Emily Shackleton's left, holding her son) declined to command the *Endurance*.

Æneas Mackintosh aboard the *Nimrod* as second officer in 1908, his first expedition with Shackleton. His right eye was destroyed in an accident.

Ross Sea party commander Æneas Mackintosh and Gladys Campbell on their wedding day in 1912.

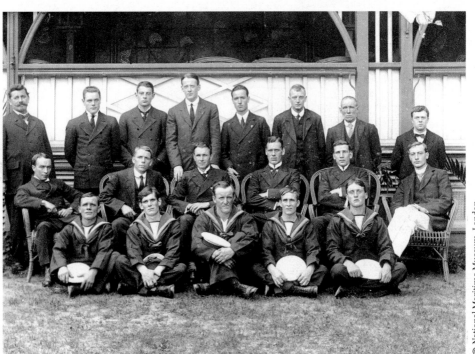

The Ross Sea party in Sydney in early November, before Mackintosh began recruiting in Australia. Rear (from left): Joyce, Hayward, Cope, Spencer-Smith, Mauger, Wise, Paton, Wallace (later resigned). Center: Mason (resigned), Leonard (resigned), Mackintosh, Stenhouse, Larkman, Stevens. Front: Warren, Mugridge, Atkin, Grady, Downing.

The incomplete Ross Sea party in Sydney in mid-December. Rear (from left): Mauger, Cope, Hayward, Gaze, Stenhouse, Mackintosh, Spencer-Smith, Jack, Stevens, Larkman. Center: Wise, Grady, Atkin, Paton, Downing, Mugridge, Gates (later resigned). Front: Kavanagh, Warren, Glidden. Not pictured: Ninnis and Wild.

The Ross Sea party during the voyage to Antarctica. Rear (from left): Hayward, Cope, Larkman, Paton, Richards, Mauger, Stevens, Spencer-Smith, Mugridge. Center: Thomson (with scarf), Ninnis (belted jacket), Warren (arms folded), Joyce, Jack, Mackintosh, Gaze, Stenhouse, Hooke, d'Anglade, Shaw, Kavanagh. Far left: Donnelly (hand on chin), Wild (holding dog). Seated: Grady, Glidden, Wise, Downing (holding dog), Atkin (behind barrel).

Reverend Arnold Spencer-Smith during his years as a teacher at a boys' boarding school in Scotland.

Londoner Victor Hayward abandoned his career in finance to sign on.

Joyce operates the motion picture camera as official photographer Arnold Spencer-Smith looks on.

La Trobe Picture Collection, State Library of Victoria

The *Aurora* in the pack ice of the Ross Sea.

Canterbury Museum, Middleton Collection

Joyce and Keith Jack in camp during the first season of sledging, with skis and a sledgemeter wheel for measuring the distance traveled.

Canterbury Museum, Joyce Collection

A party camped on the Barrier in early 1915. One man unloads the sledge, while another pitches the tent. The depot stores are behind the camp, marked with a flag on a bamboo pole. When completed, the depot would be secured with snow blocks.

Canterbury Museum, Joyce Collection

A dog team sledging in early 1915 to build the first two of Shackleton's depots on the Ross Ice Shelf. Rushed into the field without training and acclimatization, sixteen of the eighteen dogs died during the first season.

The hut at Cape Evans, built as living quarters for Scott's 1910 expedition. The hut was chosen as the main base for the Ross Sea party's scientific staff.

The Cape Evans hut, equipped with twenty-five bunks, science laboratory, and darkroom was outfitted with sprung mattresses, cast-iron range, and gas lighting.

Canterbury Museum, Joyce Collection

The *Discovery* hut at Hut Point, built by the first expedition (1901–1904) commanded by Robert Falcon Scott. The hut was not intended to serve as long-term living quarters and had been used primarily as a storage shed before being occupied by the Ross Sea party.

The Ross Sea party's approach to Mount Hope from the Ross Ice Shelf, the site of the last depot, 360 miles inland. (The rounded peak at far left is Mount Hope; the Beardmore Glacier is beyond the frame at far left.) The low rocky peak to the left of center is the eastern side of the Gateway.

Mackintosh, Wild, and Spencer-Smith leaving Hut Point on October 29, 1915, for the last time before marching to Mount Hope. The canvas sail eased sledging in favorable winds.

The sledging party on the retreat from Mount Hope in 1916, dragging Spencer-Smith on a sledge.

Four men of the party who returned from the journey to Mount Hope in March 1916 after sledging 1,356 miles to lay Shackleton's depots. From left: Hayward, Joyce, Wild, Richards.

The *Aurora* arriving in Port Chalmers, New Zealand on April 3, 1916. From left: Thomson, Wise, Warren, Ninnis, Kavanagh, Shaw, Grady, Mauger, Mugridge, Paton, Atkin, Downing, Hooke, Glidden, Larkman, Donnelly. Not pictured: Stenhouse and d'Anglade.

Physicist Richard Richards.

Ernest Joyce, in charge of sledging equipment and dogs under Mackintosh.

Storekeeper Ernest Wild.

Assistant Irvine Gaze.

Chief Scientist Alexander Stevens.

Physicist Keith Jack.

The seven survivors of the
Ross Sea party after rescue
on January 10, 1917.

Biologist John Lachlan Cope.

Members of the Ross Sea party aboard the *Aurora* after their rescue in January 1917. (From left): Gaze, Richards, Wild, Jack, Cope, Joyce, Stevens.

John King Davis (front row, center in uniform) aboard the *Aurora* as master and commander of the Ross Sea relief expedition, 1916–1917. Paton is the first at left in the back row; Ninnis is the last man at right in the front row.

Shackleton aboard the *Aurora* in Wellington, New Zealand after the relief expedition, 1917.

Four of the seven survivors of the shore party on the voyage home in 1917. Rear (standing, from left): Jack, Richards. Front: Gaze, Gunboat, Joyce and Towser, Oscar.

The onset had been insidious. The first signs—muscle stiffness, lassitude, and weakness—seemed the inevitable result of overexertion. All the while, the latent disease stole a march inside his body. Depleted of vitamin C, the connective tissue throughout his body began to break down. He hemorrhaged internally as the walls of his capillaries weakened and burst. The joints and surface membranes of bones began to bleed, and the scar tissue knitting old wounds deteriorated. New wounds refused to heal. By the time the chaplain's symptoms became severe, his swollen limbs purpling from the internal bleeding, the disease was already well advanced. On his seventh day alone, he could no longer summon the strength to stand. "Felt very rotten this morning," he wrote, ashamed of his weakness.

In the late afternoon of January 29, a muffled chorus of yelps announced the arrival of the sledgers. "It is strange but cheery to hear men and dogs again," wrote Spencer-Smith, elated to learn that the party had accomplished their mission. His companions were horrified by his wrecked condition. The chaplain was helpless, unable even to crawl from his soaking bag. As they peeled away his wet clothing, they discovered his legs had turned dusky black from the hip down and his joints had swelled. But Joyce refused to accept it. "It may be scurvy but I do not think so as his gums & eyes do not shew it," he wrote, dismissing Mackintosh's alarm as "his usual panic."

Joyce had tried to protect his party by breaking their journey with spells at the hut when he pressed them to eat seal, and by packing some fried meat when they left the hut for the last time on December 13. Mackintosh's party had subsisted on the inadequate sledging rations the longest. Their last sojourn at Hut Point had ended October 29, eighty-four days before Spencer-Smith's breakdown. Tellingly, Mackintosh seemed not much better than the chaplain, his knees sore and ankles grotesquely swollen to twice their normal size. Although Joyce, Richards, Hayward, and Wild were stronger, it was only a matter of time before the entire party was as disabled as Spencer-Smith. A continuous intake of vitamin C is needed to ward off deficiency, and their diet was devoid of it.

No explorer cared to dwell on the possibility of injury or death, and Shackleton was no exception. "Where would you be with a sick man on your hands?" demanded the president of the Royal Geographical

Society when Shackleton presented his plans for the expedition. Shackleton replied, "It is up to him, when he knows he is too bad to go on, isn't it?" It was vital that each member understood, and admitted, his limitations. There was a fine but critical line between doing one's duty and rash heroics that endangered the welfare of the entire group.

Acknowledging an incapacitating injury presented a new set of problems. The natural human impulse was compassion. As Shackleton told the RGS, "We would put him on a sledge." At a certain point, though, the ethical dilemmas became thornier, when the distances and the burdens were too great, threatening the lives of the entire group. Faced with a hypothetical crisis at the South Pole, Shackleton could only surmise how he would cope. "I would go on as I could, and it is up to him to do what is right," he supposed. "As regards disaster, they each have their two tablets of morphia." For Scott, the dilemma was real. As hope dissolved on his doomed homeward march from the Pole, he offered his men the choice of a merciful end, handing each man thirty opium tablets. The medical kit of the Ross Sea party, too, was equipped with opium.

For the Ross Sea party, the decision needed no debate. Spencer-Smith would be loaded onto a sledge and dragged back to Hut Point, although the addition of nearly two hundred pounds to the load would undoubtedly slow their pace. More time in the field meant more food consumed, and the rations were already short. Originally, only one three-man party had been intended to sledge beyond latitude 82° south to Mount Hope. Joyce's decision to push on with a team of six meant that an extra 120 to 130 pounds of supplies had already been consumed. There were reserves at each depot for the Ross Sea party, roughly enough to keep them supplied for five days between each depot without cutting into Shackleton's stores. The Bluff Depot was overstocked with an even greater surplus. If they made the 240 miles to the Bluff at a good clip, they could rest more easily.

To Joyce, that meant longer hours to cover at least thirteen or fourteen miles each day, starting immediately. Less than an hour later, he had the camp struck and Spencer-Smith wrapped in a dry sleeping bag. With the chaplain strapped securely on top of the rear sledge, Joyce led the caravan northward at full tilt. There were only four men heaving at the sledge: Joyce, Wild, Hayward, and Richards. Mackintosh was in such pain that he limped along with a slack trace. By the next day, a blizzard closed in, making travel unthinkable. The party was pinned down in the sagging tents, subsisting on short commons again. Only the hairs molting from the worn sleeping bags seemed in ample supply in the hoosh pot.

The delay aggravated Mackintosh's anxiety. "He seems properly scared," wrote Joyce. Mackintosh was frantic to speed their travel north and made some suggestions which Joyce dismissed out of hand, unleashing his own anxieties in a tirade. "I also told him that he ought to have known about the condition of Smith. He made the usual silly excuses," Joyce recorded. The accusation was ironic, given that Joyce himself still had not recognized the obvious signs of scurvy in Spencer-Smith and Mackintosh. The criticism cut Mackintosh to the quick. His sense of responsibility for the welfare of his men was poised against an awareness that his own vitality was rapidly ebbing away. He would soon be powerless to help anyone, even himself. He brooded over what might have been, confiding to Hayward that he had contemplated turning back at 80° south and leaving the advance on Mount Hope to Joyce's party. Hayward was unsympathetic. "Skipper is quite lame & I contend has hindered far more than he has advanced our main object of laying these depots to 83°30 S for Shacks support," he wrote.

Joyce mulled over that pivotal moment as well. "As I said after we left 80° that we should have the both of them on the sledge as they have been both useless from that date. If it had not been for their Primus I should have sent them back." Even if both stoves had been working, it is questionable whether Joyce could have summoned the nerve to order Mackintosh back to base. The Royal Navy had molded Joyce into the model seaman since boyhood. He was drilled to follow orders and discouraged from truly taking initiative. He had no trouble second-guessing a superior, but shied from seizing control and shouldering the full responsibility of leadership. As Richards saw it, "There was nobody in charge," which was partly true. While he and Joyce brokered the major decisions between them, neither man formally assumed the mantle of leadership or attempted to wrest it from Mackintosh. Richards criticized Mackintosh for not realizing "his folly in passing 81° S with Smith," but he and Joyce did not act aggressively to prevent it.

Their refusal to grasp the nettle had arguably cost the party dearly. Regret, however, was not in Joyce's nature. The angst that Mackintosh suffered over roads not taken was alien to him. Joyce was above all a pragmatist, existing squarely in the present. However desperate the situation, he found it pointless to surrender to despair, writing matter-of-factly, "Now comes one of the trials of the Antarctic and one must expect these things."

He was also a diehard optimist. On clear days, he mounted the sledge and scrutinized the southern horizon through binoculars, always expecting to see Shackleton. Excepting the disconsolate Mackintosh, he had fallen in with a band of like-minded comrades. "Turned in wet through, but cheerful," the refrain in his diary went. Wild also accepted the setbacks with impassive calm. Richards, whom Joyce now consulted on every move, was a cool-headed rationalist. He attacked their predicament intellectually as well as physically, taking charge of the navigation. Hayward unstintingly threw his full physical might into the job. Though Spencer-Smith was swathed in his bag and tied to the sledge, he offered encouragement whenever his comrades came within earshot. "Everyone is very kind to wretched me," he wrote in his diary.

The care of Spencer-Smith devolved to Wild, partly by necessity, because Joyce and Richards were preoccupied with logistics, and partly due to his natural inclination. Growing up in a large brood, Wild shepherded a troop of younger siblings through childhood. The caretaking instinct endured as he patiently attended to all of the chaplain's needs. "Wild is a brick the way he looks after Smithy," wrote Joyce. But Wild was no benign Florence Nightingale. "I had a fierce argument with Wild," wrote Spencer-Smith, vexed by "the strange contrast between his really fine character and his wild opinions." Wild understood that provocation was sometimes the best medicine. To escape the pain, Spencer-Smith often drifted into memories. "A rather more comfortable day—mostly hovering round Gray's Inn," he imagined one day. "If only the dream could come true." Wild was determined to keep him alert and engaged.

As the men lifted the invalid's prone body onto a sledge on February 1, he delighted in the rare blue skies and ruffling breeze. To catch the southerly, they rigged a sail. Dumping one of the two sledges lightened the load, and the mileage increased each day, from thirteen miles to fifteen miles, then topped seventeen miles. "Dogs going splendid," beamed Joyce, slipping them extra food and scrutinizing their bodies for signs of injury or illness. At midday on February 3, the party reached the 82° Depot and took only a week's worth of rations. For once, the party was banking surplus food. In the next four days, they clocked sixty-nine miles on the sledgemeter, leaving the 81° Depot behind them on February 7. The speed seemed hardly credible. Outbound, this leg had taken six days; now they raced over the same ground in four.

The unprecedented headway lifted the spirits of Spencer-Smith, who, despite his stoical good cheer, was obviously worsening. "We shall have to put our right foot forward. I am rather afraid we shall not get him in in time—still, no one living could do any more than we are," worried Joyce. Careening over the hummocked terrain, the sledge capsized from time to time, dumping the invalid to the ground. Joyce looked for ways to pick up the pace, which had peaked at just over eighteen miles in one day, surveying the gear for items to jettison. Spare shovels, ice picks, and personal items soon littered the trail behind them.

Four days after leaving the 81° Depot, the black pennants of Rocky Mountain Depot at 80°02' S 169°25' E hove into view. The Ross Sea party had sprinted across 208 miles of ice shelf in seventeen days, averaging fourteen miles each day. Joyce felt a surge of triumph. At long last, unfettered by crushing loads and contrary orders, he felt he could show how sledging was done. Due to his hard driving, Hut Point was only 150 miles to the north. This last leg of the retreat had lasted an excruciating month during the first season of sledging. Joyce had no intention of repeating the same poor performance. Nonetheless, he took extra supplies from the depot for insurance, leaving four weeks' worth for Shackleton.

He wasted no time in pushing the little band onward on February 11. He was leery of the nasty terrain and nastier weather in the seventy-four miles between Rocky Mountain Depot and the Bluff Depot. This was the "Dead Dog Trail" where, a year before, Mackintosh, Joyce, and Wild had been pinned down by blizzard conditions and had lost all their dogs in the previous season. They could not afford another such calamity. In their debilitated condition, pulling a sick man, the party had little chance of reaching Hut Point by manpower alone.

They headed north of latitude 80° south into the ill-starred no-man's-land in fine weather. Like a beacon, the familiar silhouette of Minna Bluff appeared to the west of their track on February 12 in the soft twilight. The welcome sight provoked a rush of homesickness. But by the next morning, it had vanished, curtained by mist. There was little else to spur them on. The wet snow was the worst yet encountered, "like ploughing one's way through treacle on stilts," as Hayward put it. Men and dogs floundered with every step. The going did not improve. "Rottener and rottener," Hayward wrote on the sixteenth. His trouble was not only the terrain. Joyce noticed that Hayward's movements were becoming labored, a sign he failed to recognize as the grip of scurvy tightening. Richards, now experiencing strange heart palpitations, suffered in silence.

The cost of Joyce's hellbent dash across the ice shelf was becoming apparent. Richards observed that Mackintosh appeared to be "going to pieces." The mileage plunged sharply as they marched knee-deep in drift. The party had slowed to a plod of eight miles per day, and even that would have been impossible without the sail. Desperate to take advantage of a favorable wind, they braved a snowstorm on the sixteenth, wading through powdery new drifts over a packed base some four to five feet deep. By the night of the seventeenth, the navigation book showed that they had advanced sixty miles toward the Bluff, but that was only a best estimate. When the sledgemeter malfunctioned after leaving Rocky Mountain Depot, Joyce hurled it overboard to save weight, confident that the cairns would guide them back to the coast without the time-consuming nuisance of coddling the finicky device. The navigation proved to be far more difficult than he expected. They picked up multiple lines of cairns almost immediately, laid on four separate trips, but it was difficult to tell which cairns marked the right course. Making their way into the furious whirl of snow, they soon lost sight of the cairns entirely.

Without the sledgemeter, Joyce could not truly say how near—or far—the Bluff Depot was. Judging by the rate of travel, Richards estimated ten miles. With the depot so close, they resolved to push on, even if the snow continued. But by the morning of February 18, the storm had intensified. There was no question of marching. "The wind—still howling—seems about the strongest I remember and almost carries a note of personal animus," wrote Spencer-Smith as the tent shuddered. For the sake of the chaplain and Mackintosh, Joyce could not help but be grateful for the reprieve, even though it was a luxury they could ill afford, with temperatures dropping and about eighty-five miles yet to go to Hut Point. With three days of provisions left, the men hunkered down in the two tents to wait out the storm.

The blizzard droned on, unabated, for days. Venturing outside was foolhardy, save for the essential task of tending the dogs, which Joyce, Richards, and Hayward took in turns. The other tent, housing Wild and the invalids, had fallen disturbingly silent. The snatches of Wild's singing, sometimes audible through the roar, had stopped. They had run out of kerosene and were scraping bottom on rations as well. Two biscuits and a chunk of snow made up the day's meal on February 21. "Shall have to make more holes in my belt," recorded Wild tersely. Spencer-Smith minded less; he had lost his appetite, though obsessive thoughts of food plagued him. His breathing came in irregular, shallow gasps, and he could no longer stand. Wild hoisted him to his feet and propped him up for bodily functions. He endured the pain and indignity without complaint. "He dosn't howl much like I should," wrote Wild. Mackintosh was too morose to speak. Wild could see from the red stain seeping through the snow over the latrine hole that, like Spencer-Smith, Mackintosh was bleeding from the bowels.

On the fourth day pinned in the tents, thirst became as urgent as hunger. Without kerosene, it was impossible to melt snow for water. Only a little denatured alcohol remained. Richards lit some in a mug and thawed cupfuls of snow over it to yield dribbles of water. Joyce had husbanded the rations, but by February 22, eight lumps of sugar and half of a biscuit remained. On the seventh day of the storm, Joyce could no longer tolerate the inaction. "Richards, Hayward & I after a long talk decided if possible to get under way tomorrow in any case or else we shall be sharing the fate of Scott & his party."

In March 1912, Scott's party had faced the same predicament not far from their position, struggling with weather and terrain "awful beyond words" as they retreated from the Pole. One of Scott's men had already died, another was crippled by frostbitten feet, and the three men left standing were powerless to tow the invalid, likely in the throes of scurvy. Bogged down in a blizzard eleven miles short of their depot at

79°28'53" S, 169°22'04" E, they camped and waited. Sometime after ten days, death came mercifully. Their bodies lay in a tent under a shroud of snow, somewhere along the Ross Sea party's path. Unlike Scott's party, they harbored no illusions that rescue might be on the way. There was no telling what had become of Gaze, Jack, and Cope, groping their way across the ice shelf with no compass. Stevens, alone at Cape Evans, was incapable of sledging on his own.

On the morning of February 23, Joyce, Richards, and Hayward crawled out of the tent, struggling to stand against the barrage of wind. Barely visible through the haze were surreal meringues of sastrugi rising ten feet high. Entering the other tent, they discovered that Spencer-Smith had deteriorated badly during the storm. He fainted as they lifted him onto the sledge. The party crept forward, tatters fluttering in the wind, like a procession of wretched beggars. Mackintosh followed behind in a rheumatic shuffle, tethered to the rear of the sledge. Every fifteen minutes, Richards fished his hands out of his fur mitts to check the compass, his bare fingers seared by the cold metal. To keep the course in the whiteout conditions, he stared intently at the snow streaking across Joyce's back to judge the wind direction and mentally calculated the angle between the wind and their intended direction. Marching forward, Richards tried to steer the sledge at the same consistent angle off the wind. An hour into the journey, Richards guessed that they had crept a quarter of a mile.

Even at that pace, Mackintosh was falling behind. He crabbed along in a stiff crouch, his knees locked at an angle. When the sledge jammed on peaks of sastrugi, the team yanked on the rope to drag it over and Mackintosh lurched along with it. At one point, the sledge became deadweight. Joyce called a halt. Behind the sledge, he was lying motionless in the tracks. Through the clamorous wind, they all heard him wailing piteously, "Oh my hands, my hands are gone." As the men fought their way against the ferocious gusts to reach him, they could see he was wringing his hands, repeating, "I'm done, I'm done, I don't care what happens." Joyce tried to grapple him to his feet, but he resisted, pleading with Joyce to wrap him in tent canvas and leave him behind. Richards reached out and put his arms around the delirious man, trying to comfort him. In Joyce's words, he had "completely gone." Their differences no longer mattered. He was determined to pull Mackintosh through.

"I think he has got scurvy," wrote Joyce, the first time any of them had dared to acknowledge in his diary the presence of the dreaded disease. It was impossible to ignore their own shredding and blistered gums. Richards huddled with Joyce and Wild. There was no question of leaving Mackintosh alone to die, but they no longer had the strength to pull two men on the sledge. They decided to pitch a tent and leave Spencer-Smith and Mackintosh behind while the able men made for the depot. Wild would stay to care for the invalids. "We wanted Wild to go too but he stoutly refused to desert Mr. Micawber," wrote Spencer-Smith, identifying with the hapless Dickens character. Wild had ferreted out bits of food in the tent, some chocolate, biscuits, and a little pemmican. Joyce told him to make it last until February 27. It would take at least four days, he guessed, to get to the depot and back again. As they heaved the sledge forward, Richards watched over his shoulder, thinking he would never see the three men alive again.

The blizzard seemed to redouble in force. "We could only see a few yards and even the sky was not different in appearance to anywhere else. There appeared to be no up or down," Richards wrote. It was impossible to plod much longer than two hours. The month before, the team had pulled 1,380 pounds. Now two hundred pounds was soul-destroying labor. There was no food left for the dogs. For the men, there was a half biscuit each and a cup of lukewarm tea heated with flaming alcohol, a laborious process that took nearly three hours. After the meal, Richards wrote in his diary for the first time in nearly a year. He felt compelled to leave his own record in the event that he was living his last days. Just how or by whom it might be found, he dared not speculate. Then, they lay awake in their bags, longing for the blizzard to slacken.

The three men rose at 4:30 AM on February 24 to face the ninth day of blizzard conditions. The sun shone behind the overcast, lining the clouds with a dull mercury–colored glare. They all agreed to brave the maelstrom. The wind seemed to snatch the air out of their lungs as they gasped for breath. They passed a single cairn, which could mean they were traveling across the route rather than straight down the line of markers. Joyce believed that their crude navigation might be as much as ten degrees off, raising the risk of marching past the depot without even glimpsing it. To escape the knifing wind, they pitched the tent for a meal: a half cup of tea and a quarter of a biscuit each, about fifty calories. Operating on a deficit for months, the men were wasting away, metabolizing fat, then muscle, to fuel the unceasing energy demands. Not only scurvy but a host of other dietary deficiencies were wreaking havoc on their bodies. Joyce banished from his mind the thought of slaughtering one of the dogs, knowing full well that the animals were shouldering most of the hauling. Instead, he scraped the walls of the empty dog food bin and mixed the frozen particles with water to eke out a meal for the men. Starvation made the cold feel more penetrating than the actual temperature of thirty below zero.

On February 25, Joyce and his companions were still lying in their bags with no lull in sight. The food was gone and the tent torn. "If the worst comes, we have made up our minds to carry on, on the trek & die in harness," Joyce vowed grimly. After midnight on February 26, the wind seemed to fall off slightly. The three men crawled from the tent and hitched the dogs. The animals were lethargic after four days without food. Only the hulking Oscar seemed willing to pull. According to Richards, the dog "lowered his great head and pulled as he never did when things were going well," nipping at the heels of the others to make them work harder. The men staggered onward, stopping every few yards for breath. Then, through the streaking snow, Richards sighted a bamboo pole whipsailing in the gusts. The dogs saw the depot "which seemed to electrify them & they had new life and started to run," wrote Joyce, who was dragged off his feet by the howling team with Richards.

It took hours to dig out the buried cases of provisions and set up camp. To Joyce, sitting near the Primus was "like coming out of a thick London fog into a drawing room." The depot may have saved them, but it also delivered a devastating blow: there was no message from the ship contained within. If the *Aurora* had returned, there would have been a note in the depot. Until that moment, hope had lingered that the ship had made her way safely back into McMurdo Sound in December or January. "We all think there has been a calamity, let us hope for the best," Joyce wrote.

On February 27, they woke at 5:00 AM to hurricane-force winds. "This is the longest blizzard I have ever been in," Joyce recorded. "We have not had a travelling day for 11 days and the amount of snow that has fallen is astonishing." There was no chance of steering a course, and snow covered their sledge tracks, so they crouched in the tent, waiting for the storm to abate. Each time they ventured forth in a lull, they were forced to camp again. To Richards, the delay was "awful—held up here knowing that three men are starving and worse deathly cold 10 or 12 miles back. But we can do nothing." Something else was preying on his mind: he had noticed "the dreaded black appearance" behind his own knees. Though he denied it was scurvy, he lay awake with the image of Mackintosh and Spencer-Smith in his mind, "neither able to pull much & both walking as I have now seen Hayward as I fear will shortly be myself."

On the morning of February 29, the unceasing drone of the wind dropped. They lost no time in hitching the dogs. At first, the team resisted heading south into the trackless wastes again. The terrain was abominable, the snow drifted over the team's heads in places. Shortly after starting, Richards spotted the tent in the distance. He and Joyce feared what they might find. As they drew near, he made out the bantam figure of Wild standing motionless by the half-buried tent. Stiffly, he walked toward them, drawing a harness over his head and wordlessly hitching himself to the sledge to heave along with them. Overcome by the gesture, Richards broke into sobs. A few minutes later, Mackintosh emerged from the tent and tottered toward them, hunched in a nearly squatting position. He, too, was overcome with emotion, thanking Joyce's party for saving their lives.

Mackintosh believed that the party had perished in the blizzard. A few days before, food finished, he had penned his last letters, fearing he would be unable to write for much longer. He wrote his mother, sister, and wife, and made an official record. In it, he paid tribute to Wild, "good, unselfish fellow that he is," for his sacrifice in casting his fate with the invalids. The conflicts forgotten, he praised the rest of his men, who he said had "done their duty, nobly and well." He maintained that their predicament had not been "brought about by any lack of organization," having "accomplished the difficult part of our mission, getting the depots laid for Sir Ernest Shack at Mt Hope 82, 81, 80 South and Bluff." Unsparingly, he described his own breakdown, then ended, "If it's God's will that we should here give up our lives, we do so I trust in the true British spirit as our traditions hold us in honour to do. Good bye friends. I feel sure our People my own dear wife and children will not be neglected."

"Very cold and wet and weak but not dead yet by any means, thank God," wrote Spencer-Smith. He was very much alive when Joyce's party entered the tent. When they lifted his wasted body, they found that his feeble warmth had melted a depression in the ice under his bag. Lying in the deepening puddle, he had not defecated for a week. Wild tended to all of his needs where he lay, looking after him "like a father," in Joyce's words. Doggedly, Wild had rousted the invalids into singalongs like "some picnic party," as Mackintosh put it. He hunted for bits of food and rolled them tea cigarettes. He dug out the sagging tent and built a new snow cairn flagged with black rags to ensure that Joyce's party could see it. Despondency was utterly alien to him. "My belly is singing 'Rule Britannia' now," Wild wrote as the food ran out.

Though Joyce cautioned that too much food would play havoc with their systems, the granite-hard biscuits and pemmican were manna to the three starved men. Mackintosh forsook the comforts early, hobbling out with a stick to get a head start on the team. He was determined to stay on his feet as long as possible, though the party had brought back another sledge to carry him. The fittest men struck camp and followed with Spencer-Smith tied on the first sledge. The caravan soon overtook Mackintosh, who quietly admitted defeat and climbed onto the rear sledge. Joyce, Hayward, Wild, and Richards strained to pull the unaccustomed weight of two men and an extra sledge. "For God's sake, Richy, stop or you'll bust your heart," Spencer-Smith cried out as he watched. The dogs bore the brunt. Joyce was doubly thankful that they had resisted the temptation to slaughter one of the animals for food. With a strong southerly and a sail rigged, the party arrived at the depot by the next evening.

It was March 1 and winter was fast approaching as the light faded. They were utterly unprepared. "[The snow] gets through all the holes in our clothes and the sleeping bags worse than hell," Joyce wrote, his feet frostbitten in his snow-filled finneskoe. Following the cautious route to avoid crevasses, Hut Point was about seventy-five miles north, ten or more days' travel at their present pace. In late 1915, Joyce's party had covered the distance in a record five days. With only three able-bodied men, the camp chores—pitching the tent, caring for the dogs, collecting snow for water, coaxing the erratic Primus to life, nursing the invalids, mending clothing—took hours to perform. Mundane tasks became increasingly difficult as malnutrition clouded their concentration. One night, they stopped to camp and realized that they had lost the carelessly lashed tent poles. Recognizing he was the strongest, Richards volunteered to search for them. "The hardest damn journey I ever made," he called it.

They could no longer deny that scurvy was rampant. "We are all, more or less, suffering," Hayward wrote. He and Mackintosh tried to scuttle along with the aid of sticks, "a pathetic sight to us, but we appreciated it for the brave gesture it was," according to Richards. Ultimately Mackintosh had to climb onto the sledge, but as it jolted along, he fell off repeatedly and was left lying on the trail. It was often some time before he was missed and the party turned back to find him. He told Joyce that "it would be a good job" if he was left behind.

Hayward adamantly refused to ride on the sledge and staggered along stiffly, his blackened gums so badly ulcerated that the tissue mushroomed from his mouth. The once-burly Londoner was gaunt. Crammed cheek by jowl with his tentmates at night, he was mute and withdrawn, shivering violently. The reason was hypothermia, and they were all suffering badly, dressed as they were in sodden clothing and sleeping in wet fur bags. Early in the retreat from Mount Hope, Mackintosh's body temperature was 96 degrees and Spencer-Smith's temperature registered as 94.2.

After Hayward stopped pulling and fell behind the sledge, he became even more vulnerable to the cold. When he caught up with his companions, they noticed he seemed "somewhat unbalanced mentally," speaking nonsensically and throwing his food away. On March 4, Joyce examined him again. He could not bear to be touched, evidence of how advanced the scurvy had become. His pupils were strangely dilated—not a symptom of scurvy, as Joyce assumed, but an indication that Hayward's body temperature had dropped to a dangerously low level. His irrational behavior was also a sign that he was severely hypothermic. Hayward was slipping into a twilight state, which he resisted desperately by writing to his fiancée, his script deteriorating into a barely legible scrawl. "Am now out of the team, my legs having become so swollen, stiff, painful as to make my pulling any weight in the trace impossible in fact it is as much as I can do to keep up even at this slow pace," he wrote to her. "I do not know what you will think of me, & I will not try to describe my feelings."

Joyce, Wild, and Richards were at the end of their tethers. They tried to truss their bent legs straight with bamboo poles and tight bandages. They had no idea what the pinkish tinge of their urine meant. As Spencer-Smith declined, Wild had noticed the rusty red trickle staining the snow as he supported the sick man to urinate. Richards dared not share his thoughts with the others. It had crossed his mind that it would be far easier to concede the relentless struggle and sit down in the snow to die.

The animals, too, seemed on the verge of collapse. "Poor dogs they look at one so wistfully," Joyce wrote. "They seem to ask you if you are trying to kill them with the heavy load." Throughout the day, Spencer-Smith could be heard on the sledge, muttering deliriously and reciting prayers in Latin. In the tent at night, he tried to distract himself with his diary. "'O that this too, too solid flesh would melt!'" he quoted, but unlike Hamlet, he had not given in to despair. By March 5, however, he could find no refuge from the physical torment. Reaching for the medical kit, he took a tablet of opium and aromatic chalk powder to relieve the unbearable cramps in his stomach and bowels, possibly from internal hemorrhage.

Joyce and Richards decided that the only chance of getting Spencer-Smith to Hut Point swiftly was if one of the other invalids stayed behind, along with the second sledge, to lighten the load. Joyce put it to the two men and Mackintosh volunteered to stay. On March 8, they erected the tent for him and got under way with Hayward and Spencer-Smith lashed on one sledge. A stiff southeasterly breeze sprang up, hastening their progress.

Spencer-Smith was often "wandering in more kindly climes," in Richards's words, a merciful release as the sledge jolted over the steep spurs of sastrugi. "Spent the afternoon in charge of Althorne," Spencer-Smith wrote wistfully in February, recalling idyllic summer days at his family's Sussex country house in Eden Vale. At times, he found himself back in Cambridge, the mists of the River Cam gathering around ancient college buildings the color of tea-stained linen. The party pushed on until after dark, managing eight miles. The five men squeezed awkwardly into the three-man tent. In the early hours of March 9, Richards guessed the temperature was thirty below zero. The last of the thermometers had been broken in late February. Spencer-Smith fretted in his sleep and Wild woke, accustomed to listening for his patient in the night. "Smithy had a little joke with me after 4:00 AM," wrote Wild. "He asked 'if I'd lost my bearings' because I was looking round." The others woke too, shifting uncomfortably in the stony chill. Spencer-Smith asked Richards, "I say Rich, if your heart is behaving funny, what is the best thing to do, lie down or sit up?" He didn't know, Richards answered, but he told the chaplain he thought it best to lie still.

"Woke up this morning & found poor Smithy dead at 6:00 AM," wrote Wild. Spencer-Smith's head lay exposed, his jaw slack, his eyelashes and beard glittering with ice crystals. His comrades had dragged him over three hundred miles, only to fall thirty miles short of safe haven. The last entry of his diary was dated the day before he died. "Glorious weather," he called it, his spirits lifted by the thought of his sister Fredrica's birthday. An inscription on the flyleaf of his diary came from Isaiah 66:1: "Thus saith the Lord, Heaven is my throne, the earth my footstool. Where is the House that ye built unto me? And where is the place of my rest?"

Richards and Wild clawed obstinately at the unyielding ice until they made a shallow depression and rolled the body into the hollow, "as reverently as possible," covering him with a snow cairn marked by a bamboo cross. Hayward stood apart with his back to the rites, bereft. "I am afraid he will share the same fate as poor Smith," Joyce wrote.

Forty-eight hours later, on March 11, they were threading a path across the broken sea ice around Cape Armitage with Hayward in tow on the sledge. In the open water of the sound, a colony of seals and pups frolicked. A frantic urge rose in Richards's gut, "the strongest desire to rush to one of those animals and cut its throat and drink the blood." Instinctively he knew what his deprived body needed. They pushed on to Hut Point and carried Hayward indoors, still alive but "a bag of bones." There was no sign of Gaze, Jack, and Cope. Hayward was semicomatose; it was inconceivable to leave him alone to rescue Mackintosh. For that matter, Joyce, Richards, and Wild were in no condition to brave the Barrier again. They would have to rest for a few days first, feed up on seal meat, and regain their strength.

———

On March 16, Mackintosh lay in the tent. Eight days before, he had watched the party plod north until they disappeared and he was alone. The worst of it was the terrible suspense of wondering how Spencer-Smith fared. For over five weeks, the chaplain had lain in his bag, uttering virtually no complaint and showing far more concern for his fellows than himself. Then, in the wake of the great blizzard, he had questioned why Mackintosh had allowed him to march beyond 81° when he was so obviously ailing. Though not intended as a rebuke, Spencer-Smith's plaintive question cut Mackintosh to the bone. It had been his choice to see the sledging through, no matter the cost. He was awash in regret. Alone in the tent, he thought of his wife and daughters at home in England. A year before, he had questioned his very reasons for coming to this place. "How one longs to be out of this infernal region—the dear ones at Home what are they doing?" Now he quietly gave up hope. Joyce's party had been gone far too long. Once again, he wrote last letters to his loved ones and waited for his end.

Mackintosh was outside the tent when he heard the dogs approaching from the north. He seemed in a daze when Joyce, Richards, and Wild approached. "Told Mackintosh of Smith's death and it did not seem to impress him much," Richards recounted. Joyce found him "a little peculiar." Mackintosh told them he had not been alone. There were others in the tent with him, voices he knew must be imaginary, but nonetheless he felt compelled to answer. Richards urged fried seal meat on him and they strapped him to the sledge. The trio turned in eighteen miles that day, and twenty miles the next. On the morning of March 18, the temperature plummeted to minus forty degrees. The sun had set by the time they neared the hut. They found Hayward little changed.

"As there is no news here of the ship & we cannot see her we surmise she is gone down with all hands," Joyce wrote after a search turned up no message from Stenhouse. "I cannot see there is any chance of her being afloat. I don't know how the Skipper will take it."

– 13 –

"Some Way or Other They're Lost"

MARCH 19–MAY 12, 1916

The Ross Sea party had accomplished their mission. Mackintosh had planned on departing Hut Point on September 1 and returning by March 15. They arrived just three days later, after two hundred days of sledging. For all of the improvisation and unexpected setbacks in the field, they were successful, stowing even more than the required provisions in some of the depots. Against all odds, the Ross Sea party had dragged 4,500 pounds of vital supplies across the Barrier, sledging 1,356 miles in the second season alone to lay the chain of depots to Mount Hope.

Yet the men felt little triumph. Spencer-Smith was dead, and the survival of Shackleton's party seemed doubtful. As he had told Mackintosh, "If the T. C. [transcontinental] Party is not in sight of Hut Point on the 20th March you may take it that [we] have been unable to cross or are dead." "I looked out for Shacks & his party until about the 7th [of April] & then gave them up," wrote Wild, surrendering hope for his brother Frank.

The Ross Sea party's own experience made it apparent that, in Hayward's rueful words, Shackleton's completion of the journey would be "nothing short of miraculous." Shackleton had based all of his plans on traveling an average of fifteen miles each day, but the weather and terrain from the Weddell Sea to the Pole was terra incognita. For all he knew, he would be slowed by countless mountain ranges and glaciers en route.

The Ross Sea party's tractor debacle did not bode well for the success of Shackleton's motor vehicles. Neither did their hard experience with the dogs. The notion of six novices controlling and caring for sixty-nine dogs and obtaining optimal mileage from the outset seemed improbable. His allowance of twelve days for weather delays during the four months of the crossing also seemed dangerously slim; the blizzard in February had stalled the Ross Sea party for that length of time. Drawing the inescapable comparisons with their own desperate logistical struggle, Richards was not alone in concluding that if Shackleton's party had embarked on the crossing, "some way or other they're lost."

Wild gave up his vigil for the *Aurora* as well. "I'm afraid she has gone down, or got squashed up in the ice," he wrote in his diary. The party resigned themselves to being marooned indefinitely. Mackintosh knew the food at Cape Evans and Cape Royds would hold out for more than a year, but for the foreseeable future, they were trapped at Hut Point. The limited provisions, supplemented by seal, had to last three or four months until the sea ice was solid enough for the journey to Cape Evans.

In the drafty, primitive shelter, they lived "the life of troglodytes," in Richards's words. By the flickering light of the blubber stove, they looked unlike civilized men. "Sealing in the dark is a dirty and messy business, and our faces and clothes quickly lost any resemblance to their normal appearance and it was impossible to do anything about it," Richards wrote. It took at least one or two seals to see them through each week. When a seal's throat was cut, they were sprayed with blood from the flailing creature. Handling the blubber from the carcasses mired them in oily gore. The cold was so intense that the strips of blubber froze hard in minutes. To stave off frostbite, they plunged their numb bare hands into the steaming entrails. Their clothing, hair, and beards became saturated with blood and fat, which proved impossible to wash away with water. After a second season on the Barrier, their garments were in stiff, filthy tatters. When Hayward's finneskoe fell apart, he wore his mitts on his feet, while Joyce and Wild tried unsuccessfully to make new clothes from emperor penguin skins.

Even on a steady diet of seal meat, their recovery was slow. Mackintosh's legs were still seized at right angles. He and Hayward spent most of their time in sleeping bags near the blubber stove as Richards massaged their limbs to relieve the pain. Joyce's frequent bouts of snowblindness had impaired his vision. He managed to cook and help the invalids while Richards and Wild, who suffered less, took on the strenuous chores, hunting and shoveling snow. In the evening, they took turns reading aloud to Joyce.

Joyce, Richards, and Wild counted themselves "as cheerful as school boys on a Friday" just to be alive. However bleak the outlook, Joyce was in his element, reveling in overcoming the hardships of daily existence. Mackintosh, however, found the squalor of the hut intolerable, and dreaded spending a second winter in the Antarctic away from his family. Though his conviction had flagged in the first season, he had been inspired to see the depot laying through by his promise to Shackleton and his commitment to bring his men safely back. But now their mission was complete. "To wander more I never wished to do," he wrote in a letter to his "precious loved ones."

Hayward fretted as well. His home and fiancée seemed more remote than ever. His diary tailed off to a terse litany of "blizzard" and "ditto," then the entries ended altogether in April. Dispirited, he kept a log of dates, reducing the events of days to a single check mark for each week passed and jotted lists of the garments and requisites he would need to resume his former life: "1 blue serge suit . . . 2 pr. white canvas spats . . . 2 pr. white flannel trousers . . . 1 bowler hat." He gravitated to Mackintosh, the two men bound together by their common yearning for home.

In late March, Mackintosh and Hayward began hobbling around Hut Point for exercise. Using crutches Joyce had fashioned for them, they ventured farther afield each day, testing the limits of the sea ice as it spread across McMurdo Sound. By April 18, the ice extended four miles to the north. At two inches thick, it was strong enough to bear a man's weight, but Joyce knew that the cloudy young ice, composed of briny water, could not be trusted. In time, the salt would precipitate into the sea below, leaving the ice clear and far stronger. Joyce wanted to wait until then. As he expected, high winds demolished the new ice and swept it out to sea two days later.

As soon as the ice re-formed, Mackintosh and Hayward resumed their excursions. Then on May 7, Mackintosh announced that the time was right for the trip to Cape Evans. The new ice grew to four inches thick. Joyce was dubious. "I don't know why these people are so anxious to risk their lives again," he fumed in his diary. The following morning, Mackintosh told Joyce he wanted to make the trip to Cape Evans that day. Joyce took him outside and pointed in the direction of Minna Bluff, which had served as a kind of barometer since the first season. If the Bluff disappeared, it was a sure sign that a southeasterly blizzard was brewing. The rocky spur had disappeared in the mist, and the southern sky was overcast. "I told him he could please himself but I thought it was not a day for it," Joyce recounted. He refused to go. Hayward promptly went out to see the sky and came back in, looking rattled. They would have to move quickly to stay ahead of the approaching storm. The fastest they had ever done the thirteen-mile journey was five and half hours; eight was more typical.

Joyce, Wild, and Richards argued with Mackintosh to wait for a day with fine, clear weather but Mackintosh was adamant. As Richards put it, "short of forcibly restraining him we could only urge them not to go," an echo of their vain attempt, months earlier, to talk their enfeebled leader out of continuing beyond the 83° Depot to Mount Hope. Neither Mackintosh nor Hayward seemed strong enough for the journey, both still suffering from malnutrition. Just the week before, Joyce glimpsed the dusky discoloration of scurvy lingering behind Mackintosh's knees.

To make matters worse, Mackintosh and Hayward intended to travel unencumbered to move faster. They would carry no sledge, no tent, no spare clothing, no stove, and no food. Joyce finally succeeded in forcing a bag of seal meat and chocolate upon him. Mackintosh tucked the food and his diary in the pocket on the front of his sweater. Hayward left his journal behind to save weight, first cutting his fiancée's pasted picture from the flyleaf to take with him. Joyce extracted a promise from Mackintosh that the pair would make for land at the first indication of a blow. "After bringing them back from death they seem to think they can court it again, ah well, such is life and what fools we have to put up with," he railed in his diary.

Once again, Joyce was unwilling or unable to impose his will on Mackintosh. At one o'clock, the two men said their farewells and walked down the slope behind the hut to the sea ice. Joyce, Richards, and Wild hiked up the hill overlooking McMurdo Sound and watched Mackintosh and Hayward trudge north across the ice. "I can see these two figures now quite clearly in my mind's eye," Richards wrote later. "They appeared pigmy-like as they grew fainter in the dim light against the vast expanse of sea ice to the north, and we watched them for a while in silence." Wild was characteristically blunt. "If the other two get lost, I shall be sorry we humped them back here over the barrier," he reflected. "However let's hope they get there alright."

Within an hour and a half, a stiff southeast wind came up, whipping the drift so thickly that it obscured the hill twenty yards away. By three o'clock, a blizzard had enveloped the hut. The storm raged for four days. When the wind abated on May 12, Joyce, Richards, and Wild headed out to see if there was any sign that Mackintosh and Hayward had reached safety, following the coast on the sea ice toward Cape Evans. They picked up two sets of footprints frozen in the ice. Nearer the shore, the first set, to and fro, marked the reconnaissance trip Mackintosh and Hayward had made the day before their departure. Farther from shore, a second set of prints hugged the coastline, then angled sharply northwest, as if Mackintosh had suddenly decided to make for the tip of Glacier Tongue, the usual route across the sea ice. About two miles north of Hut Point, the trail ended at open water, extending north as far as the eye could see.

— 14 —

"Drifting to God Knows Where"

MAY 6, 1915–MARCH 23, 1916

On the last night of Spencer-Smith's life, the *Aurora* was 1,100 miles northwest of the camp, beset in the pack ice of the Southern Ocean. Stenhouse had kept the battered vessel afloat for over three hundred days as it drifted with the ice. "The old ship has stood some tremendous buffeting up to the present time & one wonders exactly how much more she will take—and swim," he wrote in his diary on March 8. She was leaking badly, taking on water at a rate of three feet every day. Powerless, he could only wait and watch for open seas to the north.

_____

In the days following the breakaway from Cape Evans on May 6, 1915, Stenhouse grappled to overcome the calamity. Astonished by the colossal forces that gripped the *Aurora*, he realized that "no moorings ever made could have held her." The ice was far too dense to batter a passage back to Cape Evans, and only nine days' worth of coal remained. Stenhouse decided not to squander the short supply in a futile battle and instead allowed the ship to drift with the pack. If she was borne to the northwest, near the western shores of the Ross Sea, Stenhouse planned to disembark the crew and sledge back to Cape Evans to assist in laying the depots. If indeed disaster had befallen Mackintosh and the five men who had failed to reach Cape Evans by May, the four survivors of the shore party would not be able to do the job alone. "Stenhouse endures mental torments thinking of the Captain's party which may never have reached Cape Evans & also of those who will cross the Continent & no depôts ahead of them," wrote chief engineer Larkman.

When the blizzard died, wireless operator Lionel Hooke mounted the aerial and began transmission to Cape Evans, still less than a hundred miles away. "All well ship adrift ice pack off Beaufort Island" he tapped out repeatedly, hoping that the party had figured out how to operate their set. He also sent daily messages to the closest stations, Macquarie Island and Awarua, New Zealand, even though the two-kilowatt set was designed to transmit only two to three hundred miles. Hooke transmitted messages for weeks, but the receiver was suspiciously silent. At night, a peculiar humming sound emanated from the rigging aloft. Searching for the problem, he touched the transformer and was jolted by nine thousand volts of electricity. Sparks flaring like Roman candles from the upper spars finally revealed the problem. The grooved porcelain insulators were encased in ice and leaking the signal. Hooke suspended transmission until he could work out a solution.

By early June, it became clear that the wind and currents were driving the pack northeast, away from land. Stenhouse was forced to abandon his plan to sledge back to Cape Evans. Instead he hoped that the ship would be swept out of the Ross Sea with the pack into ice-free waters, where he could make a dash for New Zealand for repairs and coal before returning to support the shore party. During June, the ship drifted with no significant threat of ice pressure. The greatest danger was the enforced inactivity. The winter darkness and blizzards meant that the crew was largely confined to the ship, and eighteen men, cooped up in close quarters indefinitely, had the makings of a powder keg. Fortunately for Stenhouse, the men he depended upon most took the catastrophe with equanimity. Thomson decided there was "no use in being despondent about our luck [as] we might have been a lot worse off" and welcomed the chance to study the currents and ice movements. Larkman was unperturbed, working on the engines and spending evenings in the wardroom practicing astrology and palmistry. Boatswain Paton was typically phlegmatic, writing in his diary, "We came to the Antarctic to look for adventure, we are getting it, so why complain?"

The mood in the forecastle was mercurial. The veteran sailors, like able seamen Atkin and Kavanagh, kept busy "sailorizing," though they grumbled about conditions, and "Shorty" Warren did his duties with a "perpetual smile." The malcontents, the inexperienced Glidden and the three stokers, reacted badly and Thomson found it increasingly difficult to manage them. The belligerent Shaw presented the greatest threat to the peace. By day, the former policeman and heavyweight boxer was an imposing bully. At night, he sleepwalked, opening hatches and moving gear. The entire company was intimidated by his disruptive episodes. "He would be a nasty man if he only got violent in his sleep and found a fistful of opened razors handy somewhere," Thomson worried. Shaw also suffered from seizures, but Stenhouse had little sympathy, maintaining that he deserved "punishment for coming South knowing how he would be subject to such attacks."

Tensions also flared around motor mechanic Howard Ninnis. Officially a member of the shore party, Ninnis had no formal role in the well-ordered universe of the ship. His awkward position required thick skin and diplomacy, neither of which the sensitive loner possessed. He felt humiliated by the failure of the motor sledge, which Joyce had dismissed as a "useless toy," and tried to save face by boasting about his past exploits. The other men came to regard him as pretentious and "a bit of a swanker." When Ninnis was rebuffed by the seamen and the engineers, they became the targets of his caustic sarcasm.

Taut nerves were aggravated by shipboard conditions. Winter temperatures dropped as low as thirty-three below zero, and the living quarters became increasingly miserable. The men were forced to stop using eating utensils because the cold metal tore the skin from their fingers. The cracked stoves leaked noxious carbon monoxide, plaguing the officers with headaches and nausea. Twice Stenhouse was found unconscious in his cabin. Even with his faulty stove burning constantly, the temperature of his room hovered around the freezing point and the adjacent chart room was zero degrees. Paton's cabin forward resembled "a limestone cave so thick has the ice formed upon the ceiling and walls from the condensed breath."

In the face of it all, Stenhouse's confident attitude reassured the men that he would find a way through their predicament. He kept the crew busy painting, scrubbing, sewing, and restowing cargo as if the *Aurora* was a merchantman on a routine voyage. "Unremitting toil seems to be the panacea for most," he decided. But inwardly, he was troubled by doubt. "The first of July! Thank God! The days pass quickly," he confided to his diary. "Through all my waking hours one long thought of the people at Cape Evans." Although Thomson had assured him that they "had done all any human could" to secure the *Aurora*, Stenhouse wondered if he had made the best choices in McMurdo Sound. He suffered acutely as he contemplated the fact that the success of the expedition, and thirty-four lives, turned on his decisions.

"The old, old curse," he called it, as he struggled to keep despair at bay. He had battled it before, when he sailed as a third officer for the British India Steam Navigation Company in September 1913. Seven weeks into a voyage, he suffered a nervous breakdown, as the company termed it, and was put ashore in Australia. Shackleton's confidence in naming him first officer did much to salve the wounds of his last voyage, but he remained his own harshest critic and struggled to master the anxieties beneath his unflappable demeanor. "This is Hell," he wrote, then remonstrated himself to "appear to be happy & take interest in the small happenings of shipboard."

———

The winter nights were eerily quiet, the pack dampening the murmur of the waves. Shadowy hummocks loomed around the ship. There was nothing in the snowbound landscape to betray that she was adrift in the Ross Sea, moving at an average of four miles each day. By the first week of July, periodic booms sounded, like the muffled thunder of distant artillery. Around the ship, the pack grated and crackled. Stenhouse wondered if the seasonal disintegration was beginning. Moonlight revealed the towering shapes of pressure ridges to the southwest, where forces within the ice had pushed great slabs upward. Sudden spells of quaking shuddered through the ship's timbers. "We are all expectation," wrote Ninnis, as the crew watched for leads with growing excitement. But Stenhouse found it ominous. "Here one sees the impotence of man & feels a pawn in a great game," he wrote on July 13. "We have prepared for the worst & can only hope for the best . . . a release from the ice with a seaworthy vessel under us."

The crew took comfort in the *Aurora*'s redoubtable construction. Double-tiered oak beams braced her three-foot sides. The hull was sheathed with greenheart, so dense that it dulled the metal tools used to work it. The rudder was solid oak, eighteen inches thick, with iron plating. Oak and elm reinforced the bows, and the stern was built up to a thickness of fourteen feet and also sheathed with iron. Yet for all her fortification, the vessel had vulnerabilities. Stenhouse knew that the *Aurora*'s design was outdated compared with innovative vessels like the *Fram*, used by Nansen and Amundsen so successfully in the polar regions. The *Aurora*'s overhanging stern was prone to becoming wedged in the pack under pressure, and the exposed rudder and propeller could be battered by ice. Robbed of steering and propulsion, the ship would be helpless.

As cannonades of floes assailed the ship, the staunch beams creaked and moaned. Despite the assurances of the old polar hand Paton, it was unnerving for the crew to hear the timbers tested. "I doubt if anyone slept one wink," wrote Ninnis. On July 18, the men were huddled around the stove after lunch, groggy after another sleepless night. It had become so difficult to keep the crockery from sliding off the table that Hooke stuck his plate to the surface with a smear of treacle. From above, they heard a clamor "like the collapse of a high stone wall." The floes had shifted and a crack opened fore and aft, running northeast to southwest. As the men ran to the railings and watched, the crack broadened, the open water reflecting the moon.

In a matter of minutes, the *Aurora* was floating freely for the first time in months. The freedom was short-lived. New ice glazed the pool around the ship within hours and the pack closed in again. Stenhouse ordered the crew to prepare relieving tackles to ease the stress on the rudder if the ice thrust too close. They watched and waited expectantly for three days. The barometer was dropping rapidly, a sign of a brewing storm or, as Stenhouse hoped, open water on the horizon. In the early hours of July 21, Thomson's shout of "all hands" rousted the crew from their bunks. The lane around the ship had widened to a pond about a hundred feet across. The *Aurora* rotated like a compass needle until her head pointed northeast. "Our marvellous luck is 'in' I feel sure," wrote Ninnis as the pack began to break up in all directions.

Just after five o'clock in the afternoon, a deep rumbling signaled a fresh wave of pressure. But instead of liberating her, the ice crept closer and spun the ship until she was wedged diagonally across the lead. Then, "slowly, steadily, with an irresistible, menacing sort of movement," the lead closed. Clamped fore and aft, the stern took the full brunt of the pressure. The ice had found the *Aurora's* weakness. As the stunned crew watched, the massive rudder, twenty-five feet high and weighing close to five tons, was wrenched violently aside and smashed into splinters against the unyielding floe. The *Aurora* was wedged in the encroaching ice.

Stenhouse knew the loss of a rudder in the Antarctic could spell doom for the ship and the crew. Aside from rendering the *Aurora* incapable of steering, he feared the damage had undermined the structural integrity of the vessel. He decided that the time had come for preparations to abandon ship. "We are nearly 100 miles off shore, barren at that & it is a horrible situation," wrote Hooke. Even with most of the shore party's sledging equipment still on board, they were short, since the crew had not been expected to camp. There were only six tents and sleeping bags and not enough cold weather clothing for eighteen men.

By early evening, a gale howled around the beset ship, driving the men belowdecks. A few feet from the wardroom table, the massive sternpost, the last line of the *Aurora's* defenses, strained and lurched under the constant pressure as they ate in silence. The crew retreated to their bunks fully clothed, leaving the doors ajar in case the crushing pressure skewed the frames and trapped them inside as the ship sank. Stenhouse was awake throughout the night. He had not slept for four days and searched the medical kit for stimulants to shake his bleary exhaustion. Pacing the deck, he could see that the midship section was buckling upward, the bow and stern sagging as the pack pincered the vessel. Hogging, it was called, and there was only so much of the stress a vessel could take. The spine of the ship, her immense keel, was bending like a pencil in a tightening vise. "Am afraid the ship's back will be broken if pressure continues," Stenhouse wrote impassively in his diary.

If the keel snapped, she would likely go very quickly, leaving the company stranded on the floes. Each man stuffed a pillowcase with a few personal possessions and joined one of the four sledges to rehearse their assigned duties, although as Larkman confided to his diary, "God knows there's practically no hope of survival if that comes to pass." By Stenhouse's calculations, the *Aurora* had drifted two hundred fifty miles from Cape Evans. Coulman Island was the nearest land. Even if the churning pack had consolidated, towing heavy sledges over the broken surface would still take over a month. On the island, the odds of rescue would be slim. The *Aurora* was not expected in New Zealand until April 1916, and no captain would dare brave the Ross Sea until December. Then it would take months for the rescuers to systematically search the region for the stranded crew. There was little chance they would survive that long, either on land or the ice.

Amid the cacophony of creaking and whining timbers, a strange howl sounded from the sternpost on July 22. The pressure had wedged the stern in the pack, embedding the counter deeply in the ice. The *Aurora* leaped forward convulsively, then recoiled into the grip of the ice. Grasping at straws, Stenhouse ordered the men to pour gallons of battery acid on the ice in an effort to free her. It was utterly futile. Time and again, the pressure mounted, compressing the ship like a bellows until she bucked upward from the strain and fell back, hopelessly trapped.

Inching higher with each wave of pressure, the ship's bow had been shoved up onto the ice, balancing her precariously as the narrow stem was jacked higher on the floe. Then, without warning, the keel crashed through the ice and the ship rolled violently. The *Aurora* had been sprung from the vise, the stern afloat in a small pool and the bows beached on the floe. The sternpost was safely clear of the pack. By midnight, it appeared the pressure was subsiding, though few accepted that the danger had passed. "The whole crew are like a crowd of school girls, our nerves absolutely shattered," wrote Hooke. "The slamming of a door brings us all up with a start." Stenhouse and Paton examined the damage to the ship. The propeller was hidden; the rudder was smashed to wreckage; but the battered sternpost was still intact, and to Paton's amazement, "the wells and holds proved her to be worthy of her builders not many vessels of her age would have stood what she stood last night yet she has not made any water, which is the greatest surprise of all." Still, Stenhouse knew that the ship's defenses were so compromised that she would not survive pressure of that magnitude again. He also worried that the thick ice coating the bottom of the bilges had sealed gaps in the hull. Once in warmer seas, the frozen caulking would melt and the ship might leak like a wicker basket.

Against all odds, the *Aurora* had survived. For the time being, Stenhouse decided, they were safest on board and suspended plans to abandon ship. He was powerless to do much more than attempt to defend the vessel as she drifted. Stenhouse called Hooke and handed him a message for the wireless. Hooke coated the insulators with pitch and resin and climbed aloft to solder the aerial mounts. Down below, he began transmitting the distress call for King George V to Macquarie Island and Awarua, 1,100 to 1,600 miles away. "Still hoping but the chances of communicating look small," despaired Hooke on July 26.

———

The ship was at the mercy of the elements, their fate tied to the winds and currents. By early August, the *Aurora* was about four hundred miles from Cape Evans, in sight of Cape Adare. She had heeled sharply to port and crossing the deck was "like Alpine climbing," in Paton's words. Although her release appeared remote, Stenhouse asked the carpenter to jury-rig a rudder. Once in the open sea, the makeshift rudder would be slung over the stern and manually controlled with guy wires and long wooden booms, like an unwieldy outsized oar. Mauger's rudder was an immense structure, twenty-two feet long, constructed of salvaged wood and weighted with reinforced concrete and iron cut from the two-inch plating on the damaged rudder.

Hooke persisted with the wireless set, modifying the generator and the transmitter to try to increase the range. In late August, he caught occasional snatches of messages from Macquarie Island and Awarua. Then, on September 5, the mizzenmast was torn off in hurricane-force winds, carrying the attached wireless mast and aerial with it. Hooke's experiments came to an abrupt end. "Hooke dismantled wireless," wrote Stenhouse. "Heartsick."

The ice was still on the move. Lanes opened, just beyond reach, and closed again. "I cannot help thinking that if the 'Endurance' wintered in the Weddell Sea she will be worse off than the 'Aurora,'" Stenhouse wrote in his diary, wondering if sledging was under way. By late September, he calculated that the *Aurora* had been swept some seven hundred miles out of the Ross Sea and into the Southern Ocean. The drift tended sharply northwest, though he could not fix their position with certainty because the chronometers were erratic.

"This waiting for we know not what & the fact that the lives of men are at stake weighs heavy," Stenhouse wrote on November 13. The Antarctic coast was behind them, and the last glimpse of land had disappeared. As the ship neared the Antarctic Circle, summer temperatures rose into the forties. Water trickled down the frosted cabin walls and puddled in the bunks and floors, only to freeze again at night. The sodden blankets and clothing had little chance to dry. Stenhouse forbade the use of coal for the stoves, which he was determined to conserve for the ship's engines. Even with the oven fires extinguished, coal consumption in the galley was seven hundred pounds each month. There were other worrisome shortages, so he depended on seal and penguin meat as much as possible to conserve the dwindling provisions, but game became scarce in December. The stowaway rats found few scraps and turned to cannibalism to survive.

Hooke began working on the wireless once again with Stenhouse's encouragement. "Feel pretty disgusted with no communication but honestly don't think it possible under the circumstances," he wrote. Stenhouse's cabin had become a haven for troubled crew members. In the absence of a ship's surgeon, the men came to him for medical treatment, stitching cuts, pulling teeth, and excising gangrenous frostbite. Stenhouse drew Ninnis out of his shell and gave him useful duties. But there was no cure for Shaw's worsening condition. He clung to the belief that staying awake would alleviate the problem, but his seizures became more frequent. Once a menace, he was tormented by the forecastle hands as he paced the deck at night, shivering in tattered clothing. Stenhouse witnessed his sorry vigils with newfound compassion, writing, "Poor beggar he is worrying a great deal about it & gets little sympathy for'ard I imagine." He ordered a seaman to sew some clothing for the stoker out of canvas and gave him a room to himself.

His ministrations had a discernible effect. "I cannot help remarking on the happiness of the ship's company & the splendid spirit which prevails," he wrote with a touch of pride. Beneath his formal manner lay a genuine concern for the men. Even Paton found the harmony remarkable. In the throes of crisis, Stenhouse had unified the odd assortment of men and had "proved himself a thorough, reliable and energetic Seaman, no one could have been cooler than he," as Paton put it. "He has done well, and I for one am pleased to be under his command."

Mindful of morale, Stenhouse organized a Christmas dinner. The drift had slowed to a crawl in December, just eleven miles in nine days. Summer brought sleet and rain in January, but the pack remained solid. Stenhouse cut back meals to one daily to conserve the waning supply of meat, which was putrefying in the warmth. Garbage and human waste piled up alongside the trapped ship. The seamen were forced to venture farther from the ship to find seals and penguins, stumbling hip-deep into slush pools.

———

"Another month gone & still we are imprisoned," Stenhouse wrote in his diary on February 1. "The T. C. [transcontinental] party should be at the Beardmore now, if they have made good time; it will be a miserable 'homecoming' for them, after the hardships of travel, to find not only no relief ship but little stores & a prospect of another winter in the south." Unspoken was his fear that the crippled shore party had been unable to lay the depots, and finding no provisions, Shackleton's party had perished on their march to the coast. With autumn approaching, Stenhouse faced the possibility that the Aurora might be trapped in the ice for another year. "We are too late now to be of assistance to the people South or to help in any way," Stenhouse wrote on February 10.

Two days later, a fresh breeze came out of the southeast. The *Aurora* began to quake and strain against the ice bracing her sides, then rolled heavily. A swell was running from the opposite direction, heaving up against the pack. All around the ship, the solid plain of pack ice rose and fell rhythmically. The pack was breaking up. As the floes parted, revealing a dark seam of open water, Stenhouse could see that the ice was thirty to forty feet thick in places, a vindication of his decision to conserve the coal. In a widening lagoon around the stern, he glimpsed the propeller and saw that it was intact. He decided not to risk damaging either it or the jury rudder while still in the ice and gave orders to raise sail instead. To the incredulous delight of the crew, the *Aurora* began to glide slowly north, steering herself with no helmsman.

The ship nudged floes aside and made headway "like a racehorse," as Paton crowed, then became wedged in thicker ice. As she proceeded in fits and starts, the ship was taking on water, nearly four feet on the first day free of the ice, confirming Stenhouse's suspicion that the ice sheathing had caulked the strained seams until it melted. The worst of the damage was unseen: hundreds of rusted iron bolts fastening the hull planking had snapped under the pressure. The crew formed a bucket brigade to drain the bilges.

"I feel that we are taking things too easy & not trying hard enough," wrote Stenhouse in abject frustration as the ship came to another

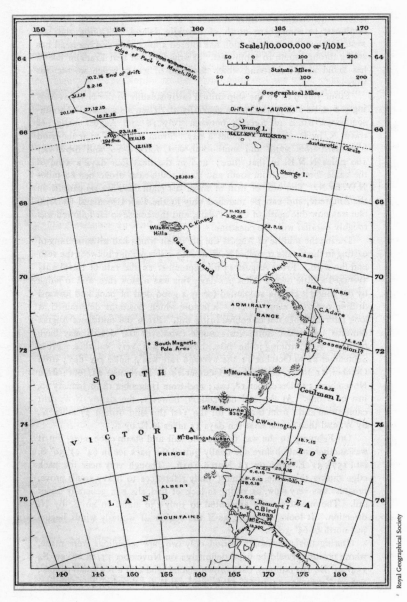

THE DRIFT OF THE *AURORA*, MAY 1915–MARCH 1916

standstill. "And yet it is no good butting against this with steam-power, for we would use all our meagre supply [of coal] in reaching the limit of the ice. And then we would be in a hole, with neither ballast nor fuel." On February 24, Paton climbed to the crow's nest and surveyed the ship's position. In all directions was a "vast white circle." Immense flat-topped tabular bergs towered out of the chalky sea, "dangerous customers to come in contact with on a dark night," as Paton put it. "Our position, in the midst of this wide expanse of pack, is not a pleasant one at any time but when a heavy swell rolls under the ice it becomes a nerve-wracking & soul-destroying ordeal," wrote Stenhouse on February 26 as the bobbing ice pounded the ship. Leaks sprang up in the propeller shaft, which the carpenter tried in vain to seal with a mixture of oakum and cement.

"The vessel cannot stand much more of this," Stenhouse wrote. On March 1, with no open water in sight, he cast the die and ordered steam on the main engines. "With the lighting of the fires, while yet in the pack, goes our last chance of relieving the people South," he wrote, surrendering the last vestige of hope for a return to McMurdo Sound before 1917. On March 3, a lane opened and the *Aurora* forged ahead. By the end of the day, they had burned through three tons of coal to gain about five miles.

Stenhouse rang off the engines and ordered the fires banked. Overnight, a mist settled over the ship as the pack crowded close. Paton patrolled the deck on watch. Beyond a ship's length, it was impossible to see anything through the opaque fog, but he could hear the rustling of the pack all around the ship. At daybreak he went aloft to the crow's nest. In the lifting fog, he sighted open water.

For the next ten days, the *Aurora* dodged and weaved, searching for a way to the perimeter. Then on March 14, the *Aurora* cleared the pack at 64°27'30" S, 157°32' E and sailed into the open sea after drifting 1,600 miles. The company was so elated to be "in the real true blue" that they relished the first wave of nausea caused by the swell. Stenhouse was guarded, bracing for fresh difficulties. The ice had protected as well as imprisoned the *Aurora*. The Southern Ocean could be fearsome sailing for any ship, much less a disabled one. Fifty-foot waves were well known in the latitudes mariners called the "screaming sixties." Stenhouse knew the ship was unlikely to ride out a gale in these waters. On March 18, about a thousand miles southwest of New Zealand, he ordered the jury rudder lowered. Woefully low on ballast, the ungainly vessel bobbed high on the massive swells as Stenhouse steamed cautiously at half speed, trying to outrace the weather without foundering. It was impossible to hold the course with any precision, not least because the thrashing booms controlling the rudder frequently pitched the helmsman bodily through the air.

In the steward's pantry, Hooke toiled away at the wireless. After fifteen months of tinkering, it seemed hopeless, to the crew if not to Hooke. Then, as he bent over the receiver during the night of March 23, it crackled to life with a burst of Morse chatter.

— 15 —

"Whereabouts Shackleton?"

MARCH 24–MAY 31, 1916

Just after midnight on March 24, the coastal wireless station in Awarua, New Zealand, received the first enigmatic radio call from the *Aurora*. Moments later, a stream of faint Morse signals formed a series of messages explaining the travails of the Ross Sea party. It continued throughout the night: "Mackintosh and party not returned . . . Lost rudder hull severely strained . . . Ship proceeding Port Chalmers New Zealand jury rudder no anchors short fuel expect arrive early April . . ."

Using a wireless set with a range of two hundred miles, Hooke's modifications had enabled him to transmit a thousand. Moored tenuously to civilization by electromagnetic waves, Hooke refused to stop sending until the sun rose and ionization swamped his transmissions. Awarua operators relayed the torrent of messages to authorities in New Zealand and Australia. Stunned officials responded, and queries flooded in for Stenhouse from the Australian navy minister, the harbormaster in Port Chalmers, and the prime minister of New Zealand. "Indicate position course speed . . . Do you need assistance? . . . How much stores did you leave Ross Base?" The naval minister released one of Stenhouse's messages to the press on March 25. By the time Hooke returned to the air after sundown, the queries had been joined by interview requests from the *New York Times* and the *Daily Mail*. All were refused because of Shackleton's *Daily Chronicle* exclusive. The burning question—"whereabouts Shackleton?"—remained unanswered.

The news flashed around the globe, colliding en route with another item, coincidentally released by the expedition organizers a few days earlier. Seeking to revive interest in the expedition, they had released a

1914 bulletin from Shackleton at South Georgia. Details of the outdated report and news of the *Aurora* became garbled and fueled fresh editorial conjecture. In London, the *Morning Post* reported that Shackleton was "expected daily to return" to Buenos Aires, while the Associated Press confidently announced Shackleton's triumphant arrival in Sydney, "news of achievements withheld for the present." A London medium claimed to have "met Shackleton's Spirit," who said that he and his party had died two months before. Most accurate was the *Daily Chronicle*, which published a message from Stenhouse under the banner headline, "The Shackleton Expedition, Bad News from the Antarctic 'Aurora' Breaks Loose and Drifts Away in the Ice, the Shore Party Stranded."

In England, Emily Shackleton endured the wild speculation in silence, shielding herself from the intrusions of Fleet Street. In the fifteen months since the expedition's departure, she had submerged her private worries in the placid routine of family life. On March 27, after three days of waiting for conclusive news, Lady Shackleton hastened to the Gray's Inn offices of Shackleton's solicitors, Hutchison and Cuff. Expedition secretary Frederick White and Ernest Perris of the *Daily Chronicle* were waiting. Alfred Hutchison confirmed what she had already sensed: The stories trumpeting her husband's imminent return stories were false. There was no word of Shackleton or the *Endurance*. She had not heard from him since his last letter from South Georgia in December 1914, in which he confided his misgivings about the iceworthiness of the *Endurance* and the reliability of Captain Frank Worsley. There was little she could do, short of writing to the Australian high commissioner, Andrew Fisher, "Is it too much to ask to come to our aid and send the *Aurora* for her rescue work in the Ross Sea?"

For Gladys Mackintosh, there was more certainty but far less reassurance as she sat in the solicitors' office. Her days of waiting were measured by her youngest daughter's life, born just a month before the *Aurora* left for the Antarctic. As the wireless messages from Stenhouse clearly stated, her husband had been counted as missing with five other men when the *Aurora*'s moorings snapped. It did not bode well that they might have been on the Barrier in early May in the darkness and subzero temperatures of the Antarctic winter. Ten men were stranded ashore— sixteen if Shackleton and his party had arrived safely from their transcontinental crossing. They had been marooned for ten months.

As the pall of grim conclusions settled on the gathering, messages continued to flow from the crippled *Aurora*. As well as coping with the hazards of the open seas, Stenhouse grappled with crew problems. The constant demand for steam required the firemen to stoke coal around the clock. Shaw reached his breaking point and refused duty. The mental strain of the seizures and ostracism had become unbearable. The steward d'Anglade quit work on the same day. The mercurial Frenchman, who had alternated between bouts of malingering and industrious activity throughout the voyage, decided he had nothing further to gain by working. After two days, Shaw returned to the boiler room. No manner of threats or persuasion, however, could induce d'Anglade to return to the galley.

Stenhouse feared that the *Aurora* could not reach port without assistance. On March 26, the bunkers held sixty tons of coal and the engines were burning through six to eight tons each day. Stenhouse resisted calling for help. Since the ship first made wireless contact, he had refused all offers of a rescue tow from New Zealand, and not merely out of pride. Admitting that the vessel was imperiled laid her open to a salvage claim. A convention of maritime law since Byzantine times, marine salvage entitled a rescuer to a percentage of the value of a disabled ship. Stenhouse faced losing the *Aurora* to either the sea or an Admiralty court forcing a sale to pay off a salvor. By March 30, the *Aurora* was 330 miles south of Awarua. If she could hold her own against the seas, it would be about three days to port.

Stenhouse resisted calling for assistance even as the glass dropped and snow squalls buffeted the ship. The engines raced and whined as the ship plunged over the crests of rollers. The listing *Aurora* bowled northward, her stern lanterns glowing red to signal "vessel not under command" as she veered in the direction of New Zealand's south island. The rudder had proved useless with a heavy sea running, so it was hoisted. With the rocky Snares Islands on the starboard bow, Stenhouse ordered the engines cut and the ship's topsails taken in, knowing that a sudden gust on a rudderless ship under sail could knock her down or wreck her on the rocks. Reluctantly, he accepted the offer of a rescue tow from New Zealand authorities. A tugboat left Port Chalmers, the city of Dunedin's harbor, to rendezvous at sea. The *Aurora* drifted through the night with engine fires banked, rollers mounding under her hull, as her company watched and waited through a howling storm. In the early hours of April 2, the tug *Dunedin*'s searchlights swept across her starboard bow as her captain hailed the *Aurora*.

The *Dunedin* stood by until daylight, when the tug's crew sent over cases of porridge, bacon, and eggs from well-wishers. Then came the newspapers and sacks of mail. The headlines confirmed one of the first wireless messages that Hooke had received from Awarua: "War still going strong havoc with civilization." Under tow, the *Aurora* reached Otago Heads on April 3, when Stenhouse asked the tug to cast off so the *Aurora* could steam into port unassisted.

Stenhouse had already begun planning the rescue of his stranded shore party. For days, he had been sending entreaties to Frederick White in London for funds to organize the rescue. His questions about money were studiously ignored. Instead, White responded with a barrage of messages forbidding Stenhouse to speak to anyone in the press other than the *Daily Chronicle*. Finally, after repeated queries, Stenhouse received a reply. "No news *Endurance* . . . Unable send sum required as no funds available." In a reprise of his wires to Mackintosh fifteen months before, White continued, "Can only find five hundred pounds . . . Secure free facilities for ship and any financial gifts. Following publication your trying experiences might be possible discover Australian benefactors. Do all your power obtain assistance."

In fact, the expedition coffers were nearly empty. Bills from 1914 were still in arrears and the organizers could not meet new obligations. With no finances available, the organizers turned to the Royal Geographical Society. While not an official sponsor, the RGS had traditionally assumed the role of "keeper of the conscience of the geographical world," as Shackleton once put it, and had contributed to the expedition. RGS president Douglas Freshfield summoned Perris to the society's Gothic Revival mansion overlooking Hyde Park to account for the expedition.

A tough mountaineer known for legendary first ascents in the Caucasus and the Himalaya, Freshfield was two years into his term as president. A lawyer by profession, he was intellectually uncompromising, a man intolerant of evasion who had acquired a reputation as a "verbal chastiser." With the barrister's gift for economical fact-finding, Freshfield cut swiftly to the chase and extracted the story from Perris. Before departing South Georgia for Antarctica on December 5, 1914,

Shackleton had sent Perris a letter. The *Endurance*, he wrote, was "a great disappointment." Her bunkers were so small that he expected to run short of coal on the return voyage and use seal blubber to stoke the engines. In any event, he did not plan for the ship to return until sometime in 1916. "Shackleton knew perfectly well he could never cross the first season," according to Perris, although Mackintosh had not been informed of the change of plan. Perris admitted that Mackintosh's instructions were "to be corrected by a long cable" before the *Aurora* left Australia on December 24 "which was never sent."

The outlook was bleak. Perris's admission explained why the Ross Sea party had rushed into laying depots in the first season and had remained on the Barrier so late in the season. There was no question in Freshfield's mind that a relief expedition for both parties was necessary, but financing, men, and a ship, possibly two, would be scarce during wartime. Moreover, there were precious few vessels suited to braving the Antarctic. On April 7, the RGS Council passed a resolution that while "unable to offer financial assistance or accept any responsibility, [the RGS] is prepared to give its advice to any responsible body which may be appointed to control the expedition." The implication was clear; the RGS did not consider Shackleton's own organization to be that body.

Finding a ship's master capable of carrying out the relief operation was also a formidable problem. As one insider put it, there were "very few men fitted to take charge of such an expedition." Few navigators knew the unique hazards of the Antarctic seas, and fewer still had sailed into the Ross Sea. "In fact, Davis is the only man," advised a council member. Freshfield was relieved to learn that John King Davis, the man who had refused command of the *Endurance*, was due to arrive in Britain in early April.

Ten years younger than Shackleton, Davis had five Antarctic voyages to his name, three as captain and expedition second-in-command. Since joining the *Nimrod* as first officer he had measured his "trial by ice" with telltale precision, counting "one year, nine months and eleven days" on voyages south of the Antarctic Circle. For he felt the burden of command in the ice keenly, the intense vigilance demanded by the knowledge that at any moment, the fate of the ship and the lives of her crew might hang in the balance, stayed only by his hand. Tall and lean, with the face of a temperance parson, Davis was austere of physique and disposition, his remote manner on the bridge earning him the nickname "Gloomy Davis." "If you are a charlatan the ship will find you out," Davis once declared. "It is a disquieting reflection." He had made up his mind not to go south again, but Freshfield was equally determined to change his mind when he invited Davis to the RGS on April 10.

It was a meeting of two like-minded realists. Davis spoke plainly. He was unsparing in judgment: "There are all these people who have friends; the *Endurance* was full of them—people without any qualification or knowledge just pushed in there because they thought it was an adventure." He was not surprised that the *Aurora* had survived the pack because "unlike the *Endurance*, the *Aurora* was a stoutly built old Dundee whaler." He harbored "grave fears of the fate" of the *Endurance*. The bleak ice reports from whalers returning from the Weddell Sea dimmed hopes that Shackleton's ship had even reached the Antarctic continent. At best, Davis reckoned, the ship had burned through her entire supply of coal in a losing struggle with the pack ice. Powerless and unable to steam north, Shackleton might have tried to run with the currents for Cape Town under sail. More likely, Davis thought, the *Endurance* had succumbed. "I think they have got in and then the sea has frozen and they thought it was alright," Davis told Freshfield. With mounting seasonal ice pressure, "the ship would be crushed." The doubtful prospects for the crew and Shackleton's transcontinental party remained unspoken, but understood. Both men were keenly aware of the odds against finding six men lost on the vast continent had they embarked on the crossing. The search area would be thousands of square miles. Shackleton himself had predicted "it would be no good looking."

Freshfield asked Davis to take command of the Weddell Sea relief effort while the *Aurora* was fitted and dispatched to the Ross Sea. Davis agreed, with one condition: He would be given "absolute and undivided control." "I will have nothing to do with the thing unless they allow me to decide who is necessary and who is not," he told Freshfield. Davis insisted on overseeing the staffing and outfitting of the ship, including a wireless set. Shackleton had organized his expedition in an eight-month cyclone of preparation. Somehow, an equally complex venture would have to be equipped in less.

Hutchison and Perris also offered Davis command of the Weddell Sea relief, failing to grasp that a major funder would likely demand control of the expedition. Their paramount aim was to rescue Shackleton's party. "There is no misapprehension in our minds as to the relative importance of the two expeditions," they wrote to Davis on April 10. "A ship will have to be sent down to the Ross Sea to bring off the men there, but as this is a matter of far less importance we need not trouble you about it." In the attorneys' view, it was not certain that Stenhouse would assume command. If substantial financing was offered, they were prepared to acquiesce to the funders' nominee for a new master of the *Aurora*.

Private funding was improbable in wartime; government support would be needed to mount a rescue. However, Prime Minister Henry Asquith's office and the Admiralty declined to take the initiative. Asquith's secretary suggested that the RGS sponsor the relief and approach the Treasury. Freshfield resisted assuming responsibility for the expedition. He had been one of the chief skeptics when Shackleton appealed to the RGS for support in 1914. "In my opinion they ought never to have let the expedition start after the war began," he thundered, and a series of correspondents joined in to air their views on the matter in a raft of memoranda and dispatches.

In fact, much of this doubt had arisen in hindsight. In August 1914, when the *Endurance* sailed from the West India Docks, politicians and populace alike thought a few decisive battles would win the war by Christmas. They could not have predicted that by April 1916, the conflict would evolve into total war, monopolizing the economic and industrial resources of the combatant nations. Nor could anyone have anticipated the staggering human cost: 1,154,822 British casualties by 1916, five out of every nine men sent to France alone. The popular mood shifted as the nation mourned and struggled to grasp the chaos in Europe. Tabloids rallied to Shackleton's cause in a patriotic vein. "Shackleton's ideal is that of the Allies," the *Daily Graphic* opined. "He is laying siege to darkness in the interests of truth and scientific conquest."

Indignation percolated in other quarters. "Fancy that ridiculous Shackleton & his South Pole—in the crash of the world," Winston Churchill, then in command of a battalion in Belgium, wrote to his wife, Clementine. "When all the sick & wounded have been tended, when all their impoverished & broken hearted homes have been restored, when every hospital is gorged with money, & every charitable subscription list is closed, then & not till then wd I concern myself with those penguins." Grudgingly, he added, "I suppose however something will have to be done."

The government ultimately mobilized support for the rescue. An advisory committee was convened on May 13, chaired by Arctic admiral Sir Lewis Beaumont. The members included former RGS president Major Leonard Darwin, Admiral Sir John Franklin Parry, Sir Douglas Mawson, explorer William Speirs Bruce, representatives of the Board of Trade and the Treasury, Perris, and Hutchison. The committee resolved to buy the *Discovery* or the *Terra Nova* and equip her for a relief voyage to the Weddell Sea in December. Search parties would scour the coast four hundred miles east and west of Shackleton's proposed landing site at Vahsel Bay. A reward would be offered to encourage whalers to come forward with clues to the expedition's fate. No plans were discussed for the relief of the Ross Sea shore party.

In the daily deluge of mail arriving at the RGS on May 24, a letter came from the mother of William Mugridge, one of the *Aurora*'s firemen: "Seeing by the papers that they have arrived at New Zealand, I should be very thankful if you could give me any information if they are being sent to England." The organizers had not notified the families of the men's welfare.

In New Zealand, the eighteen officers and crew of the *Aurora* struggled to come to grips with the world they had longingly imagined for fifteen months. Automobiles, trees, horses, insects, all seemed bewildering and delightful. Ninnis received a parcel of forwarded letters from his sweetheart, Ethel Douglas. For some, though, the joy of homecoming was riven by sadness. A telegram informed Paton that his mother had died in Scotland, and Hooke's brother Frank, a cavalry captain in the Australian Imperial Force, had died of wounds sustained at Gallipoli in June 1915.

"Now that we are in touch with civilization once more the past events seem unreal," wrote Ninnis. The men lingered in Port Chalmers and Dunedin in limbo, their fates bound up in the wrangling in London. They had been liberated from captivity, yet they were unable or unwilling to resume their lives. For some, particularly the men of the lower deck, there was nowhere else to turn. Only as secure as the berth on their next ship, the best prospect seemed to hang on for the back pay owed by the expedition. Other members of the *Aurora*'s crew agonized over the fate of the shore party and waited in the hope of returning to rescue their stranded mates.

The onus of providing for them fell upon Stenhouse. The burden was considerable, even before he considered the cost of a refit. To pay for living expenses and back wages, Stenhouse estimated that he needed at least £2,500. The London office sent £500 on April 7, then communications stopped altogether. Stenhouse made anxious inquiries and learned that White had parted company with the organizers in late April. Perris finally responded from London, sending £250 and the news that the expedition was flat broke. The crew was attempting to attend to duties aboard the *Aurora* in the tattered clothes they had worn in the Antarctic. Stenhouse used his personal savings and borrowed small amounts from friends and supporters to pay the creditors and the crew.

After two months in port, the crew began to disperse. Mugridge, carpenter C. C. Mauger, and three of the seamen, Arthur Downing, Sydney Atkin, and "Shorty" Warren, joined the New Zealand Expeditionary Force. The two mutineers, Emile d'Anglade and Harry Shaw, were hauled into criminal police court. Stenhouse relented and moved for the charges to be dropped, though he intimated that the battlefield or deportation were their only options. D'Anglade volunteered and Shaw was discharged as "mentally unsound" and returned to England.

On May 28, John King Davis received a cable when he arrived in New York City on a war transport voyage: "You are appointed Captain of the relief ship *Discovery* and to command expedition. Report at Admiralty, London, without delay." The vessel that had conveyed Shackleton to the Antarctic for the first time was to be his deliverance. Three days later, a cable arrived in London from the Falkland Islands, the southernmost outpost of the British Empire:

> *31st May have arrived Falkland Islands Endurance crushed middle Weddell Sea October 27th 1915 drifted 700 miles on ice until April 9th 1916 landed Elephant Island 16th April left 24th April leaving 22 men in hole ice cliff proceeded South Georgia with 5 men in 22 foot boat for help at time of leaving island all well but urgent need immediate rescue.*

It was signed Shackleton.

— 16 —

Port Chalmers

JUNE 1–DECEMBER 20, 1916

Shackleton never set foot on the Antarctic continent. On December 24, 1914, as the *Aurora* sailed south from Australia, the *Endurance* was threading her way through the pack ice of the Weddell Sea. By January 18, Shackleton's progress was arrested, the ship frozen fast in the ice. He had come within a hundred miles of his intended Antarctic landfall when it slipped from his grasp. "What the ice gets, the ice keeps," he warned the ship's captain, Frank Worsley. The *Endurance* was beset until November 1915, when she was crushed by the ice. Her company of twenty-eight was stranded on an ice floe "drifting along under the caprices of wind & tides, to heaven knows where," in one man's words. It was an uncanny echo of the plight of the *Aurora* in the Ross Sea, "drifting to God knows where."

Shackleton and his men dodged the hazards of the polar seas for months, finally launching three lifeboats in April 1916 to search for land. After a punishing eight-day journey, they found refuge on uninhabited Elephant Island, northeast of the Antarctic Peninsula. Shackleton knew there was no hope of rescue in a place where ships rarely passed. In a desperate bid to save his men, he and five companions navigated a twenty-foot open boat, the *James Caird*, seven hundred miles to the Island of South Georgia. They landed on the unpopulated coast and were forced to traverse twenty-six miles of uncharted mountains and glaciers to reach safety at a manned whaling station. There, Shackleton raised a ship and sailed south to rescue the twenty-two castaways on Elephant Island. A barricade of impenetrable pack ice repelled the rescuers. Defeated, Shackleton made his way to the Falkland

Islands to seek support for another attempt, where he learned about the Ross Sea party's predicament.

"I have had a year and a half of hell: and am older of course but so no lives have been lost, though we have been through what no other Polar Expedition has done," he wired to his wife. Other lives still hung in the balance. "I am rushed to death with cables and arrangements for the relief of our people." For the moment, that meant Frank Wild and the men on Elephant Island. Half a world away, the ten men stranded in the Ross Sea region, including Ernest Wild, would have to wait.

The news of Shackleton's reappearance electrified the press. "There are some failures almost as glorious as successes," extolled the *Edinburgh Evening Dispatch*. "Sir Ernest Shackleton's is one of them." The glass plate negatives Shackleton had stowed in the *James Caird* survived the ocean voyage. Photographs of the mangled wreck of the *Endurance,* heeled over in the chokehold of the ice, displaced the images of battle on the front pages of newspapers around the world. Demoralized by a war that seemed to be no closer to an end than when Shackleton left, the public celebrated his triumph: "This mission is one of sheer science and idealism—an enterprise undertaken in the cause of pure knowledge and pure achievement without any ulterior attachment to the great objectives that now convulse mankind," observed the *Pall Mall Gazette*. But there was disapproving commentary as well. "The British public has always had a tender spot for a sportsman, and [polar] expeditions provide a strong sporting element," noted the *Outlook*. "Nevertheless it may be questioned if the purpose they serve is in any way commensurate to the long toll of gallant lives they have exacted and the anxiety and suspense they cause to relatives at home."

The Royal Geographical Society greeted the news of Shackleton's return to civilization coolly. "Congratulate safety. Sympathize disappointment," ran the terse response to Shackleton's cable from Port Stanley. The relief committee in London now deemed it unnecessary to commit precious national resources to rescuing the men of the *Endurance*. On June 2, Davis received a cable in New York advising him that his services were no longer required. He took the abrupt about-face in stride and resumed his command of the war transport ship *Boonah*. Shackleton was on his own, combing the southern hemisphere for an iceworthy vessel.

———

In New Zealand, the *Aurora* languished at the Port Chalmers docks. Stenhouse champed at the bit to begin the refit. If the repairs had only started promptly, he reflected, the ship might have been available to relieve Shackleton's men on Elephant Island before sailing to the Ross Sea. But until he could raise the funds, even the Ross Sea relief was in doubt.

Stenhouse was fighting a battle of attrition. Only three forecastle hands remained on board and they frequently disappeared for days without leave. He found it hard to reprimand them as they were yet to be fully paid. The cantankerous Paton grumbled over the lack of discipline as he patrolled the ship on his customary night watch. Fortunately, most of the men of the wardroom remained resolutely at Stenhouse's side, although Larkman had returned home to England in June. In his stead, Donnelly stood by for word to begin overhauling the engines. Thomson supervised work aboard the *Aurora* and Ninnis served as Stenhouse's clerk. Hooke left to visit his family in Melbourne, but pledged to return for the rescue voyage.

In early June, circumstances were looking particularly bleak when Stenhouse received an unexpected letter from Leonard Tripp offering help. An old friend of Shackleton, the Cambridge-educated attorney was a partner in a prestigious Wellington firm and scion of a founding family on the north island. He counted a legion of friends in high places in the capital. The quiet, dapper lawyer became the expedition's champion, arranging for Stenhouse to meet several members of parliament. Then he personally pleaded Stenhouse's case with Prime Minister William Ferguson Massey, who sympathized with the plight of the Ross Sea party. He urged the cabinet to approve a grant of £500 and contacted the secretary of state for the colonies in Britain, Andrew Bonar Law, about taking action. The British government began negotiating with New Zealand and Australia to contribute to the relief expedition. They agreed that Britain would finance half the cost, and proportional to population, Australia and New Zealand would contribute roughly 40 percent and 10 percent respectively.

The dominions mirrored Britain in setting up advisory committees to coordinate the rescue effort. The Australian committee was headed by Rear-Admiral Sir William Creswell. Two distinguished scientists were invited to join, Orme Masson, Keith Jack's former professor, and Griffith Taylor, a geologist on Scott's last expedition. Captain J. R. Barter and Captain J. B. Stevenson contributed naval expertise. Since the *Aurora* was moored in Port Chalmers, direct oversight would fall to the two New Zealand committee members, Joseph Kinsey and John Mill. The proprietor of a flourishing shipping and insurance firm, Kinsey had a long association with Antarctic exploration, acting as an agent for Scott, Mawson, and Shackleton.

On June 28, Kinsey arranged for the *Aurora* to enter dry dock. The bent chain cable links were testament to the battering the vessel had suffered at Cape Evans. Badly hogged by the ice pressure, her stem was strained, the post of the missing rudder damaged, the sternpost and keel loose, the iron stem plating wrenched away, and the planking gapped apart. The jib boom, fore topgallant mast, mizzen topmast, main derrick, and bow anchors were broken. The marine surveyor estimated £6,000 to set the "exceptionally bad and damaged condition of the vessel" to rights, about twice what Shackleton had paid to purchase her outright in 1913. With government support for the relief expedition, cost was no longer an obstacle.

By mid-July, a hundred workmen thronged the ship. Gangs of carpenters began systematically dismantling the ravaged vessel. Yet Stenhouse was still unable to rest easy. Beneath the efficient flow of activity, he detected an odd undertow of tension and "many influences at work." Kinsey dropped occasional waspish comments about Shackleton. They had been close at one time, but in the aftermath of the *Nimrod* expedition, Kinsey came to take a stern view of Shackleton's off-the-cuff organizational style. As he put it, Antarctic exploration was "not all beer and skittles." Recent events had hardened Kinsey's attitude. As work proceeded on the *Aurora,* he openly disparaged Shackleton. Stenhouse felt bound to defend his leader and challenged Kinsey. Thereafter, he perceived "a marked difference in [Kinsey's] attitude" toward him. Kinsey assumed control of the refit and Stenhouse felt that his role had been diminished to that of "an interested inspector with little to say."

Kinsey was not alone in his disenchantment with Shackleton. The circumstances of the *Aurora* in 1914 left an acrid aftertaste with many influential Australians. Some blamed him for the Ross Sea party's poor organization and subsequent misfortunes. In 1914, one contributor had been so incensed that he told Stenhouse "Shackleton should perish." Although Mackintosh and Stenhouse protested that the difficulties originated with the organizers in London, many observers were unconvinced.

The memory of those problems was still fresh. In early May, an Australian parliamentary committee concluded an inquiry into the backwash of expenses left in the wake of the *Aurora*'s rushed departure for the Antarctic, totaling £3,938. Crippled by insufficient funding, Mackintosh had channeled a slew of unauthorized orders through the dockyard, well beyond the £500 approved by the government. At the time, Sir William Creswell of the Naval Board acknowledged that, with lives at stake, there was "no other course but to accept the position of fitting out and defer the question of payment for future settlement." With Shackleton beyond communication, the Naval Board was forced to absorb the costs. The outcome of the affair smarted. The final report accused Mackintosh of "taking unfair advantage of the situation." A month later, Creswell was presiding over the Australian Advisory Committee charged with bailing out the expedition once again.

Shackleton's Australian associates from the *Nimrod* expedition, including Edgeworth David and Douglas Mawson, were critical as well. David, who had assisted Mackintosh in 1914, had been hauled before Creswell's inquiry to account for the shipyard overages. Mawson disapproved of the slipshod organization of the Ross Sea party and the lack of provision for a relief expedition. He believed that Shackleton's improvidence set an unfortunate precedent. "All that explorers in future will do is to raise enough money to get away to where they want to do their work then call out to the Government to complete the job," he wrote to Andrew Bonar Law, concerned that ill-conceived expeditions would jeopardize government support for future scientific exploration of Antarctica.

Bad feeling aside, the moment was hardly opportune for an appeal to public largesse in Australia. The Commonwealth was committed to supporting Britain in the war and the demand for capital and lives was seemingly insatiable. On July 1, General Douglas Haig ordered the Big Push against the western front north of the River Somme. By the end of the first day, there were 57,000 casualties. As the offensive dragged on, reinforcements were ordered from fresh Australian troops, and by

August, more than 23,000 Australia and New Zealand Army Corps soldiers had been killed or wounded, one of every two soldiers sent into battle. The incongruity of spending thousands of pounds to save ten men in Antarctica, measured against the thousands of Australian fathers, husbands, sons, and brothers risking their lives in the war, was ever present in the minds of the committee members.

By August, the estimated cost of refitting, provisioning, and wages had escalated to £20,000. Since the governments were footing the bill, they were determined to have their orders carried out. They had no intention of allowing costs to spiral beyond their control again. The question of command was thus critical. Kinsey was open to Stenhouse retaining command, provided his loyalty to the committees could be assured. Some members wondered if he was the best man for the job, so Kinsey sought John King Davis's opinion. When Davis arrived in Wellington in early August on the *Boonah,* he was "disturbed and anxious" to hear the details of the *Aurora's* troubles. While he commended Stenhouse's conduct as meriting the highest possible praise, he flatly stated that he thought their predicament stemmed from the acting master's "inexperience and lack of judgement." Though Stenhouse lacked firsthand knowledge of the region, "even a slight acquaintance with the literature of the Ross Sea should have warned them of the risk involved," since it was well documented that the ice north of the Glacier Tongue migrated during blizzards. Davis knew that "no matter how strong her moorings and no matter how solidly frozen in," a ship anchored off Cape Evans would "inevitably" be swept away. "An experienced commander, with knowledge of ice conditions in that area, would never have attempted it," he concluded, astounded that "such a fatal policy had been followed."

Even though Shackleton's orders had expressly forbidden mooring south of Glacier Tongue, Davis faulted Stenhouse for following them. "That man obeyed instructions and consequently the ship ran adrift. If I had been there I should have simply said 'Those instructions are absolutely nothing: I have got to protect the ship. [Hut Point] is the only place where I have any knowledge that the ship can be safe and I must go there." In Davis's eyes, Stenhouse had not yet developed the judgment and conviction essential to command. Davis's assessment bolstered the sense that Stenhouse should not lead the relief. The committee members also questioned his allegiances. Stenhouse had made it clear he was Shackleton's man and the governments wanted a commander answerable only to their authority. Kinsey believed Stenhouse would agree to revert to the position of first officer if it was "tactfully explained to him." As Kinsey assured the government, "So keen is he in the work that he offered to go South again under anyone whom we might appoint and have more confidence in than himself."

———

By August, Shackleton had failed in his third attempt to rescue the men on Elephant Island, repelled each time by impassable pack ice. He returned to the Falklands with Captain Worsley at his side, where he learned that the Admiralty was sending the *Discovery*. Shackleton was relieved, but nagged by a suspicion that the Admiralty meant to push him aside. His fear was well founded. Planning had proceeded without him, and they had appointed Lieutenant James Fairweather as commander, intending to limit Shackleton's role to that of an adviser with no authority over the relief expedition. In short, he would be deprived of the privilege of rescuing his own men. As a condition of joining the ship, Shackleton would be required to concede the "rights of publication and copyrights of all matters which result from the Relief Expedition" to the British government. The Admiralty publicity effort would ensure public humiliation and also jeopardize Shackleton's future financial prospects, since his ability to repay the 1914 loans for the expedition was tied to the newspaper, film, and book exploitation of his story.

Shackleton had no intention of becoming a bystander in the Admiralty's heroic pageant. By late August, he had borrowed a ship from the Chilean navy. On August 30, the *Yelcho* ran the gauntlet of pack ice and made for Elephant Island. Through the mist, Shackleton recognized a windswept spit of land. He scrambled into a boat to row to shore. Figures appeared on the rocks and he shouted, "Are you all well?" "All safe; all well, Boss," came the reply from Frank Wild. By September 3, the reunited company was marching through the streets of Punta Arenas to a rousing heroes' welcome.

"I have done it," he told his wife in a triumphant cable. "Damn the Admiralty: I wonder who is responsible for their attitude to me. Not a life lost and we have been through Hell." Before sailing home, he planned to go to New Zealand to take command of the *Aurora*. As he told an expedition backer, "I cannot settle down to anything whilst my mind is on those ten men of mine in the Ross Sea."

Once again, official plans had galloped ahead of him. The advisory committees had resolved to appoint a new commander over Stenhouse. Davis received a coded wireless message aboard the *Boonah* on September 5 offering him command. Davis accepted. Then, in the wake of newspaper reports that Shackleton intended to command the relief, Davis tendered his resignation, making it clear that he "had no wish to seem to vie with [Shackleton]." The committee members refused to accept his decision, insisting that Davis, arguably the most qualified man in the world to command the ship, had a moral obligation to do so. Davis felt duty-bound to accept the appointment, reiterating that he would only do so if he was granted "absolute and undivided" control. The committees agreed.

Stenhouse was kept in the dark. On October 4, he opened the local newspaper, the *Otago Daily Times* and was shocked to see an article announcing Davis's appointment. Stenhouse had assumed that Shackleton's rights as owner of the ship, including the appointment of officers, were inviolate and had only accepted the committee's control until Shackleton declared his intention to lead the relief. Thomson, Donnelly, and Hooke were outraged. They wrote a letter of protest to the mayor, published in the newspaper, taking "exception to the discourteous manner in which this alteration was conveyed to the captain," who they respected as "a worthy leader in danger and difficulty, after 16 months of hardship in the Antarctic."

The committees did not anticipate the furor that ensued. The heroic exploits of Stenhouse and his men had been in the news since their first wireless plea for help, capturing the imagination of the public. The people of Port Chalmers and Dunedin had taken the men to heart, and the officers became part of the community, joining various charity clubs and the local Masonic lodge. "No man other than one endowed with the greatest courage and resource could have brought that ship back to port," read a letter to the editor of Dunedin's *Evening Star*. "I trust, therefore, that the powers that be, backed up by public opinion, will see to it that British fair play will prevail, and thereby save a great injustice being done to a brave man."

Tripp was appalled by the treatment of Stenhouse and arranged for him to meet with the acting prime minister, James Allen, and Robert McNab, minister of marine. McNab was supportive but pragmatic. Since the Australian and British governments were shouldering about 90 percent of the relief costs, he acknowledged their right to insist upon Davis. The 10 percent stake carried by New Zealand bought only a minority influence in the conduct of the expedition. McNab suggested a compromise arrangement, with Davis commanding the ship and Shackleton heading shore operations assisted by Stenhouse. Tripp saw that this was the best they could realistically expect. He advised Stenhouse to put his feelings aside and agree to sail under Davis.

Stenhouse cabled Shackleton in Valparaiso for instructions. Shackleton replied, ordering him not to "sign any documents official or otherwise relating to the Ship" and advising him to retain a lawyer. With that, Stenhouse prepared to square off with the man legendary for his "unapproachable dignity." On October 10, Davis had arrived at the Port Chalmers shipping office and signed on as master of the *Aurora*, noting on the articles that Stenhouse officially reverted to chief officer and Thomson to second officer. On October 12, the two men met. Davis listened impassively as Stenhouse declared, "I will not relinquish command," and rejected "the authority of anyone, other than [the] owner or his nominee, to appoint another to the command of the *Aurora*." Davis kept his stony composure and gave Stenhouse three days to reconsider or face discharge.

"My position here is an almost impossible one," Stenhouse wrote to a perplexed Tripp, who, ignorant of Shackleton's secret messages, could not understand why Stenhouse refused to cooperate. The day of reckoning passed. On October 17, Stenhouse refused to hand over the ship's keys and instructed Thomson, Donnelly, and Hooke to ignore Davis's summons to the shipping office. When they failed to appear, Davis discharged them immediately. The following day, Stenhouse received a letter informing him that he had been discharged as well.

Davis intended to sail by December 15. Forging ahead, he assembled his crew around a core of Antarctic veterans from Mawson's expedition. Only Paton, able seaman Kavanagh, and Ninnis remained

from the Aurora's original complement. Ninnis had sidled away from
Stenhouse when he scented the change of wind and aligned himself
with Kinsey, acting as the committee's clerk. The defection came as a
personal betrayal to Stenhouse. He had mentored Ninnis aboard the
ship and depended upon him in New Zealand, entrusting him with
the Ross Sea party's diaries, papers, and film negatives. Stenhouse discovered that his confidant had delivered all of Shackleton's property to
Kinsey. Stenhouse was shut out of the relief effort and alone in Port
Chalmers. Although much of the opinion in the press was sympathetic, some wondered if the Aurora's plight was due to his actions.
The implication cut him more deeply than the "discourtesy & double
dealings" of the authorities. His drive to save the stranded men had an-
imated him every day since he watched the lighted window of the hut
disappear in May 1915. Now there was little he could do until Shackleton
arrived.

The explorer's journey from southern Chile to New Zealand by train
and ship was tortuously slow. Ernest Perris continued negotiations on
his behalf from London. On October 16, Perris received a cable from
the Australian prime minister stating, "Sir Ernest Shackleton should
be informed that there is no reason for him to go to Australia." In a
burst of indignation, Shackleton cabled back, "I am dead tired and very
lonely but all my fighting blood and spirit of endurance is alive at this
last damned impertinence from Australia and the cheek of Davis and
that bloody old fool Kinsey."

On November 10, Shackleton boarded a steamer in San Francisco
with Worsley for the long voyage across the Pacific. Before leaving, he
cabled Davis to announce his imminent arrival and instructed Tripp,
"if necessary put embargo on departure Aurora if attempts sail before
arrival but be tactful." When he arrived in Wellington on December 2,
Tripp ushered him to McNab's office. McNab immediately warmed to
the explorer and reassured him that he would not allow the Aurora
to leave without him. The minister ordered activity on the ship to cease to
buy time.

Despite the sympathetic hearing, Shackleton and Tripp let the minister know that they would not hesitate to take aggressive action. Tripp
informed McNab that Shackleton was prepared to take the matter to
court to defend his legal rights as owner of the Aurora. It may have

been a bluff; a legal action would likely have favored the government and delayed the rescue efforts, possibly endangering lives. However, Tripp's hint that Shackleton might reveal the quarrel and provoke "a howl against the British government" in the press was a potent threat. He shrewdly supposed that the governments feared the adverse publicity of the Empire running roughshod over a British hero. Britain could ill afford cracks in the imperial unity. Australian enlistment was critical and volunteer numbers had fallen sharply, as the bloody combat at Gallipoli and the western front had already claimed so many eligible men.

When Shackleton faced his former comrade-in-arms John King Davis, he was defiant. "I found him somewhat changed," Davis wrote, assuming from Shackleton's imperious manner that "popular adulation had spoilt him a little." The exuberant explorer he once knew seemed careworn and out of touch. "His failure, though glorious in its way, had imposed on him grave financial and personal worries," Davis observed. "He had emerged, moreover, after nearly two years' isolation, into a strange, embattled world that bore small resemblance to the world he had left behind in 1914."

Davis felt compelled to share some home truths with his friend and former leader. He explained the resentment of the Australian government over being forced to bail out the *Aurora*, not once but twice, at a time when the conflict abroad was draining the resources of the British Empire. It gradually dawned on Davis that Shackleton was "unable to realize, yet, that the war was engrossing the thoughts and emotions of the majority of civilized man and that, consequently, people were apt to be impatient with polar exploration." The battles had made too many heroes.

After the meeting, Shackleton relented and told McNab he would yield command to Davis, on one condition: Stenhouse and Worsley would sail with him on the *Aurora*. Davis rejected the idea. Sailing with four qualified masters on board with divided loyalties would have courted disaster. It appeared to be an attempt to lay the groundwork for a takeover of the ship by Shackleton's cadre after she sailed. In fact, Worsley had suggested to Shackleton that they bypass the formalities and pirate the ship. The Australian committee refused to consider Shackleton's proposal, calling it "either childish or criminal," and reaffirmed Davis. Davis, weary of the machinations, was ready to surrender command. As he told Mawson, "The only thing that matters is the party down below, I am quite prepared and will hand over to Shackleton or the devil himself as long as they get the job done."

McNab suggested a solution: returning control of the relief to Shackleton, provided he would repay the costs later. It was out of the question. The expedition was deeply in debt. McNab sensed that a financial offer might break the impasse and proposed to guarantee Shackleton that he would not be obligated to repay the costs of the *Aurora*'s 1914 refit or the relief and would retain ownership of the vessel if he stepped quietly aside in favor of Davis. The Australians and British objected to his plan. By the morning of December 18, the *Aurora* was ready for departure, but the negotiations were deadlocked. Shackleton had not yet accepted Davis's legitimacy, and the British government refused to free him of liability for the costs of the relief. Marine Minister McNab saved the negotiations by persuading the New Zealand prime minister to agree to a side deal. If Britain made a claim against Shackleton in the future, the New Zealand government would assume the cost of his legal defense. They also agreed to pay the wages of the discharged men and buy first-class passage to England for Worsley and Stenhouse. Shackleton acquiesced and signed on as a subordinate under Davis's authority, reserving his "rights as owner to the *Aurora*."

At Bowen Pier in Port Chalmers, just after seven o'clock in the morning of December 20, the crew cast off and Davis steered the *Aurora* out into the stream and waited. Shackleton arrived, flanked by McNab, Worsley, Stenhouse, Thomson, and Donnelly and boarded a waiting tugboat with his entourage. As the tug made its way to meet the *Aurora*, a cheer went up on the pier.

Two years after the *Aurora* first sailed south with the Ross Sea party aboard, she was bound for Antarctica once again. "With every degree of latitude that we crossed on our southerly course, the period of daylight increased and the temperature decreased," Davis wrote. "And as we retraced our steps over this old, familiar road to the last continent, we seemed to have embarked upon a voyage where, instead of advancing into future time, we were receding into the past."

— 17 —

Rescue

The *Aurora*'s southing was rugged. Spirits were subdued, as most of the seasick company retreated to their bunks off duty. But for Davis, ever watchful on the bridge, the turbulence was behind him. Sailing a ship he knew inside out, he felt he had "come home to an old friend."

Shackleton had never before set foot on the *Aurora*. On his first voyage aboard his own vessel, he had been reduced to the role of supernumerary officer. Both Shackleton and Davis handled the highly charged situation with courtesy and diplomacy, setting the bewildered crew at ease. Assigned to room with shore party leader Morton Moyes, Shackleton told astounding stories of his *Endurance* adventures, deftly concealing the raw reality behind the ripping yarns. He and his men had endured an ordeal of unspeakable privation that had threatened to rob some of their reason and humanity. Shackleton quieted the ravings and quelled the rebellions, denying himself even the solace of a confidant, sensing that to give voice to his fears and doubts would admit the seeping corrosion of pessimism. "All cheerful," Shackleton repeatedly noted in his diary, repressing any hint of despair.

It was this experience, not official rank, which set Shackleton apart from the *Aurora*'s company as he paced the decks. As the *Aurora* pushed south, he dreaded what he might find—or not find. "I just pray that those men on the Ross Sea Side are alive," he wrote to his wife. "I don't care then twopence about the loss of the whole Expedition, plans and ambitions."

Eight days later, the peaks of Ross Island appeared on the horizon. Davis ordered a wireless message sent to Cape Evans: "Aurora is now within 100 miles of Cape Evans, hope you are all well and that ice will allow us to communicate with you shortly." There was no reply. Throughout the night, Davis braced for the expected blockade of pack ice crowding the entrance of McMurdo Sound, but the ship cruised unopposed through open water. Just after ten o'clock in the morning on January 10, the *Aurora* lay at anchor at the pack edge off Cape Royds. Farther south, McMurdo Sound was frozen fast. From the crow's nest, Cape Evans was visible, about seven miles away. Davis ordered a distress rocket fired. The expectant crew vied for position on deck to watch for the reply, but there was no hint of activity on the black slopes. Shackleton, Moyes, Ninnis, and the ship's surgeon, Frederick Middleton, disembarked and made their way across the ice to search the Cape Royds hut first.

The hut seemed suspended in time since the day in 1909 when Shackleton's men had rushed to board the *Nimrod*, leaving a batch of fresh bread and ginger snaps on the table. But the graffiti scrawled boldly across one wall was new: "Joyce's Skining Academy," it read, and the paint was still wet. The searchers found a cryptic note stating that the Ross Sea party had been living at Cape Evans in December 1915. "It increased our fears," Ninnis wrote. "For why had they not left later news if all were alive?"

As they were en route back to the *Aurora*, a party was sighted sledging across the ice from Cape Evans. Shackleton, Moyes, and Middleton hurried out to meet them. The crew watched through binoculars as the parties came together. One by one, three men walked apart from the huddle and lay down on the ice. The strange rite was a prearranged signal from Shackleton to Davis on the ship to report the number of dead. Seven had survived, "almost more than we sometimes hoped for," Ninnis admitted.

The united group began moving toward the ship. The crew crowded the railings as Ninnis readied the motion picture camera to capture their arrival. Davis disembarked to greet the survivors: Joyce, Wild, Richards, Gaze, Cope, and Stevens. Jack was so overcome that he remained behind at the hut. Davis was appalled by their feral condition: "They were just about the wildest-looking gang of men that I have ever seen in my life. Smoke-bleared eyes looked out from grey, haggard faces; their hair was matted and uncut, their beards were impregnated with soot and grease." But he was most distressed to find that "their great physical and mental privations went deeper than their appearance. Their speech was jerky, semi-hysterical and at times almost unintelligible, [and] their eyes had a strained and harassed look."

At close quarters aboard the ship, the stench of their bodies, redolent of the seal slaughter, was overpowering. The men were hustled below for their first real bath and change of clothing in two years. Jack arrived late and, wrapped in "wonderfully white woolens" by the crew, "ate & ate & wished I could eat more." They smoked their fill of coveted tobacco, but to Joyce's disgust, the ship was dry. In his view, "any exploring ship ought to carry stimulants." Given the frantic state of the survivors, it was impossible to obtain a coherent account of what had happened to Mackintosh, Hayward, and Spencer-Smith. The men were reeling from their encounter with Shackleton, who they believed had perished, and the joy of rescue. "A good deal of emotion present and torrents of excited talk," recorded Ninnis. "Questions that one did not wait to have answered before another sprang up." The rescuers fielded anxious queries about the war and home and the *Aurora*'s original crew, long ago given up for dead by the party. Then, exhausted, the castaways fell silent. "I just want to sleep & sleep & dream," wrote Jack. "Everything seems so strange." In the days that followed, Davis and Shackleton coaxed snippets from the overwrought men, piecing together their story.

They had found no trace of Mackintosh and Hayward since the fateful day of their departure in May 1916. The break-up of the ice began while they were en route to Cape Evans. If they had escaped drowning, the pair had covered two or three miles at best before the full onset of the blizzard. The question was whether they were able to negotiate the mosaic of broken floes in whiteout conditions to reach Cape Evans, or the safety of any land, for that matter. Between Hut Point and Glacier Tongue, the slopes of Mount Erebus met the sea in vertiginous cliffs and icefalls daunting even to highly skilled mountaineers. Not far from Turtle Rock, a short stretch of coast moderated into steep snowy slopes accessible without gear, but from there, the route south to Hut Point was riddled with crevasses. Even if fortune had been on their side, the seven- or eight-mile trek would have taken hours to negotiate. They would have succumbed to hypothermia sooner.

"Whether they got there or no they deserved to be badly frostbitten or lose their lives," Joyce wrote in his diary on the day they left Hut Point. It would be some time before he, Wild, and Richards knew how the two men fared. Joyce was determined to delay the trip to Cape Evans until the ice was thicker and illuminated by a full moon. The conditions were finally right on July 15. In the still midwinter darkness, they trudged across the sea ice, a lunar eclipse shading the moon to a bright sliver. As they approached the Cape Evans hut, the dogs raised a commotion. Jack came out and asked, "Where is Captain Mackintosh?" Joyce, Richards, and Wild were incredulous, unprepared to hear the worst. The grief was tempered by the anger they all felt, including the tolerant Wild, who reflected that if the pair had "slung their lives away" earlier, then there might have been some chance of getting Spencer-Smith to Hut Point in time to save him.

The Cape Evans party was astounded to hear of the "hardships endured & the heroism displayed by all," in Jack's words. Gaze and Stevens emerged from the hut and welcomed the new arrivals. Then Cope approached, shouting, "Hello Richy, I'm not good, I've got appendicitis—I'm shitting nanny-goat turds." Later, Cope's companions confided that he had been "off his head" since arriving at Cape Evans in January. Withdrawing as he had at Hut Point the previous winter, the biologist spent most of his time in his bunk obsessing about his health. But he did not have appendicitis, and his most intense pain was psychological. He refused all meals, instead eating stealthily at night. Richards passed near his bunk one night and heard Cope's paranoid mutterings about "thieves and murderers." His behavior so alarmed his companions that they hid the harpoon gun and other potential weapons and avoided leaving him alone.

The news of his cousin's death stunned Gaze. "It seems to me so hard that he was so near safety and then just to miss getting in," he wrote to his aunt. They all mourned, but for Stevens, the loss of his closest friends was devastating. After living on his own at the hut for six months while the others were sledging, he was desperately lonely. His dedicated meteorological work during his solitary vigil at Cape Evans could fairly be called heroic. Grappling with his physical limitations, he had slogged up Wind Vane Hill every four hours, day and night, for six months, to monitor the weather instruments. With Mackintosh gone and Cope non compos mentis, the chief scientist was estranged from the entire party and crippled by his disillusionment and poor health.

As expected, the Cape Evans party had heard no word from the ship. The *Aurora* had not been expected back in port until March or April 1916 at the earliest. It would be some time before the alarm was raised. If Europe was still in the throes of war, they realized that Britain would be hard pressed to mount a relief operation. Grasping at straws, Joyce wondered if the Americans would send a ship. Even so, no vessel could penetrate the ice before December. He estimated that there was enough food to supplement the seal for at least eighteen months, and perhaps another year's worth at Cape Royds, but how long they could mentally bear the conditions was another matter.

The dark winter months of 1916 had seemed one unending blizzard. During the lulls, Wild and Gaze joined Joyce in training the puppies. The progeny of Nell and Bitchie were nearly ready to pull in a team. They were the lucky survivors. Seven pups had died in accidents, and the overstressed Nell had attacked and eaten several. Sickened by the relentless savagery, Gaze shot her. Joyce skinned any suitable carcasses to make mittens, reserving his sentiment for Oscar, Gunner, Towser, and Conrad. "If ever persons owed their lives to anyone, we owe ours to the dogs," he wrote in the aftermath of sledging. The canines felt no such tenderness for each other. They fought rancorously, except the Greenland husky, Conrad, who managed to keep fairly aloof from the fray. On July 27, the three mongrels ambushed him, and, mortally wounded, Conrad bled to death as the men struggled to save him.

There was little respite from their stark existence. Most of their time was taken up by hunting. The scarcity of seals in winter forced them to roam as far as twenty miles in a day. Jack and Richards found solace in their scientific work, devising new experiments and making weather observations. Salt and alcohol had run out. After reading a novel in which a penal colony convict extracts salt from seawater, Wild tried the same technique successfully. A batch of home brew made from alcohol and malt yeast produced "a high degree of conviviality," as Jack put it, but triggered vicious hangovers.

On August 22, the sun edged above the horizon. The early days of spring were colored by melancholy. Wild injured his foot while skiing and was confined to his bunk. Then, in September, Richards suddenly fainted and collapsed. Cope diagnosed a heart attack. Joyce was shattered. "He had been my constant companion for ten months, and a better pal amidst toil and trouble never existed," he wrote. When two more episodes of arrhythmia followed, Richards quietly despaired of recovery. Lying in his bunk, he faced the graffiti he had recently scratched on a plank beside his head—"Losses to date: Haywood Mack Smyth Shck(?)"—and wondered about Shackleton's fate and his own.

Cope surprised them all by nursing Richards solicitously, sleeping in the adjacent bunk and tending to his patient's needs around the clock. He assured Joyce that Richards would pull through. "All the same, I felt a broken man," Joyce later wrote. He seemed anxious to leave the hut where his closest friend lay bedridden. In October, Joyce made for Cape Royds to collect zoological specimens as ordered by Shackleton. Others came from Cape Evans and stayed for short spells, but Joyce preferred the solitary life there with reminders of happier days on the *Nimrod* expedition.

On his return to Cape Evans in early December, Joyce organized another journey, this time a return to the Barrier to pick up depoted geological samples. Wild and Gaze joined him to visit Spencer-Smith's grave. They solemnly rebuilt the eroding mound that had been hastily erected on the morning of the chaplain's death. The new cairn stood twelve feet high, surmounted by a cross etched with the epitaph, "SACRED TO THE MEMORY OF REVD. SPENCER-SMITH, WHO DIED ON MARCH 9TH, 1916. A BRAVE MAN." Gaze photographed the grave for his aunt and uncle, and the group headed north. They made for Cape Crozier to conduct the postponed emperor penguin study and found it impossible when they bogged down in waist-deep drift. Instead, Joyce mounted another fruitless search for traces of the missing men. Back at Cape Evans, Richards was still confined to his bunk. The first day of 1917 passed with little notice. Though speculation turned inevitably to the probability of rescue, the men prepared for another year in the Antarctic.

In time, Richards found his feet again. Too weak for hunting, he became the lookout. Each morning, he stepped outside the hut to watch for seals. Raising the binoculars on January 10, he saw the *Aurora*. He rushed back inside and shouted, "Come on, the ship's here." The men sat, dumbfounded, until Richards repeated himself. Jack, like the others, was incredulous, his feelings "too deep to describe in his diary":

> It was not a time for words, our hearts were too full for this & I am not ashamed to say that tears of joy forced their way into my eyes in spite of myself. To think that our long time of waiting was over at last & that relief had come—no more hunting for seals, no more reeking blubber fumes, no more sledging, no more blizzards & frost-seared feet & hands, it was too good to be true. Was it to be wondered that anyone should be overcome at a time like this?

"Ship O," Wild recorded simply, the last entry in his diary. He and the others hitched the dogs to sledge toward the approaching rescuers. As they closed in, Joyce recognized the solid figure of Shackleton. "Joyce old man, more than pleased to see you, how many of the party are alive?" he said, reaching out to clasp hands. Joyce asked for a smoke and news of the war. They were all astounded to learn that the ship lying off the ice edge was the *Aurora,* afloat after all. Davis's customary reserve thawed when he greeted the survivors, many of whom he knew. He was anxious to learn the fate of the missing men. After patient questioning, Davis felt certain that Mackintosh and Hayward had lost their lives on May 8, 1916. "As neither man had any equipment, I consider it impossible for them to have survived more than a few hours, and now eight months have elapsed without news," he concluded.

It was the outcome that Shackleton had dreaded since first learning of the Ross Sea party's plight. Davis, too, was devastated. Mackintosh was a close friend and it was "a great blow to learn of his loss under such tragic circumstances," he wrote. The only possible consolation to offer the families of the dead was to find the bodies or some clue to their fate. Davis asked all of the men to write statements detailing the circumstances of their disappearance and suggest areas for fresh searches. He intended to leave no stone unturned, though some crew members, like Ninnis, saw "little hope in this wilderness of finding any traces of the lost men."

The Ross Sea party had already scoured the few accessible low-lying slopes between Hut Point and Glacier Tongue, then ventured briefly north of the floating glacier and to the Dellbridge Islands. Shackleton decided to make a more thorough northern search, from Glacier Tongue to Cape Evans. On the slight chance that the two men had floated across McMurdo Sound on a floe and scrambled ashore, he and Davis decided to examine the west coast of the sound as well.

On January 12, Davis steamed for Butter Point on the western shore of McMurdo Sound, where one of Shackleton's *Nimrod* parties had laid a large depot. Mackintosh knew of the cache, and it would have been logical for the stranded men to go there had they landed nearby. After the ship anchored, Shackleton, Joyce, and Wild sledged for the depot and found it untouched. Davis headed back across the sound to Cape Evans, where Shackleton disembarked with Joyce, Wild, Jack, and a dog team to resume the search on January 13. Richards, Gaze, Cope, and Stevens had no desire to leave the *Aurora* again.

As they sledged away from the ship, Shackleton gave the men news of home. Wild was eager to hear more about the doings of his brother Frank. Despite the bonhomie, there was an undercurrent of tension in the air. The Australians were leery of Shackleton and "rather hostile," as one man put it. They blamed him for the expedition's misfortunes and the futility of their sacrifices, though their indignation was unspoken. While Shackleton and Joyce headed out to search Little and Big Razorback islands, the introspective Jack quietly sorted his feelings as he packed up the expedition property in the hut with Wild. The scientific notebooks, diaries, documents, and photographs, were all loaded into boxes for Shackleton. The personal effects of the dead were collected as well, including Spencer-Smith's communion vessels.

By his own account, Joyce felt no qualms about lashing out at Shackleton when the two men were alone. Joyce accused him of "playing of chances in equipping the ship in Australia" and recruiting men "only fit for drawing room tea parties." He demanded to know "where [Shackleton's] eyes were when he engaged them." Joyce had joined the expedition brimming with expectation, then his hopes were dashed when it became clear that Mackintosh regarded him as nothing more than an underling. He had cooperated with Mackintosh's schemes, however wrongheaded he believed them to be, and given his all, but "owing to the unfortunate breaking up of the *Endurance,* the main object, that was, the laying of the depots, was after all of little use." The failures he laid at the door of Shackleton, the man he had once called "the Boss."

The sortie was cut short when a southeasterly swept in, driving Joyce and Shackleton back to Cape Evans and pinning the four men in the hut for two days. Huddled around the stove, Shackleton told them of the *Endurance*'s travails. Slowly, the men warmed to him. For Jack, the last days in the hut were "amongst the most enjoyable spent in the South, Shackleton having the three of us in fits of laughter hour after hour relating his experiences." It was more than the breezy banter; he was an attentive listener, gently drawing out the details of their ordeal. He understood because he had suffered himself. Grudgingly, Joyce conceded that Shackleton had "nearly put up with as much hardship" as themselves and granted that the journey in the *James Caird* was "simply miraculous."

The storm subsided on the sixteenth. Shackleton and Joyce rose at three in the morning for their last planned search on the north side of Glacier Tongue. The glacier itself was heavily pocked with crevasses and drifted fifteen feet high in snow near the coast. Shackleton concluded that any remains there would be hidden and called off the search.

At Cape Evans, Jack and Wild were performing a last ritual. They erected a cross on Wind Vane Hill and attached a copper tube containing a paper scroll with a message written by Shackleton:

ITAE

1914–1917

Sacred to the Memory

of

Lieut. Æneas Lionel A. Mackintosh RNR

V. G. Hayward

and

The Rev. A. P. Spencer-Smith BA

who perished in the service of the expedition

"Things done for gain are nought

But great things done endure."

I ever was a fighter so one fight more

The best and last

I should hate that death bandaged

my eyes and bid me creep past

Let me pay in a minute life's

arrears of pain darkness & cold.

Shackleton turned, as he did so often in turbulent times, to poetry, invoking Swinburne's exhortation to a life well lived in the first two lines. The last six lines were from Browning, slightly paraphrased. It was Browning who spoke to Shackleton most directly. Years before, he had come to terms with the immediacy of his own death as he went about the business of exploration, but he had never accepted the possibility of losing men under his command. Shackleton sacrificed his own food and clothing and peace of mind rather than see his men do without. When some talked of suicide on Elephant Island, he yanked them back from the brink, substituting his will for their own. But he had been powerless to impose his will across a continent.

Davis watched for the returning party, away from the ship for three days. He admitted to himself there was "no chance of finding anything." He knew the ice dynamics of McMurdo Sound well and he believed the ice had broken up quickly, leaving "no doubt that these men were drowned before the blow ended," a quicker, kinder end than the other possibilities. In 1909, Davis had kept an anxious vigil aboard the *Nimrod* when Mackintosh and a sailor went missing on the sea ice and had been haunted by "the harrowing picture of two helpless figures on a diminishing ice-floe, drifting northward to certain death." Now he grieved for Gladys Mackintosh and their daughters, for whom there was "so little to comfort."

"Just why he was so set on making the crossing is not clear," Richards later wrote as he tried to divine Mackintosh's motivations. He believed that both men had found the primitive conditions at Hut Point intolerable. It did not occur to him that the fact that their actions seemed so senseless was telling in itself. Mackintosh and Hayward had not fully recovered from the intense physical trauma of sledging, and the possibility remained that their judgment was literally impaired by disease and deprivation.

The search party returned to the *Aurora* on January 16, empty-handed. Shackleton drafted a formal report for Davis, concurring with his conclusions. Davis asked Dr. Middleton to examine the survivors and contribute his medical opinion to the report. Considering the living conditions, he judged their health to be good. Ten months of fresh meat and a varied diet had resolved the scurvy and malnutrition. Save for abscesses, frostbite damage, and sunburn, he found no significant problems. The nutritional deficiency diseases that they had undoubtedly suffered were beyond the ken of medicine of the time.

On January 17, Davis gave the order to weigh anchors. "As these men cannot be alive, it is no use risking further lives in trying to recover their bodies, or records," he wrote in his diary. But the ice was not inclined to release the *Aurora* without a fight. Guided by the steely glare of ice blink on the horizon, Davis dodged and weaved to outflank the pack and work the ship north.

As the *Aurora* plied her homeward course, the band of Robinson Crusoes avidly read every newspaper and magazine on board. All were dominated by the war. The sheer scale of the human loss in Europe was unspeakable. The bloody stalemate at Verdun, which the Germans had conceded a few weeks before, had claimed about seven hundred thousand casualties to contest six square miles of wasteland. In Richards's words, the men of the Ross Sea party were "utterly dazed and stupefied" to see the horrors of modern warfare illustrated in the tabloids. The artillery barrages, poison gas, aerial bombardment, barbed wire, and tanks seemed to belong to a "world of veritable devils," as did the daily litany of atrocities. Along with the crew of the *Endurance,* Richards reflected, they were "the only civilized beings to have experienced such complete isolation" during this time of momentous upheaval.

The strange news provoked other emotions, pangs of remorse for being absent in Britain's time of need. The conflict was in its early days when the Ross Sea party left British shores in the autumn of 1914. At the time, the scoldings of a London dowager that they "should all be fighting and not playing the fool in the snow" held little sting. Now, they all vowed to enlist upon reaching home.

On January 20, Shackleton gathered the men on deck for official photographs. The seven survivors obligingly shaved, slicked, and trussed themselves in suits, with the exception of Joyce, who sported a seaman's sweater, full beard, and unshorn mane. Ninnis filmed the proceedings. Emerging from his paranoid gloom, Cope was once more "the life and soul of the party," blowing kisses to the camera. Joyce was in constant motion, promenading arm in arm with Cope. Oscar, Gunner, and Towser romped about the deck. The others shuffled nervously in the background, smoking. Their smart attire failed to mask the effects of two years of suffering. In still photographs, the men drew the barrel-chested dogs protectively toward them. Richards averted his eyes, looking older than his twenty-three years, while Stevens appeared twitchy and anxious. Joyce and Gaze stared straight into the camera, weary and deflated. Only Wild seemed unscathed in his sharp pinstriped suit, arms akimbo and grinning like the man who broke the bank at Monte Carlo.

The Ross Sea party presented Davis with a gift one evening. It was the set of *Encyclopedia Britannica* Davis had given to Mackintosh for the expedition. As Stevens explained, "There was not a man of the party who did not make frequent and eager use of the volumes." The grubbiest pages, blackened with blubber soot, betrayed the most popular topics. From engine repair to cooking, the encyclopedias had been the party's survival guide and the last word in settling bets. The Ross Sea party and Shackleton signed a volume "as a remembrance of his kindness." Davis recalled the words of Robert Service:

> *Do you recall that sweep of savage splendour,*
> *That land that measures each man at his worth.*
> *And feel again in memory half fierce, half tender,*
> *That brotherhood of men who know the South?*

A week after turning north, the *Aurora* was still blockaded by the pack. She had been driven so far south that Davis contemplated returning to Cape Evans to resume the searches and wait for the ice to ease. Then a northwesterly opened a passage on January 29 and the *Aurora* sailed into open seas. As the vessel plied the Southern Ocean, the wireless operator regained contact. Davis transmitted a report and asked the relief committee to notify the next of kin while Shackleton composed messages for the British press. On February 8, the *Aurora* neared New Zealand's north island. "This strange voyage, that had seemed almost to set time at nought and be a voyage into the past, was over," Davis wrote, sensing that an era had ended.

— 18 —

"The Men That Don't Fit In"

FEBRUARY 9, 1917–MAY 8, 1985

The first sign of the life they had left behind was invisible. Wafting on the morning breeze of the antipodean summer, the lush fragrance of New Zealand's coastal rainforest overwhelmed the senses. After two years in an antiseptic wilderness, the survivors of the Ross Sea party stood on deck and inhaled the pungent scent of eucalyptus as the *Aurora* approached Wellington. It would be their last chance to catch their breath before returning to civilization.

The pilot boat met the *Aurora* in the stream on February 9, jammed with a supercargo of dignitaries. The mayor of Wellington led a chorus of cheers as the tug glided alongside the battered barquentine. On board, Shackleton and the Ross Sea party ran the hand-pumping and back-slapping gauntlet. When the *Aurora* finally eased into her berth, cheering throngs jostled on the quayside. Reporters were out in force to report the spectacle: "Sir Ernest Shackleton back again, so proud of the men he has brought back from the Antarctic, so deeply regretful for the men who have died, and ready now to go Home and give a hand where it is most needed." Shackleton spoke plainly: "I deeply deplore their deaths. They succumbed after great privations. They displayed heroism and fortitude in a continuous series of misfortunes."

Dazed and unsteady on their sea legs, the seven survivors were whisked away to the Grand Hotel and lavished with unaccustomed luxuries. Local merchants clamored to fit them with new suits. Their own clothing, left on board the *Aurora* in 1915, had been sold to raise cash during the 1916 refit in Port Chalmers. Invitations flooded in as the local gentry vied for the privilege of hosting the expeditioners as house guests. "Everyone is very kind to us—too kind in fact," wrote Jack.

However heartfelt the welcome, the rescued men were overwhelmed by the commonplace of their former lives. Beneath the surface, profound changes stirred. "We had left the civilized world in 1914 with the rather naïve outlook that prevailed in the very early years of this century," said Richards. "We returned in 1917 to a world that could never again be the same as before." The Ross Sea party wandered like stunned refugees from another time.

Sensing their disorientation, Shackleton steered Jack and Gaze clear of the well-wishers and ushered them off to the cinema for an evening, where he entertained them with a store of riotous anecdotes and antic impersonations. Jack and Gaze were won over. Whatever anger they had once felt had been exorcised. It was the same for most of the Ross Sea party, who gave him their unreserved admiration. He was "a man like Churchill—a great man," in the words of Richards. "His charm and persuasiveness were not insincere . . . He was a big man in every way. He had his faults—some say he was unscrupulous. If so it was probably due to his overriding passion for his task."

On February 12, the citizenry was out in force as Mayor John Pearce Luke hosted a civic reception at the town hall with the acting prime minister, James Allen, and a platoon of the great and the good. "Were positively lionised & we all had to stand up for inspection & be individually named by the mayor in front of all the people," recalled Jack. The Ross Sea party had "played the game," according to Luke, demonstrating the "spirit of adventure, the courage and the self-sacrifice that made the British race what it was."

―――――

One by one, the men left for home. On February 20, Stenhouse boarded a steamship for England, along with Cope and Frank Worsley. Adrift in the Southern Ocean a year before, Stenhouse had known utter isolation. Now, Allied escort ships shadowed protectively to fend off German raiders. In the lamplight of his cabin, porthole blackened, Stenhouse wrote to Æneas Mackintosh's widow and steeled himself for the train journey to Bedford, where he would visit Gladys Mackintosh and their two young daughters. He carried the letters

Mackintosh wrote to his family when he was alone in his tent on the Barrier.

Cope arrived home to learn that his brother, Gerald, had been killed on the western front and resolved to join the Royal Naval Air Service. In August 1917, he resigned his commission to become commandant of the British American Overseas Hospital, although he had yet to complete his formal medical training.

Ernest Wild returned to England in March, raring to "get into the scrap." He reported for duty in Malta and joined the minesweeper HMS *Biarritz,* which patrolled the Mediterranean. The usually acerbic Stevens reserved his highest praise for Wild. "There are some things that have great value but no glitter. Consistent throughout the whole period in the south, long-suffering, patient, industrious, good-humoured, unswervingly loyal, he made an enormous contribution to our wellbeing." For Stevens, Wild's presence was one of the few redeeming aspects of the expedition. After the rescue, Stevens wrote a frank report for Shackleton, tapping away on a typewriter lashed to the wardroom table during the rough voyage across the Southern Ocean. The report was even-handed for the most part, leaving aside, as he expressed it, "most of the unpleasantness which cannot be blinked any more than it can be righted." With that, he put the experience behind him and boarded the S.S. *Medina* for England.

On April 28, he glimpsed home for the first time in nearly three years. As the steamer plied the waters of the English Channel fifteen miles off the Devon coast, German submarine UB-31 fired a torpedo that tore through the ship's starboard side, killing six crewmen instantly. In a matter of minutes, the *Medina* sank out of sight as Stevens and 411 souls shivered in a flotilla of lifeboats. Gone were Stevens's personal scientific notebooks and diaries, as well as most of Spencer-Smith's personal effects.

Stevens arrived at the Spencer-Smith home bearing their son's diary. The last letters the chaplain had sent home before the *Aurora* left civilization in 1914 had been full of expectation. "By the way, Stenhouse & I have a favourite song when employed on such a job," Spencer-Smith wrote to his mother after a day of loading cargo. "'I wish my mother could see me now!' I wish you could—for you wouldn't be sorry for me at all: I'm awfully fit & happy." The grief was too intense for his father;

bereft, he refused to see Stevens. The chaplain's youngest brother, Martin, had been killed on the battlefield in France in September 1916, and the eldest, Philip, was a prisoner of war. In late 1917, another brother, Charles, would die on the western front.

Stevens also carried Hayward's effects to his family. In the packet was a letter to Victor from his younger brother, Stan, a machine gunner preparing to deploy with the Royal Highland Fusiliers to France in early 1915. "Well, Vic I hope you have a good and successful voyage & that you come back in the best of health & we meet again," he wrote. "Guess the war will be all over except the shouting by the time you are back, I for my part believe the end is in sight now, but there have been very heavy casualties on either side." Stan Hayward was killed in action at Ypres on June 13, 1916, a month after his brother set out to cross McMurdo Sound and disappeared. Bundled with the letter was Hayward's diary. A neatly trimmed hole in the flyleaf remained where he had cut out his fiancée's photograph to take with him to Cape Evans.

The sorrow and disillusion were more than Stevens could bear. "On reaching home I was disgusted with the whole affair, with the general behaviour and attitude of Shackleton, and glad to shake free of the whole thing and get on to other matters." However solicitous Shackleton seemed, Stevens could not forgive him for the organizational failures that had engendered so much hardship and lost opportunity. "I went out as a geologist, with, I think, a keen interest and a desire to make good," he explained to a colleague. "All the way out I found my attempts to study Antarctic geology more thoroughly thwarted in absurd ways." Two weeks after his arrival in England, he joined the Royal Engineers as a wartime surveyor.

Military service was still voluntary when Richards returned home to Australia. His health seriously compromised, he assumed a post teaching mathematics and physics at the Ballarat School of Mines and Industry and convalesced. Jack's expertise as a chemist was in high demand and he was quickly recruited for explosives work in a government munitions facility. For the moment, his passion for pure research was deferred and he put his Ross Sea data notebooks away. His personal journal, however, was officially the property of Shackleton. He managed to put Shackleton off and later sent him an abridged transcript with the most forthright comments omitted.

His fellow Australians Richards and Gaze had also been brutally frank in their diaries. Unquestioning deference to authority did not come naturally to the Australians on the expedition. Throughout their time in the Antarctic, they debated, advocated new ideas, and criticized. In trying times, they blew off steam in their journals. As John King Davis later observed, "Mackintosh was a sahib and the rest excepting Stenhouse apparently did not understand the rudiments of playing the game." Davis, an Englishman, found their ethic incomprehensible, but, arguably, their outlook had enabled them to persevere and find purpose in the face of tribulation, much like their countrymen at war. This culture clash was enacted in parallel, far from the polar wilderness, on the battlefields abroad. "Obedience is the supreme virtue," as Prime Minister David Lloyd George declared. But the British gentlemen's code did not sit well with the Australians. Their valor in combat was exemplary, yet for every thousand Australian volunteers, field punishment was meted out to nine, as compared with one British soldier. British officers were astonished that Australian troops would not simply follow orders, instead debating strategy and tactics with their superiors. Similar conflicts were recorded in the diaries of the Ross Sea party. In the end, Gaze managed to keep his journal, while Richards submitted only summary reports to Shackleton.

Joyce lingered in New Zealand, savoring the comradeship and acclaim. Before Shackleton departed, he paused to write a letter of commendation for Joyce:

> *I consider that your strenuous work throughout the Expedition; the care you took of your parties under most adverse circumstances; your optimism when things looked black indeed: all go to prove that my judgement was correct in obtaining your services for the Expedition. . . .*
>
> *Lastly, and most important of all, your conduct on that long, trying southern journey—especially after Mackintosh broke down—ranks in my mind, and will, when the other men know of it, rank with the best deeds of Polar Exploration. You had charge of the party, and through your instrumentality as leader, and with the help of your loyal comrades, all except Spencer-Smith reached Hut Point in safety. I, as Commander of the Expedition, feel this to be of paramount importance, for, though I was thousands of miles away, the responsibility still lay on my shoulders.*

It was the affirmation that Joyce had always craved. In 1915, his boast may have sounded like overweening bluster to his companions—"I suppose it will be really the biggest thing ever been done"—but he had been right after all. What had begun as the prosaic chore of supporting an historic expedition had eclipsed it, for they had succeeded even though Shackleton had failed. And in the completion of their mission, the Ross Sea party had made one of the longest continuous land journeys in the history of Antarctic exploration to date, 1,356 geographical miles, placing them in the same rank as Shackleton, Scott, and Amundsen. No other expedition had come close to the two hundred days they spent sledging on the Barrier.

Basking in the warm glow of Shackleton's approbation, Joyce was riding high. He was a celebrity, recognized on the streets and besieged with invitations. He felt sure his strivings for fame and fortune were about to be fulfilled. After a whirlwind courtship, he married a New Zealand lady named Beatrice Evelyne Curlett, the daughter of a gentleman landowner from Christchurch. She was thirty-six, Joyce was forty-one, and for both marriage was a fresh adventure. After the wedding, Joyce and his new bride decided to begin their life together in Sydney, taking the dog Gunner with them.

Leonard Tripp attended the ceremony and reported to Shackleton. "His affairs will be practically the end of the Antarctic Expedition here," he wrote. The indefatigable Tripp had freed the expedition from a thicket of legal obligations and debts dating back to 1914. He sold everything he could, including the *Aurora*'s chronometers and the puppies, then approached influential associates for cash to cover the liabilities. The members of the Ross Sea party were paid in full—Richards, for one, received £93—though there was arguably no adequate recompense for the ordeal they had endured. Finally, he made arrangements for Oscar and Towser to take up residence at the Wellington Zoo.

The modest attorney was an impeccable organizer and prudent manager. When funds became available, he duly paid back every pound donated in 1917 with interest, and for the most beneficent, enclosed a stuffed emperor penguin from Joyce's zoological collection. With the books balanced, he was able to send bonuses to most of the Ross Sea shore party at Shackleton's behest. The Weddell Sea party did not fare as well; some never received their full due from the London organizers. In retrospect, the expedition would arguably have unfolded differently with Tripp at Shackleton's side from the beginning. His patient attention to detail was the ideal complement to the explorer's improvisational style. Shackleton put great stock in Tripp's counsel, which was crisply frank. "It seems to me that it will be impossible for you to do any exploring anyhow for many years," he advised. "It would be unwise for you ever to take on another expedition unless you not only had sufficient money to pay your way, if everything went alright, but you would have to have money in hand to provide for accidents." Above and beyond the outstanding debts handled by Tripp, the governments of Australia, New Zealand, and Britain had borne the entire cost of the relief expedition, totaling £21,000.

The committee was as good as their word and handed the *Aurora* over to Shackleton, who promptly sold the ship for £10,000. After forty-two years of polar seafaring, she was destined for a far less romantic trade, hauling coal from Australia to Chile. Once there, her holds would be filled with explosives and fertilizer. In the twilight of the square-rigged ships, sail still reigned supreme on this route, making the long passage across the Pacific with ease when steamships were unable to stray so far from coaling ports. It was a dubious triumph. The cargo could spark into a ferocious blaze and reduce a wooden ship to charred wreckage in minutes. The terrible conflagrations were the dread of every seaman who sailed aboard the nitrate ships.

On June 20, the *Aurora* left Newcastle, New South Wales bound for South America. Too old for battle, Scotty Paton had a family to support and a hankering to return to sea, so he joined the voyage as boatswain. The grizzled veteran was characteristically phlegmatic about his Antarctic odyssey, saying, "although we had our little trials, we suffered none of the real hardships that I expected." The *Aurora* never arrived in port. She was presumed to have fallen victim to an enemy raider or foundered in a gale.

———

A month after his return from the Antarctic, Shackleton was still in New Zealand. Observing the explorer's electric effect on audiences in Wellington, Tripp urged him to lend his voice to boosting wartime morale. He believed the gesture would quiet criticism that Shackleton and his men had not done their part in the war. Shackleton traveled around the country lecturing to packed houses, yielding the podium to Joyce to tell the story of the depot laying. The proceeds were divided between the Red Cross and a fund for Mackintosh's widow. Buoyed by a rising tide of goodwill, Shackleton prepared to leave New Zealand in an "odour of sanctity."

Australia was a tougher nut. Sentiment against Shackleton had been far stronger there, due in large part to the hostility of the Australian relief committee. Some believed that Shackleton was to blame for the plight of the Ross Sea party and the lives lost. Undeniably, the party's task was made immeasurably more difficult by Shackleton's haphazard preparation, a product of his refusal to delegate the planning for the Ross Sea party far earlier in 1914. His staffing decisions, based on expedience and loyalty rather than proven polar travel and leadership ability, had handicapped the party. The unsuitable dogs and inadequate supplies had caused needless misery. Moreover, his instructions for mooring the *Aurora* had nearly doomed the ship and crew and condemned the shore party to twenty months of suffering and uncertainty. And leaving Mackintosh reliant upon an expedition organizer in London, Ernest Perris, whose chief concern was newspaper coverage for the *Endurance*, did a grave disservice to the Ross Sea party. Yet the argument for Shackleton's culpability in the loss of the three men was emotional rather than logical. The deaths did not result directly from flawed planning or negligence. Remarkedly, Mackintosh had overcome the expedition's hobbled beginnings to accomplish its stated aims. He and Hayward had crossed the unstable sea ice of their own volition. Even if Shackleton's best-laid plans had come to fruition, and the innovative diet devised by his nutrition expert had been available, Spencer-Smith and the rest of the party would have been vulnerable to scurvy and other nutrient deficiencies due to the heavy reliance on preserved foods. Their record-setting journey tested the limits of human endurance as well as medical knowledge.

The question for many observers was one of Shackleton's moral accountability, an issue that seemed to trouble him as well. The death of three men was a blow to his ethic of leadership, as he had always prided himself in his selfless care for his men. At first, he rebelled against the burden of guilt, criticizing the Ross Sea party in letters to his wife. "I am afraid they made a hash on the other side the way they moored the ship," he wrote to Emily before embarking on the relief. On his return, he called it "a mercy" that he had gone. "Everything was in a state of chaos," he told her. "Mackintosh seemed to have had no idea of discipline or organisation and it required all the tact I possessed to square things up. He had made a very bad impression for the Expedition in Australia by the way things were done at the start, but for that he is dead through his own carelessness and I won't say more now."

Upon reflection, he told Worsley that he believed the depots would not have been laid without Mackintosh's "supreme efforts" in Sydney and on the Barrier. "Honour to whom honour is due," he said in tribute. Publicly, he spoke mainly of the *Endurance*. "This is the story of the Weddell Sea side of the expedition. There is no time now to tell of the other side," he told a lecture audience in 1917. Ultimately, he held himself to account: "My name is known as an explorer, as the leader of the Expedition. With the leader lies the praise or blame, and rightly so."

Australian audiences responded to his candor. At the Sydney Town Hall, one lecture attracted 3,500 people who "wavered from wild enthusiasm almost to tears." Emboldened, Shackleton insisted on meeting with the Australian relief committee in Melbourne. He was intent on a showdown, accusing them of conspiring against him. "If I have an enemy, I would like to meet him face to face and I feel I have got one in this room," he told the surprised members, including Sir William Creswell and Orme Masson. "We went at it hammer and tongs for about an hour and there was plain speaking on both sides," Masson recalled. Shackleton left the room. But before the dust settled, he returned, grinning and offering a handshake to every man. Bygones, he decided, were bygones. As he wrote to Tripp, "Have had committee on carpet for very frank talk now feel better and buried hatchet."

Shackleton said his farewells to Davis. Privately, Davis criticized the expedition as "badly organized and injudiciously manned." He felt a growing skepticism over the value of record-seeking Antarctic exploration. As he later wrote to a member of the Ross Sea party, he believed "the whole motive which prompted Shackleton in the first place was wrong."

In late March, Shackleton sailed for America. The United States had remained aloof from the conflict in Europe, and some saw Shackleton as an ideal good will ambassador to rally support for the British cause. Ninnis was at Shackleton's side as Man Friday, having changed allegiances again and dissociated himself from the relief committee. Ninnis predicted that "the tail of his comet would be as full of unfinished affairs as ever" as he grappled with mounds of correspondence. Shackleton awaited orders from the Admiralty, expecting he would be tapped for duty in Arctic Russia, where his expertise in polar travel could be put to good use. In the meantime, Shackleton and Ninnis crisscrossed the United States on a lecture tour, fêted in grand style from coast to coast. Ninnis occupied his free time poring over Russian grammar books, anticipating the call to the front with Shackleton.

———

By autumn, word came that the *Aurora*'s carpenter, C. C. Mauger, and able seamen Sydney Atkin and Ginger Kavanagh had been seriously wounded in battle. Joyce, too, was determined to do his part and resolved to rejoin the Royal Navy. New recruits were so desperately needed that conscription had been introduced in May 1916, mandating the enlistment of all men from the age of eighteen to forty-one. As Tripp dryly commented, army doctors would pass anyone for service, "even a man with a wooden leg" provided it was free of termites.

Joyce was rejected because of his impaired eyesight, which he attributed to his frequent bouts of snowblindness. He wrote to Tripp and Shackleton in July, insinuating that Shackleton was responsible. "I don't know what he is up to; he says in his wire which I had to pay for!!! That his eyes are bad," Shackleton replied to Tripp. "I think this is a playup. I am not responsible for his eyes he signed on at his own risk. I don't believe they are so bad." The expedition agreement signed by Joyce clearly stated that all members joined at their own risk, absolving Shackleton of liability for injury or death and explicitly denying the possibility of any compensation. Nonetheless, Shackleton had given Joyce an open-ended offer of help, saying, "If ever I can assist you, please call upon me; and if it is in my power it will be done." While he never hesitated to put a man's needs above his own in a crisis, he could be irregular about fulfilling promises once back in the hurly-burly of society.

In the past, Shackleton had striven to do right by Joyce. Knowing Joyce had turned his back on a navy pension to join the *Nimrod*, Shackleton took special pains to compensate him, pledging a bonus of at least £1,000 if the expedition ended in profit. In the event of his death on the 1908–9 South Pole attempt, he had left a will ensuring that Joyce received the bonus, goods, publishing royalties, and assistance in finding a job. A select group of other intimates, including Frank Wild, were rewarded as well. But for Joyce, the money, opportunity, and praise never seemed enough. The special perquisites only reinforced his feeling that he was owed.

A few weeks after his disability plea, Joyce changed tack. "Joyce troubling about Albert Medal," Shackleton wrote in exasperation to Tripp. The award, created by Queen Victoria, paid tribute to lifesaving beyond the battlefield, honoring firemen, mariners, and others who had selflessly risked their own lives. Annoyed by Joyce's presumption, Shackleton cabled Tripp: "Shut him up his proposition impossible he not right to make recommendations."

In fact, Joyce did not have to wait long for recognition. In February 1918, King George V awarded the Polar Medal to the members of the Imperial Trans-Antarctic Expedition. Joyce received a gleaming silver medallion anchoring a stepladder of bars commemorating his upward progress from lowly able seaman of the *Discovery* to hero of the Ross Sea party. Though the award was conferred by the monarch for Arctic and Antarctic exploration, Shackleton had nominated the recipients. All but four members of the Weddell Sea party were honored. Of the Ross Sea party, only the insubordinate Shaw and d'Anglade were excluded. The relief party was not honored, apparently excluded from Shackleton's recommendations. The former relief committee members in Australia and New Zealand were outraged, and a highly charged stream of correspondence to the British government ensued. Davis took no offense, having decided long before that "gratitude does not seem to be a weakness of the exploring fraternity," and lobbied for the medal for his men, which they ultimately received in 1920.

Shackleton joined "the Great Adventure," as he called the war, and by the autumn of 1918, he was en route to Murmansk as an army major. He requested the transfer of Worsley and several other *Endurance* members with visions of "driving a dog-team over the snow into Berlin." Shackleton gladly complied when Worsley asked for his "old pal Stennie" to join them, and they were "together again & happy as sandboys."

Joyce was not among them. In September 1919, he was down on his luck, convalescing after a car accident. From his hospital bed, he spun a woeful tale in a letter to Edgeworth David. After being rejected by the Royal Navy, he had been hired by the Broken Hill Proprietary Company to run a marquesite mining operation in Bowen, Queensland. The job had lasted less than a year when the mine proved a bust. Joyce moved on, becoming a traveling salesman. His new career was derailed when the accident landed him in the hospital with a crushed kneecap. Joyce unburdened himself to David, detailing the Ross Sea party's trials at length. Between the lines of the thirty-one-page missive was a plea for help and a wounded rebuke of Shackleton, despite the fact that the explorer's invisible hand was evident throughout Joyce's postexpedition career.

The November 1918 armistice closed a chapter for Shackleton. In January, he left the Army and returned to England. He pinned his hopes on his book and film to pay off the expedition's debts. By his own reckoning, the ITAE had cost £80,000. His chronicle, entitled *South: The Story of Shackleton's Last Expedition 1914–1917*, was published in November 1919. He devoted five chapters of the book to the Ross Sea party, with quotes from their diaries and reports, and described their achievements in glowing terms. "I think that no more remarkable story of human endeavour has been revealed than the tale of that long march," he wrote, calling the expedition "a record of dogged endeavour in the face of great difficulties and serious dangers." But he also found fault. "The ground to be covered was familiar, and I had not anticipated that the work would present any great difficulties," he wrote in *South*. Shackleton concluded that the inexperienced members of the shore party lacked "any sign of the qualities of leadership" and asserted that "though there was a good deal of literature available, especially on this particular district, the leaders of the various parties had not taken advantage of it." He also criticized Mackintosh's decision to rush the dogs into sledging in January 1915.

South was reviewed widely and sold well, though Shackleton saw no income from the book. The executors of a deceased benefactor demanded repayment of a loan and Shackleton signed over the rights to the book and motion picture. *South*, the film of the *Endurance* expedition, premiered at the Royal Philharmonic Hall in London in December 1919. Six days a week for six months, Shackleton appeared on stage to tell the story of his "spectacular failure." Scarcely a month into the engagement, audiences tapered off. "Are we growing a little blasé about our heroes?" asked the *Daily Mail,* decrying the half-full houses. "Here is one of the great pioneers of our history, a super-man of adventure . . . Can it be that the war has made heroism commonplace?"

In February 1920, Shackleton peered out into the audience, recognized the piratical figure of Joyce, and saluted him from the stage. Recently arrived from Australia, Joyce was in London to join a new Antarctic expedition as second-in-command to John Lachlan Cope. Cope was planning a grand five-year expedition to assess the potential for future economic exploitation of Antarctica, at an estimated cost of £150,000. Due to depart in June 1920, it soon became apparent to Joyce that they would be delayed. The Royal Geographical Society declined to support Cope's endeavor, publicly calling it "ill-considered in detail and unsound in principle." His case had not been helped when Shackleton read the Ross Sea party diaries and learned of the biologist's erratic conduct. When asked by the RGS to comment privately on Cope's prospectus, Shackleton had called the plan "thoroughly impracticable, absolutely suicidal," and evidence of "a complete ignorance of Antarctic experience and travel." The Royal Society followed suit in withholding its scientific imprimatur, and Cope's expedition collapsed.

Joyce was at loose ends again and hit upon the idea of lecturing, even though his contract with Shackleton had forbidden it. He contacted Tripp and asked for permission to organize a tour using photographs and motion picture film of the Weddell Sea and Ross Sea parties. Shackleton relented and allowed Joyce to lecture, writing, "You have a great story to tell, and with the assistance of the film and pictures it should prove of vital interest to the people of this country—a people that have always taken an interest in the wide spaces of the World and in the work of the pioneer."

Shackleton had tired of the lecture circuit himself, and he chafed at the humdrum of everyday existence. "Alaric, King John, pearls in the South Seas," he had jotted in his *Endurance* diary, daydreaming of the lost treasure troves he would one day seek. There were entrepreneurial notions as well: "Siberian trade, Paraguayan tea, electric tin opener." By the spring of 1920, Shackleton was promoting another polar expedition. The company was familiar, a contingent from the *Endurance* and Stenhouse from the *Aurora.* However grateful Shackleton may have been for Joyce's heroic efforts, the relentless pestering had soured their twenty-year friendship. Shut out of Shackleton's inner circle, Joyce was not invited to join the expedition.

The *Quest* sailed from London on September 18, 1921, on a voyage to circumnavigate Antarctica. At forty-seven, Shackleton was headed south again. In January, the *Quest* neared South Georgia. The first sighting of the island's snowy peaks conjured glimpses of past triumphs, when Shackleton crossed the forbidding mountains to rescue his men in 1916. His arrival in Grytviken was a kind of homecoming, and he celebrated with old friends late into the evening of January 5. Back on board the ship a few hours later, his recurrent heart trouble flared. He died and was laid to rest in the whaler's cemetery.

––––––

Not long before Shackleton's death, Joyce had received a letter from one Sir William Jury, demanding that he desist from showing the motion picture film of Shackleton's expedition during his lectures. Shackleton had transferred ownership of the film to Jury to pay off the expedition's loans. Pained by his fall from grace, Joyce bitterly concluded that Shackleton had "double-crossed" him.

"I am very sorry to hear that Sir Ernest Shackleton's estate is in such a muddle," Joyce wrote to Shackleton's friend and biographer Hugh Robert Mill. With more than a trace of schadenfreude, he gossiped that Shackleton had died insolvent. Joyce asserted that Shackleton still owed him a considerable sum, a contention not borne out by financial records. Bemoaning the loss of the income in question, he told associates that "under the circumstances I had to forego my claim."

Though his trials and achievements needed no embellishment to beggar belief, Joyce increasingly exaggerated both. He persisted with his campaign for honors, writing to the Admiralty and making a dubious claim that Shackleton had promised to recommend members of the Ross Sea party for the Albert Medal, an intention unfulfilled due to his untimely death. Though one Admiralty official sensed that Joyce's account was "calculated to give an incorrect idea of the real facts," the Ross Sea party's story impressed the first lord, Leopold Amery. Amery advocated for Joyce, asserting that his diary should be required reading for every man and boy in naval service. On July 4, 1923, at Buckingham Palace, King George V pinned the gold medal on Joyce and Richards in recognition of their efforts to save Spencer-Smith and Mackintosh on the Barrier. Wild and Hayward were awarded the medal posthumously. Joyce, the errant petty officer, was paraded as the model Jack Tar for the Royal Navy. But his hunger for recognition and tangible tokens of his heroism was unsated. A few months later, he made an unsuccessful pitch for a Royal Geographical Society medal, an award that had been bestowed upon the likes of Amundsen, Shackleton, and Scott.

Though Joyce had managed to impress a parade of influential figures, Shackleton's circle had gotten wind of his curdled relations with the explorer and took a dim view of him. "I do not like to write you what I think about 'Captain' Ernest Mills Joyce but I do not think he will 'cut much ice,' as they say," sniffed John Quiller Rowett, the wealthy backer of the *Quest* expedition, commenting on Joyce's appropriation of a title without benefit of official promotion. Joyce routinely wrote long letters to prominent acquaintances when he was "out of a job and up against the wind," enclosing copies of Shackleton's commendation letters and increasingly strident criticism of the disposition of the Ross Sea party. The recipients began to suspect his version of events was not entirely reliable.

People were talking, among them Stenhouse, who confided his knowledge of Ross Sea party politics to Frank Worsley. Apparently, Mackintosh had resisted when Shackleton assigned Joyce to the *Aurora* in 1914, but Shackleton reportedly said "he didn't want him & he had to go somewhere." It was secondhand, but had the ring of truth. Shackleton might well have been reluctant to take the overweight Joyce on the transcontinental crossing, yet, loath to exclude him from the expedition, particularly since he had previously left Joyce behind on the *Nimrod* Pole attempt. Whatever Joyce's sins, Shackleton attempted to honor his obligations.

Undeterred by, or perhaps unaware of the whispers, Joyce applied to join the British expeditions to Mount Everest in 1921–22 and the Antarctic in 1924. He was rejected: "did not want an old fogey like me:—(not experienced enough)," he concluded. Blocked from lecturing, Joyce next set his sights on a book. Shackleton had already exercised his right of first publication of excerpts from Joyce's diary, so Joyce was free to write his own account. In 1928, he landed a deal with Duckworth, the prestigious house founded by the family of Virginia

Woolf and publisher of such literary lights as Woolf, Evelyn Waugh, and D. H. Lawrence. The manuscript was to be drawn from Joyce's expedition diary. Hugh Robert Mill had apprehensions about Joyce's possible revelations. "I understand that the full story of the Imperial Trans-Antarctic Expedition is one which had better remain unpublished for some time to come," Mill wrote to Alexander Stevens. In his 1923 biography, Mill had taken pains to insulate the explorer from the troubles of the Ross Sea party.

The South Polar Trail was published in early 1929. Mill, as it turned out, had little to fear. For the most part, Joyce resisted the urge to air his differences with Shackleton in print. He shrewdly realized he was cruising the thermals of Shackleton's fame and resisted blaming his former benefactor. Diary passages reflecting poorly on Shackleton's leadership were scrupulously withheld, though traces of resentment seeped into the book. Elaborating on the party's struggle to improvise their own clothing and gear, he archly quotes Shackleton's *Heart of the Antarctic:*

> It is true that the explorer is expected to be a handy man, able to contrive dexterously with what materials he may have at hand, but makeshift appliances mean increased difficulty and added danger. The aim of one who undertakes to organise such an expedition must be to provide for every contingency.

Knowing his old comrade to be a fabulist, Stevens anticipated that Joyce would apportion the lion's share of credit to himself. However, it was with regret, not indignation, that he wrote, "I should be sorry if Joyce gave himself away in his book." The temptation proved too great, and Joyce's mates got short shrift in his self-aggrandizing epic. Stevens was pained that Joyce had been especially critical of Mackintosh. In spite of their quarrels, Stevens admired Mackintosh, finding "something steadfast and reliable about him which would make one pause in attributing deficiency of judgement to him." He firmly believed "the Ross Sea party would have achieved much less than it did but for his unwearying drive in circumstances which were pretty well impossible. He showed a devotion to Shackleton quite out of measure to Shackleton's treatment of him." John King Davis, too, admired Mackintosh's dedication. In his opinion, the "great march south to lay the depôts for Shackleton may not have been as spectacular as others that have attracted more attention but it was undoubtedly a magnificent achievement, accomplished for a purpose other than his own safety." Tragically, his shining accomplishment was overshadowed by his unhappy end.

By casting himself as the prime mover and de facto leader of the Ross Sea party, Joyce downplayed the contributions of others as well. Richards confessed he was "astounded" to discover "how much his account differed from reality." It was not merely the numerous factual errors and exaggeration, but outright invention. Among other dramatic flourishes, Joyce falsely claimed that the party had seen the tent where Scott and his perished companions lay, and implied that he had euthanized the dying dogs in the first season. More seriously, Shackleton's letter appointing Joyce to the expedition, reprinted in its entirety in the book, had been altered to misrepresent the explorer's intentions. A line defining the command structure was deleted—"The shore party for the Ross Sea would be in command of an officer, to whom you would be responsible"—to give the impression that Joyce was solely "in charge of all equipment, store[s], sledges, clothing, dogs, etc." Moreover, Shackleton had made clear to Joyce that he was not under consideration for the main transcontinental party, yet Joyce's book quotes the letter as stating, "I can promise you a position on that party."

Richards forgave Joyce for the "inaccuracies and false impressions" he detected in *The South Polar Trail*. "Indeed it is what I would have expected of him. He was bombastic—a bit of a swashbuckler & saw everything—quite genuinely I think—more colourful and larger than life—but true-hearted and a staunch friend." Stevens was similarly generous. "He is a good fellow in his own sphere, and it is not fair to expect more of him than his fortune in life warrants. Like the rest of us, or at least as much as the rest of us, he is entitled to his 'foibles,' and the critic ought to have some background to give perspective to views of him." They knew that, whatever his faults, Joyce possessed the will to literally drag men back from certain death, even at the risk of his own life.

The reviews of the book were kind though not sensational. The *Times Literary Supplement* called the expedition "a wonderful performance." London's *Daily Express* praised it for exemplifying "the guilelessness of unadorned gripping truth," investing the account "with truer and fuller romance and merit than if it were 'written up' after the manner of film sub-titles, as is so common." Author J. B. Priestley was similarly impressed. "The prose epics of the Antarctic have long been familiar to us, but let me recommend, as appendage to them, *The South Polar Trail,*" Priestley wrote, calling the book "a plain blunt chronicle, very finely illustrated." James Wordie, geologist of the *Endurance,* reviewed the book in *The Geographical Journal* as "first-rate and most thrilling."

The modest reception in the press and marketplace failed to match Joyce's soaring expectations. On the strength of the book, he floated a proposal for a new Antarctic expedition in partnership with a South African university, which ultimately collapsed. Over the years, he launched schemes to explore the North Pole and hunt fur seals by airplane, hawked seal pelts at trade shows, and lost his savings to a shady investment scheme. "I know you know me well enough to say that things ought to be brighter for me," he wrote to a former shipmate, Charles Royds, since elevated to vice-admiral. The cavalier tone of Shackleton's criticism in *South* still rankled. "Talking is easy and everyone is wise after the event," Joyce told Royds. "The record sledging journey; the hard pulling on the trail, what man suffered for science was all sunk into oblivion. for why: it is difficult to understand."

He soldiered on with literary efforts, penning dozens of articles about polar exploration, and even fiction, searching for the words to express the profound effect the expedition had wrought on him. He favored tales of adventure and derring-do at sea, sometimes writing under the pen name "Draycot Dell, An Old Shell-Back." Some of his stories reached the pages of the boys' story paper *Chums.* Joyce's hero and alter ego, Hard-Fist Jemsen, was a tough old salt left behind by modern times and values. "I've found it to be a world where few people keep their promises," Jemsen proclaims. Joyce's vision of himself had not changed. He was ever the maverick adventurer and vagabond of the Robert Service poetry he was so fond of reciting:

There's a race of men that don't fit in,
A race that can't stay still;
So they break the hearts of kith and kin,
And they roam the world at will.

By any measure, Joyce had achieved remarkable success in life: traveling to the ends of the earth, receiving honors from the king, publishing his memoirs, and collecting a scrapbook bulging with news clippings of his exploits. For a boy from the Royal School for Navy Orphans, it was astonishing. The Antarctic had seemed at once an escape from the stifling hierarchy into the wilds and, back at home, a means of springing up the ladder in society. Scott and Shackleton had held the same expectations. Following their example, Joyce believed the hidebound rules no longer applied in the Antarctic, and indeed, on Shackleton's ships, every man had a voice. There were unmistakable signs of social transformation after the war, as if the expedition had sailed from one world and returned to another. But what Joyce craved—acceptance in the rarefied circles inhabited by Shackleton—would never come, bound as he was by his roots in the Victorian age.

His literary output tailed off to countless anonymous letters to London newspapers signed "An Interested Observer." "Whilst crossing the Western Ocean," wrote a supposed passenger on a Cunard ship, "we on board had the pleasure of listening to one of the finest lectures that I have ever heard. The lecturer was the famous Polar Explorer Ernest Mills Joyce." Inevitably, the letters digressed into a recounting of "the world's greatest polar effort" and the litany of his accomplishments.

There was still a living to be earned for himself and his wife. Among other ventures, scuttlebutt put him at the helm of a rumrunner on the Thames, which, true or not, the raffish Joyce would have relished. But it was his post as concierge for London's Eccleston Hotel that paid the bills. Every new guest was a fresh audience, and they could not fail to be amazed by the stories he had to tell. Sobered by the workaday life, he mellowed from the swashbuckling Joycey into a "middle-aged, stockily built, grey-haired and quiet-spoken man," as one newspaperman reported. Joyce attended the annual dinners of the Antarctic Club, relishing the reunion with other polar veterans, many now laureled with knighthoods and academic honors, and savored the limelight when it chanced his way.

"At 64, He Still Wants to Explore," read a May 1939 headline. The reporter found the forgotten explorer in an autumnal mood. "Given the chance to join an expedition, I'd be off like a shot," Joyce declared.

"I suppose I shall have to be satisfied with my memories." He watched the Antarctic exploits of Mawson, Wilkins, Byrd, and Ellsworth from afar, keeping his scrapbooks assiduously up to date. In the early days of a wet London spring, Joyce fell ill with bronchitis, complicating his battle with emphysema. On May 2, 1940, his heart failed as he lay in bed in his basement flat at the Eccleston, tucked up in his old reindeer fur sleeping bag.

——

"Not a rotter in my book," said Richards. "'Warts and all' I liked old Joycey." Theirs was an unlikely friendship, forged by necessity and tested by hardship. The tolerant, steady Richards was unfazed by his mate's bumptious temperament, as he explained, because he believed everyone has "plenty of faults if one goes looking for them & very few haloes."

Despite their bond, their lives diverged after the expedition. Richards remained in Australia and Joyce returned to England in 1920. "After the trouble with my heart, my health was never the same," said Richards. "My days of travelling were over." However, life brought other fulfillment. Richards had found his calling as a science teacher and, like his fellow Australians, Jack and Gaze, he was flattered by the honors and recognition that followed the expedition, but had no desire to dwell on the past. Letters came from Joyce, inviting him to join one scheme or another. Richards always declined, gently. Joyce's yearning for the Antarctic was unfathomable to Richards. Neither could he imagine anyone else returning to the continent. "We did think strongly [in 1917] that we were going to be the last people that were down there," he recalled. "The Pole has been found and the transcontinental journey may have been accomplished or may not have been accomplished, but nobody is going to send down another expedition."

At first, his prediction seemed prescient. A decade after their rescue, no exploring expedition had ventured into McMurdo Sound. It was not until 1928 that American Richard Byrd sailed into the Bay of Whales, where Amundsen had once established his base. His main ship was a traditional wooden sealer, but a steel-hulled vessel followed in its wake bearing three airplanes, snow vehicles, and a battery of communications gear. Byrd's team became the first to fly over the South Pole, ushering in a new era of technological exploration in the Antarctic which promised to shift the advantage to humankind in the hostile polar environment.

"What is left for exploration?" asked the *Daily Mail* in 1933, lamenting the end of "exploration in the large old sense which inspired the Elizabethans and Victorians." When the United States launched Operation High Jump in 1946, the aims were strategic. The U.S. Navy dispatched an invasion force of thirteen ships, thirty-three aircraft, amphibious tractors, and 4,700 personnel to Antarctica. The modern explorers regarded the huts with curious wonderment. For all the seeming impermanence, the weathered wooden buildings appeared unchanged since the survivors of the Ross Sea party rushed to meet the *Aurora* three decades before. At Hut Point, their last meal of seal meat sat in a skillet on the stove, and the sledging biscuits were still edible. Outside the Cape Evans hut, the preserved body of a chained dog stood frozen upright. Nearby, the newcomers found the copper tube containing the scrawled epitaph of Mackintosh, Spencer-Smith, and Hayward tumbling over the slopes.

Richards watched avidly as this new era of exploration unfolded. "The modern Antarctic adventurer receives a very substantial salary, is covered by insurance, and has all the blessings of the Welfare State in leave credits of various kinds," he noted with bemusement. Henceforth, Antarctic exploration would be dominated by national enterprises rather than driven by the vision and will of one man. Government-backed expeditions launched calculated assaults on all sides of the continent, mapping the coastline and plunging deeper into the uncharted interior than ever before. Though strategic goals were ever present, scientific research once again came to the fore. In 1955, the United States founded a permanent research base at Hut Point called McMurdo Station. In 1957–58, the International Geophysical Year (IGY) focused on Antarctica. As part of the effort, Everest pioneer Edmund Hillary and scientist Vivian Fuchs fulfilled Shackleton's dream and made the first crossing of the Antarctic continent. Scientists from sixty-seven countries participated, founding fifty bases on the continent, including one at the South Pole. Not far from where the Ross Sea party routinely clambered from Ross Island onto the Barrier, New Zealand founded a scientific station called Scott Base.

Some of the newcomers were mesmerized by the past. They rummaged through the huts, sifting the relics of the bygone era they called the Heroic Age. The traces of the Ross Sea party's time there were omnipresent but enigmatic, their story obscure beside the legendary exploits of Scott and Shackleton. Some sought out Richards to hear the story. He had by then retired as principal of the Ballarat School of Mines and Industry and he welcomed the chance to reminisce. The outward evidence of the experience was gone, save for eccentricities like his lifelong aversion to frozen foods. He had forgotten what had lured him to Antarctica in the first place, but memories surfaced as he was plied with questions. Remembering was not always easy. "Cape Evans was 'home' to us, but I never recollect Hut Point without a touch of mental distress," he once said. More happily, he was moved to renew his correspondence with the other survivors of the shore party: Jack, Gaze, and Stevens. Unlike Stevens, Jack and Gaze never regretted joining the expedition. Jack kept the poetry he wrote in the Antarctic, inspired by "the beauty untold" he felt privileged to have seen, even in adversity. Gaze's enthusiasm was undimmed since the day in 1915 when, at the end of a day of backbreaking labor, he wrote,

> I don't wonder at people's being drawn back and back to the place in spite of the hardships and risks—there's a fascination about the life that would appeal to most anybody. It's astonishing too, how one forgets about the outer world. No doubt there are times when one simply longs for civilization again with its attendant comfort and luxury but these fits don't last and you thank your lucky stars that you're down here, living a real life.

Richards was inspired to write a book, *The Ross Sea Shore Party 1914–17.* The slim volume told the story plainly in forty-four pages. It was as modest as Joyce's chronicle was flamboyant, and without bitterness. Richards was untroubled by disappointed hopes. After a lifetime of reflection, "weighing the credits and debits—and there were very weighty debits," he, too, had "no regrets." Richards never regarded the struggle as futile. "It was something that the human spirit accomplished," he said. "It was something you tried to do." He died on May 8, 1985 at ninety-one, the last of the men who laid Shackleton's depots to Mount Hope.

Epilogue: "The Brotherhood of Men Who Know the South"

Ernest Wild

Wild dreamed of manning the bar in his own English country pub, but he put his aspirations on hold until after the war. Back among his kindred Jack Tars in the Royal Navy, serving as a petty officer aboard a minelayer in the Mediterranean, he seemed unaffected by his Antarctic ordeal. His breezy good nature endeared him to his new mates and his outstanding performance impressed his superiors. In February 1918, he was struck down by a raging fever. The unmistakable symptoms of typhoid followed. An aggressive vaccination campaign had fortified troops against the deadly contagion, but Wild, arriving late to battle, had not been vaccinated. He died on March 10, 1918, and was buried with full naval honors in Malta, where his shipmates contributed a marble cross in his memory.

Irvine Gaze

After reuniting with his family in Melbourne, Gaze sailed for England to join the Royal Flying Corps. Aerial warfare had come into its own since 1914, and the RFC had grown from sixty-three to twenty-two thousand aircraft. Though lionized by the public, fighter pilots led a precarious existence. During "Bloody April" 1917, when Gaze was deciding to join, the life expectancy of a front line pilot was eleven days. Gaze was flying over occupied France on October 13, 1918, when his Bristol fighter was shot down. Only slightly injured, he and his crew were taken prisoner by German soldiers. Summoned for interrogation, Gaze was asked what his Polar Medal signified, and when Gaze said he had been a member of Shackleton's expedition, the adjutant exclaimed that he had attended Dulwich College with the explorer. He invited Gaze to dinner with the squadron, where Gaze met flying ace Hermann Göring.

Eleven days later, armistice was declared. Gaze remained in the RAF in England into 1919 and married Freda Sadler of the Women's Royal Air Force. They returned to Australia to settle in Melbourne and raise two sons, both of whom later became RAF pilots. Dubbed "the local millionaire" by the Ross Sea party for his ever-ready cache of unlikely luxury items, he proved to be a natural in business. Gaze enjoyed a prosperous career as a footwear company executive and later tried his hand at farming. He cherished lifelong friendships with his Australian mates from the expedition, Jack and Richards. He died on April 22, 1978, at the age of eighty-eight.

Andrew Keith Jack

In June 1917, Jack took charge of the Guncotton and Nitric Acid Section of the Commonwealth Cordite Factory in Maribyrnong, Victoria. After the war, he devoted himself to his family and continued his career as an industrial chemist in munitions, ultimately rising to senior assistant manager of the Maribyrnong facility during World War II. From 1944 to 1946, he was the Australian munitions representative to Britain. At his retirement in 1950, he was chief safety officer and secretary of the Operational Safety Committee of the Australian Department of Supply and Development. He was a fellow of the Institute of Chemistry and the Royal Australasian Chemical Institute.

After the expedition, Jack carefully preserved the scientific notebooks of the Ross Sea party but did not publish any results. Jack later shared the data with meteorologist Fritz Loewe. In 1961, Loewe published a paper on Jack's tide ablation experiments in the *Journal of Glaciology*, and later authored the first publication of the Ross Sea party's meteorological results. Loewe's analysis provided fresh corroboration and notable contrasts with the results from earlier expeditions to the Ross Sea region. Jack died on September 26, 1966, at eighty-one.

John Lachlan Cope

While serving as commandant of the British American Overseas Hospital in 1917, Cope married Norah Robinson, daughter of Lord and Lady Rosmead and granddaughter of the First Baron Rosmead, Sir Hercules Robinson, who was during his illustrious career the governor of Hong Kong, Ceylon, New South Wales, New Zealand, and South Africa. The following year, Cope resigned his post and joined the Royal Air Force. After the war, he lectured on his polar exploits and worked as a teacher and journalist. In 1919, he launched the British Imperial Antarctic Expedition. Cope's extravagant program dwarfed Shackleton's effort in both scope and cost. Sailing on Scott's old ship, the *Terra Nova*, with a sixty-strong crew, including Joyce and Larkman, he planned to circumnavigate the continent for five years, establishing multiple scientific stations with a view to future commercial enterprises including mining and whaling. The ship would also be equipped with a plane, intended for reconnaissance and a first flight to the South Pole.

When he failed to garner establishment support and funding, his grand plans collapsed. There was no ship and no plane, and the crew dwindled to four men. Cope arranged passage on a whaling vessel sailing into the Weddell Sea in 1921 to disembark his party on the Antarctic Peninsula. One man, George Hubert Wilkins, quit the expedition on reaching Antarctica. Cope decided to return home to drum up more funds and persuaded Maxime Charles Lester and Thomas Wyatt Bagshawe, the latter aged nineteen, to stay in Antarctica until he could return. The two young men spent a year making scientific observations, until early 1922, when a whaling captain returned to pick them up.

Resolute in his ambition to become a doctor, Cope resumed his medical studies while supporting his family, a formidable effort with four young children at home. Twenty years after he had been forced by finances to withdraw from his medical courses at Cambridge, he qualified as a general practitioner in 1933. He practiced in London until the outbreak of World War II and then moved to the Midlands. In March 1947, his wife Norah died, and eight months later, John Lachlan-Cope passed away in his sleep at the age of fifty-four.

Alexander Stevens

After completing his wartime service with the Royal Engineers, Alexander Stevens returned home to Scotland. In 1919, he joined the Scottish Spitsbergen Syndicate as a geologist and traveled to the Arctic islands of Norway for field research. Afterward, he married and retreated to the still waters of academic life, resuming his post at the University of Glasgow. Stevens took on the challenge of developing a geography department at the university, becoming its first professor when the chair was established in 1947. Graduate students could be intimidated by his "brittle sharpness of intellect and trenchant criticism," but discovered in time that the austere, blunt Scotsman could be an amiable adviser. They often found him in his office late at night, peering at documents by the dim illumination of a blue light bulb— necessitated, he explained, by the damage done to his eyes by snow-blindness in the Antarctic. In 1953, he lost his sight entirely. He retired that year and lectured at St. Andrews University and Johns Hopkins University. Stevens died on December 20, 1965, at the age of seventy-nine. His notebooks lost in the sinking of the *Medina* in 1917, he left no record of any geological work in the Antarctic, published or otherwise, though he had contributed to the surviving meteorological reports.

— SELECTED MEMBERS OF THE *AURORA*'S COMPANY —

Joseph Stenhouse

As he himself would have put it, Stenhouse had a good war. With Stenhouse as gunnery officer and Frank Worsley in command, mystery ship PQ61 cruised the Irish Channel, accompanying merchant vessels as a decoy to flush out German U-boats. On September 26, 1917, PQ61 forced UC33 to surface with a combination of strategy and sheer audacity, then rammed and sank the submarine. Worsley was awarded the Distinguished Service Order (DSO), and Stenhouse received the Distinguished Service Cross (DSC) and command of the HMS *Ianthe*, where Leslie Thomson joined him as first lieutenant. In October 1918, Stenhouse reunited with Worsley and Shackleton in Murmansk with the North Russian Expeditionary Force. By armistice, Stenhouse had been decorated again, this time with a DSO, and he was awarded the Order of the British Empire (OBE) in 1920. The sojourn in Arctic Russia sparked a business enterprise, Stenhouse, Worsley and Company, a shipping concern established for trade with the Baltic states. Their assessment of the market proved overly optimistic, and business dwindled. Stenhouse left to join an expedition in Brazil.

Stenhouse was deeply affected by the tragic plight of Mackintosh's widow and two young daughters. In 1923, he married Gladys, and the first of two daughters was born the following year. From 1925 to 1927, he sailed in command of the Royal Research Ship *Discovery* as scientists studied whaling and fishing grounds and conducted oceanographic surveys in the South Atlantic and the Antarctic.

When his time on the *Discovery* ended in 1928, he tried his hand at a series of entrepreneurial ventures, including a sealing company and an Antarctic passenger cruise line. His motivation was not merely wanderlust. After a brief respite of postwar prosperity, the European economic downturn of the late 1920s depressed trade and curtailed opportunities in the mercantile marine. In the dismal climate of the Great Depression, all of Stenhouse's enterprises foundered. In 1934, he rejoined Worsley in a new adventure, sailing for Cocos Island in the Pacific in search of pirate treasure. Like Mackintosh before them, they failed to find the booty. Casting about for work, he outfitted a sailing ship for a wealthy adventurer's round-the-world cruise and served as nautical adviser to the 1937 British film *Mutiny of the Elsinore*, doing a turn in front of the camera as well. In 1938, he reunited with Worsley to form a sailing cruise company with the yacht *Westward*. He also wrote a rollicking account of his days as a young apprentice in sail, *Cracker Hash*.

During World War II, Stenhouse served with the Royal Navy in the Gulf of Aden. In 1940, he risked his own life to rescue one of his crew when his ship struck a mine, killing half of the crew. During operations aboard the *Tai Koo* on September 12, 1941, Stenhouse was killed when his ship was sunk after an explosion, presumably caused by a mine.

Leslie Thomson

In 1916, Thomson paid his own fare to sail to England and join the Royal Naval Reserve. In late 1917, he was appointed to the mystery ship

Ianthe under the command of Stenhouse and later assumed command of the Q ship *Margaretha*. By war's end, he was awarded the Victory Medal, the Mercantile Marine Medal, and the Service Medal. After armistice, he served as master on merchant ships sailing from Britain to Australia. During his London lectures in 1920, Shackleton brought Thomson to the podium, along with veterans of the *Endurance,* to acknowledge his role in the expedition. Returning to Australia, he married in 1922 and contented himself with a shore job to be with his wife and two sons. In the early days of World War II, at fifty-five, he volunteered for the Royal Australian Navy, but was rejected due to his age and heart problems. After persistent efforts, Thomson signed on as third officer on a Dutch merchant ship and served in dangerous waters, for which he was awarded the Africa Star. He was discharged due to ill health in January 1942. Undeterred, he signed on to the Australian Civil Construction Corps, charged with building facilities to support the war effort. His health worsened, and Thomson died on June 20, 1946, at the age of fifty-nine.

Lionel Hooke

Like the other Australian expedition members, twenty-year-old Hooke demonstrated a dynamic optimism that propelled him beyond the Antarctic ordeal. He thrust the intrigue and disappointment of the relief expedition resolutely behind him and joined the Royal Naval Volunteer Reserve in late 1916. His bond with Stenhouse endured. Reunited in London on leave, Stenhouse and Hooke looked eagerly forward to the day "when the war will cease & we shall go into the Wilds again." In early 1917, Hooke was posted to a shore job inspecting rescue tugs and chafing at the bit. He had no intention of sitting out the war on the sidelines and applied for transfer to the Royal Naval Air Service. He was selected for training as a pilot of the RNAS airships employed escorting convoys and patrolling for enemy submarines and mines. On his last flight of the war in September 1918, his airship was downed by friendly fire and crashed in the English Channel.

Recovered, Hooke returned to civilian life in 1919, and rejoined Amalgamated Wireless Australasia Ltd. (AWA) as the company's manager for Melbourne. Wartime advances had propelled the technology beyond Morse code to the broadcast transmission of voice and music.

With evangelical zeal, Hooke set out to literally spread the word. In 1920, he rigged a speech transmitter at his home with an aerial twelve miles away, delivering a radio program to the astonished legislators in Parliament House. In 1921, Hooke launched the AWA wireless concert service, which hosted the first theater broadcast in Australia.

As AWA blazed the trail for broadcasting and high-speed telecommunications in Australia, Hooke was in the vanguard, demonstrating his flair for leadership when he oversaw the upgrade of Australia's coastal radio stations. Shy of his thirtieth birthday, he was appointed deputy general manager of the company. He kept a hand in the laboratory, inventing the Automatic Distress Transmitter to allow endangered vessels without a wireless operator to signal for rescue, a forerunner of EPIRB. As a champion of research and development, he was a visionary advocate of beam wireless, television, and FM radio. During World War II, he redirected all of AWA's energies to developing and supplying communications equipment to Australian and American forces in the Pacific theater. As AWA's managing director from 1945, he guided the company to its preeminent position in electronics manufacturing in Australasia. In 1957, Hooke was knighted for his services to industry. He assumed the chairmanship of AWA in 1962, a position he held until his death on February 17, 1974. Throughout his career, he kept a cablegram from Shackleton framed in his office: "Am not in position yet to return your services which were invaluable. I appreciate your loyalty and work."

Alfred Larkman

Chief engineer Larkman returned to sea as a merchant mariner in 1916 and was awarded the wartime Mercantile Marine Medal. Profoundly affected by the generosity of New Zealanders during the *Aurora's* time in Port Chalmers, Larkman emigrated to New Zealand in 1920. He settled in Wanganui and became a teacher, married, and raised three children. He taught engineering at Hawera Technical High School and later at Wanganui Technical College. At his retirement in 1954, he was head of the college's engineering department. Over the years, he also worked as a consulting engineer and was a member of long standing in the Wanganui Astronomical Society. Still the quirky eccentric his expedition mates remembered, Larkman devised an innovation on Daylight

Savings Time which required that clocks gain a minute each day for the first six months of the year, losing the same amount for every day from July to December. His system fixed the sunrise at 7 AM, according to his preference to wake at first light. Larkman Time was never adopted. Two weeks before he died in 1962, the Larkman Nunatak of Antarctica's Queen Maud Mountains was named in his honor, one of a number of features named for members of the Ross Sea party as their story became known among explorers and scientists. An American Hercules aircraft was diverted to scatter his ashes over his namesake peak, fulfilling his last wish.

James "Scotty" Paton

Scotty Paton was among the twenty-one men lost at sea when the *Aurora* disappeared en route to Iquique, Chile. The vessel was officially posted as missing by Lloyd's of London on January 2, 1918, and no trace of the crew was ever found. An inquiry established that the German raider *Wolf* was laying mines in Cook Strait and the Tasman Sea in June and July 1917 and concluded that the *Aurora* likely ran afoul of a mine. Shocked by the tragic news, Shackleton searched for Paton's orphaned daughters to contribute to their support.

Aubrey Howard Ninnis

After parting with Shackleton in 1917, Howard Ninnis returned to England, where he cut his ties with longtime sweetheart Ethel Douglas, the baker's daughter from Surrey whose memory had sustained him during the *Aurora*'s drift. Ninnis reported for duty with the Royal Naval Volunteer Reserve as a chief petty officer. He was appointed to a shore post in Portsmouth, where he served for the rest of the war. Both of his brothers died before armistice, the elder at Passchendaele and the younger in a riding accident. Ninnis was later involved in the repatriation of troops to Australia and New Zealand.

But it was love, not war, that brought Ninnis back to New Zealand in 1924, lured by an Auckland singer named Constance Howard. In his adopted home, Ninnis parlayed his wartime skills as a radio operator into a broadcasting career. His greatest success, much to his chagrin, was as the costar of *Uncle Bob and Aunt Betty*, a children's radio show he performed with Constance.

During the 1920s, Ninnis fixed his aspirations on the Antarctic once more. Like Joyce, he saw the path to riches in Ross Sea whaling and mineral wealth and conceived a string of schemes based on his expertise and other people's money. The efforts failed. So, too, did his relationship with Constance. Ninnis moved to Dunedin, not far from Port Chalmers, and joined station 4YA. He lectured frequently at schools on his Antarctic adventures. In later years, he bought a smallholder's claim in Otago and spent his holidays prospecting for gold. In 1956, he died of heart failure at the age of seventy-three. Like Joyce, he relived the experience in the written word until the end of his life, leaving a prodigious body of unpublished work about the expedition.

Harold Shaw

Fireman Harry Shaw, the former tough-talking bully, returned to England in 1917 and rejoined the Manchester City Police Force, where he had been a constable for five years before moving to Australia in 1913 to seek his fortune. Five days later, he resigned for reasons unknown. His seizure disorder would have disqualified him. His "roving disposition" won out, as his father put it, and he drifted from job to job and angled for handouts. In 1919 he applied to the National Relief Fund, intended for victims of the war, claiming to be the sole survivor of the *Aurora*. He was unsuccessful. After the award of the Albert Medal to Joyce, Richards, Wild, and Hayward in 1923, Shaw renewed his pursuit for "recompense for his sufferings." He entreated members of Parliament for assistance in obtaining the medal for himself. Shaw's lobbying attracted the support of Fabian socialist Sidney Webb, then president of the Board of Trade. Shaw's appeals failed when it was determined that he had not saved lives on the expedition.

— THE RELIEF EXPEDITION —

John King Davis

Davis and Shackleton parted friends in Sydney in March 1917, the last time they would ever meet. "A great explorer" is how Davis remembered him after his death. "One feels how much one misses him now that he has gone." After the war, Davis returned to Australia and in 1920 was appointed commonwealth director of navigation, a post he held until he retired in 1949. Early in his tenure, he became convinced that a weather station in the Coral Sea was crucial for early warning of cyclones headed toward the coast, and he championed the establishment of a wireless outpost on storm-swept Willis Island. When detractors insisted the project was too dangerous, Davis volunteered to help build and man the station himself. To this day, it plays a vital role in Australia's meteorological monitoring.

Davis succumbed to the siren call of the Antarctic again in 1929, when he joined Sir Douglas Mawson on the British, Australian and New Zealand Antarctic Research Expedition (BANZARE), sailing as second-in-command and master of the *Discovery*. After his retirement from public service, he consulted on Antarctic exploration and wrote extensively, including *High Latitude* (1962). He was a seminal force in the planning of the Australian National Antarctic Research Expeditions (ANARE), and the Antarctic base, Davis Station, and the Davis Sea were named in his honor. In addition to the Polar and RGS medals, he was appointed a Commander of the Order of the British Empire (CBE) in 1964. Remembered as a "free thinker and plain speaker" and a "deepwater sailorman of the old school," he died a confirmed bachelor at eighty-three on May 8, 1967.

ACKNOWLEDGMENTS

I am so grateful to the descendants of the members of the Ross Sea party, who generously shared their memories of the men, personal papers, family histories, and artifacts, and have kindly permitted me to publish excerpts and photographs. I spent a memorable day with Elisabeth Dowler, Æneas Mackintosh's daughter, who was born just weeks before the expedition left civilization, and his granddaughter, Anne Phillips. Patricia Stenhouse Mantell recounted stories of her father, Joseph Stenhouse, and generously opened his papers at the Scott Polar Research Institute, which brought a new perspective on the heroic efforts to keep the *Aurora* afloat and rescue the stranded men. I enjoyed many conversations with John Hooke about his father and wireless technology. He also kindly made his father's diary and log available to me, as well as the fruits of his own detective work. Anthony Lachlan-Cope and Anne Magee shared their father's life story, as did Malcolm Thomson and Tony Gaze. Allan Mornement kindly offered background and photographs of his great-uncle, Howard Ninnis. Cicely Douglas contributed a copy of Ninnis's diary for Ethel Douglas. Clifford Smith and Dorothy Beglin are first-rate genealogists and shared their research on the Spencer-Smith family. Anne Wild Fright painted an engaging picture of her godfather Frank and uncle Ernest, known to his young nieces and nephews as a fun-loving prankster, which was a marvelous complement to the memoirs and diaries. John Sanderson, Stenhouse Martin, Colonel Denis R. Stenhouse, Peter Wordie, and Pat Bamford also contributed valuable information. I am indebted to Richard and Peggy Krementz for sharing Ernest Joyce's important original diary. Jenya Osborne and Ingrid Davis kindly allowed me to quote from John King Davis's *High Latitude* and unpublished papers and letters. My thanks go to Rosa Peacock and the Tripp family for quotations from the papers of Leonard Tripp.

Alexandra Shackleton is a dedicated advocate of Antarctic research, scientific and historical. As well as making her grandfather's papers available, she brings scholars together with common research interests. She presciently introduced me to Brenda Hudson of the British Film Institute, who headed the superb restoration of *South*, which led to the discovery of the Ross Sea party film footage, which had been presumed lost in 1917. Erich Sargeant, Jan Faull, and Simon Brown at the BFI were unfailingly helpful.

The lion's share of my research was conducted at four institutions with remarkable polar collections. I am grateful to the Scott Polar Research Institute and its director, Professor Julian Dowdeswell, for welcoming me as a visiting scholar from 2002 to 2004. The late William Mills, the gifted librarian and keeper of collections, offered much encouragement and aid. His successor, Heather Lane, as well as Caroline Gunn and Lucy Martin, were unstintingly helpful with maps, manuscripts, and photographs, often bringing new treasures to my attention. I also appreciate the assistance of Shirley Sawtell, Liz Crilley, and Bob Headland. Tim Lovell-Smith steered me to little-known gems in the Alexander Turnbull Library of the National Library of New Zealand. The Canterbury Museum in Christchurch, New Zealand, holds an exceptional Antarctic collection, and Jo-Anne Smith, Kerry McCarthy, Paul Hobson, and Fran Pashby were instrumental in making it accessible. Royal Geographical Society archives officer Sarah Strong and picture librarians Joanna Wright and Justin Hobson were unflappable in the face of frequent visits and requests.

New facets of the story emerged from important material in the Mawson Collection, thanks to Mark Pharaoh, assisted by Tim Tolley. Gerard Hayes, Mark Showalter, and Lois McEvey were unfailingly helpful at the State Library of Victoria. Martin Beckett, Allen Ventress, Mark Hillenbrandt, Rosie Block, and Jennifer Broomhead offered ever-ready aid at the State Library of New South Wales. The great pains taken by these archives in Australia and New Zealand to make their collections accessible to scholars in the northern hemisphere are impressive.

The research took me far and wide to unexpected places, and archives and archivists all over the world were very helpful: Kiri Ross-Jones of the National Maritime Museum; the late Jim Wilson of the Port Chalmers Museum; Anna Blackman of the Hocken Collections; the National Archives of Australia; Yvonne Snider-Nighswander of the Hudson's Bay Company Archives; Johannah Massey of the Southland Museum, Invercargill; David Tetlow of the Greater Manchester Police Museum; Ian Pierce and Jenni Wright of the Archives Office of Tasmania; Paula McKenzie of the Southland Post and Telegraph Museum in Awarua; British Library Newspapers at Colindale; the National Library of Australia; Radio New Zealand; the University of Sydney Archives; Rosemary Shivnan; and Moira Raulin at the Glasgow University Archives.

I am indebted to Gordon Johnson for welcoming me as a senior member of Wolfson College, Cambridge. I am immensely grateful to the National Science Foundation's Artists and Writers Program for supporting my research in Antarctica in 2002, and to program manager Guy Guthridge for his continuing enthusiasm. Andy Young and Elaine Hood were instrumental in the intricate logistics, and Brian Johnson, Brennan Brunner, and Phil Austin were great guides in the field. Bob Marstall was a boon companion during many trips to the huts. I appreciated the time that numerous NSF-supported scientists took to share their expertise and shed new light on the expedition, including Rick Aster, David Bromwich of the Byrd Polar Research Center's Polar Meteorology Group; ornithologist David Ainley; Steve Arcone of the Cold Regions Research and Engineering Laboratory; medical anthropologist Lawrence Palinkas; and Lester Reed. Volcanologist Phil Kyle also graciously invited me to visit the Mount Erebus Volcanic Observatory and learn firsthand about Ross Island's origins and his group's research. An incidental benefit was spending a week camping at altitude in temperatures hovering between twenty and forty below zero and several days in a raging blizzard—all in a routine day's work for them, but for me, a small taste of the conditions endured by the Ross Sea party. Nelia Dunbar was an excellent guide to the geography, the science, and the art of subzero camping.

Desmond Lugg, chief of medicine of extreme environments at NASA, advised me on the Ross Sea party's health issues. Lugg was formerly the head of polar medicine for the Australian Antarctic Division and interviewed many polar veterans, including Richard Richards. Henry Guly of Derriford Hospital in Plymouth, England, authored a seminal paper about the Burroughs Wellcome medical kits used by several Antarctic expeditions, including the Ross Sea party. He offered insights into the pharmacology of the era and medical issues of the party. John W. Hare of the Joslin Diabetes Center, Boston, reviewed the evidence of Alexander Stevens's physical examination and symptoms which indicated undiagnosed diabetes. Orrin Devinsky, director of the New York University Medical Center Epilepsy Unit, confirmed my suspicions about Harry Shaw's ailments and discussed his symptoms in depth. Joel D. Howell, director of the Program in Society and Medicine of the University of Michigan, interpreted the cardiac symptoms and diagnoses contained in the diaries and cardiac medicine of the era. Andrew Weinberg, author of an incisive paper on hypothermia, answered myriad questions and sparked my research into the early-twentieth-century understanding of the condition. For stimulating conversations that shed light on other issues, I also owe my thanks to glaciologist Charles Swithinbank, glaciologist Andy Smith of the British Antarctic Survey, Peter McDowell of Antarctic Logistics and Expeditions, Steve Wheeler of the U.S. Coast Guard, Roger Mear, and oceanographer Elizabeth Hawker. Sue Hamilton and Geneviève Montcombroux of the Inuit Sled Dog International, Geoff Somers, Ben Hodges, Ken MacRury, and Andrew Bellars, a former British Antarctic Survey veterinarian, spoke with me about sledge dogs. Hill's Pet Nutrition shared data arising from their involvement in Antarctic and North American dog sledging.

One of the joys of this project has been exploring unexpected byways with people like Frank Barlow, a former radio operator at the Awarua station in the 1930s and 1940s, who shared a wealth of knowledge about the history of his craft. Bill Deverall, a radio operator on Ross Island in the 1960s, contributed as well. Nigel Watson of the New Zealand Antarctic Heritage Trust, whose mission it is to preserve the Ross Island huts, shared inventories of the hut artifacts. Peter Chapman shared interviews with Lionel Hooke and Irvine Gaze. Thomas Dixon of the Cambridge University Faculty of Divinity elucidated many aspects of Reverend Arnold Spencer-Smith's life in the church. I valued all of these fascinating interviews and conversations; any errors in expression of these ideas are my own. Likewise, the views expressed here are my own.

I am also grateful to these individuals and institutions for permission to reprint material: The Alexander Turnbull Library Manuscripts and Archives Collection, National Library of New Zealand (Ernest Joyce papers and manuscript of *The South Polar Trail*, the Antarctic Petrel, and the S. Y. *Aurora* agreement); the Scott Polar Research Institute (the Arnold Spencer-Smith diary, the Fisher Papers, and the Edward Wilson papers); the Canterbury Museum (the Richards, Spencer-Smith, and Gaze diaries; the Quartermain correspondence); the La Trobe Australian Manuscripts Collection, State Library of Victoria (Andrew Keith Jack diary); the Royal Geographical Society (the diaries of Eric Marshall and A. H. Ninnis, the *Discovery* expedition papers, and correspondence relating to Shackleton's expeditions); the Mitchell Library, State Library of New South Wales (Lionel Hooke diary and log); the University of Otago, Hocken Collections (diary of James Paton and papers of A. H. Ninnis); the Mawson Collection, South Australian Museum (correspondence of Douglas Mawson); the National Library of Australia (Richard Richards oral history interview); the Institution of Professional Engineers New Zealand ("An Engineer's Antarctic Log" by A. H. Larkman); the *Geographical Journal;* and PFD (www.pfd.co.uk, on behalf of the Estate of J. B. Priestley for a quotation from a review by Priestley in *The Evening News,* March 1, 1929). Excerpts from *The South Polar Trail* are by permission of Gerald Duckworth & Co. Ltd., and those from *The Life of Sir Ernest Shackleton* by H. R. Mill are reprinted by permission of The Random House Group Ltd. Quotations from the poetry of Robert Service, including "L'Envoi," "The Men That Don't Fit In," and "The Quitter," are reprinted by kind permission of Anne Longepe. Excerpts from the letters of Sir Winston Churchill and Lady Churchill are reproduced with permission of Curtis Brown Ltd., London on behalf of the Estate of Sir Winston Churchill (copyright Winston S. Churchill). Quotations from Crown Copyright documents appear by permission with record numbers. For permissions advice, I also thank William Krasilovsky. Every effort has been made to trace the copyright holders of material reproduced here. The author and publishers apologize for unintentional omissions and would be pleased to correct any oversight in future editions.

Many thanks are due to polar historian Beau Riffenburgh for our enlightening conversations on topics of mutual research interest. Beau, Mike Sharp, Mimi Berlin, Karen Barss, Dan Weinstein, historian

Alan Gurney, and Louise Crossley, a scholar of John King Davis, were astute readers of early draft sections. Dan also made me the precious gift of a first edition of Dick Richards's rare book. I benefited greatly from conversations with polar researchers and aficionados: Bob Burton, Jan Piggott, the late David Yelverton, Stephen Venables, Leif Mills, Jim and Sue McCarthy, Bruce Buchan, Cyril Halstead, and Alistair Cruickshank. Henry Worsley kindly permitted me to examine material in his personal collection. I am grateful for the legacy of the late L. B. Quartermain, who interviewed many Ross Sea party survivors in the 1960s after participating in early hut conservation efforts, and A. G. E. Jones. Both men recorded their wide-ranging research in various articles and deposited their papers at archives. All scholars of polar exploration owe an incalculable debt to Roland Huntford, a historian of surpassing insight and a writer of incomparable grace. I am grateful for his body of work and his personal generosity in sharing his knowledge with me—and for translating the words of Nansen from the Norwegian in a pinch.

My gratitude goes to my editor at Viking, Rick Kot, and associate editor Hilary Redmon. Publisher Clare Ferraro and Bloomsbury Editor-in-Chief Alexandra Pringle offered gracious support of this ambitious undertaking. I am so grateful to my Bloomsbury editor Rosemary Davidson, for her encouragement. I owe warm thanks to my agents, Theresa Park and Abner Stein.

For lending research support for a project that spanned four continents: Glenn Coster provided meticulous archival transcription in New Zealand, and Sandra Jackson and Keely Cousins helped in Britain. David Rootes of Poles Apart was kindly at the ready for special document photography. Mapmaker Kelly Brunt used a GIS to merge coordinate information and bearings from the expedition diaries with Mackintosh's charts. These maps were loosely georeferenced based on their background latitude and longitude grids to produce these definitive maps of the Ross Sea party's depots, with design by Elles Gianocostas. No detail escapes the gimlet-eyed scrutiny of Lisa Roberts, and for that reason I am most grateful that she was by my side during production. By the same token, logistics were an ever-present and often daunting fact of life. Dear friends including Tom Koch, Lisa Roberts and Kit Buckley, Jackie Mow and Patrick Cox, Ellen Dockser,

Gareth and Nicola Ward, and Michael Roth offered home away from home and countless kindnesses. For encouragement and advice, I also have Shelley Bennet, Alex Chronis, Delyth Lewis, and Richard Gregory to thank.

And finally, my gratitude to my husband, Nick Lewis, as inspiration, support, and discerning reader is inexpressible. For me, he is living proof of T. E. Lawrence's words, "All men dream: but not equally."

APPENDIX: UNITS OF MEASUREMENT

Temperature

Temperature is indicated in Fahrenheit.

Length, Distance, and Position

One fathom equals six feet.

Archival references to distances, both land and sea, are given in British standard nautical miles, unless otherwise indicated. Used in the United Kingdom until 1970, one British nautical mile is 6,080 feet or 1.15 statute miles.

A geographical mile is the length of an arc on the surface of the Earth subtended by one minute of latitude; one degree of latitude is 60 geographical miles. However, because the sphere of the Earth is slightly flattened at the poles, the geographical mile varies slightly with latitude, being shortest at the equator and longest near each pole. The mean is 6,076.12 feet, the length of the international nautical mile used in modern navigation (1.852 km or 1.15 statute miles).

The Ross Sea party depots were originally intended to be spaced roughly one degree of latitude apart at 79°, 80°, 81°, 82°, and 83° south, up to Mount Hope, at about 83°30', and are referred to as such in the text for simplicity's sake. In reality, the placement was approximate. The recorded coordinates for the depots, determined by the party and reported in John King Davis's official log of the relief expedition, are shown on the maps.

The navigation records are fragmentary, making it difficult to reconstruct positions with absolute certainty. Even these coordinates are not precise, due to the navigational methods and error in the instruments. There was some debate in the party about the positions. For example, by Spencer-Smith's reckoning, the 80° depot was placed at 80° 08' S and the 82° depot at 82°22' S.

The Mount Hope depot was laid in the Gateway. Judging by diary descriptions of the terrain, it was probably constructed within a two-mile radius of latitude 83°31'30" S longitude 171°00' E. Modern surveys place the summit of Mount Hope at 83°31' S 171°16' E. Cope's depots and the camps were plotted on the maps using recorded bearings and diary notes on mileage.

Chart (grid) bearings are cited, not magnetic.

Rate of Travel

One knot is one nautical mile per hour, or 1.15 statute miles per hour.

Wind Speed (Beaufort Scale)

References to wind speed are based upon the Beaufort Scale, which assigns values of 0 through 12 to signify conditions. Gale-force winds of Force 8 are 34 to 40 knots per hour (17.2 to 20.7 meters per second) and storm-force winds of Force 10 are 48 to 55 knots per hour (24.5 to 28.4 meters per second). At the top of the scale, hurricane-force winds are defined as Force 12, sustained winds of 64 knots per hour (32.7 meters per second) or higher.

Time

The year 1916 was a leap year. Some later writings of the expedition members neglected to take account of this fact, and thus included errors in dates and time spans. The errors are corrected here. The time kept by the shore party and on board the *Aurora* was Greenwich mean time (GMT or UTC) plus ten hours and ten minutes, the time observed in December 1914 at Macquarie Island, the *Aurora*'s last port of call.

Currency

Comparable values of modern to historic monetary amounts are based upon data provided by Economic History Services. Different methods are used appropriate to the commodity. When discussing wages, income, and wealth, the retail price index (RPI or CPI) method is used to obtain the relative purchasing power of money. For investment and capital equipment, the gross domestic product (GDP) deflator is used. The rate of exchange at the time of writing was £1 = US$1.80 and the exchange rate for £1 was $2.35AU or $2.58NZ. During the period 1914–1917, the average US dollar price per British pound was $4.80.

NOTES

— A NOTE ON THE SOURCES —

Shackleton devoted five chapters of his account of the Imperial Transantarctic Expedition, *South: The Story of Shackleton's Last Expedition 1914–1917*, to the Ross Sea party. The book contains some factual errors and revised quotations. There are two published books about the Ross Sea party written by members of the expedition. Ernest Joyce's book, *The South Polar Trail*, has been regarded as the authoritative account since its publication in 1929. A detailed exegesis of Joyce's book is a work in itself; in short, his biased account contains many errors of fact. "A strange mixture of fraud, flamboyance, and ability," historian Roland Huntford once described Joyce in his sterling biography, *Shackleton*, and it is an apt description. It is important to remember Joyce's very real accomplishments, however, which stand on their own merits.

It is telling that the reason Richards gave for resuming his diary in 1916 was that he wanted to leave an independent account of events because he believed Joyce's would be skewed. Richards's *The Ross Sea Shore Party 1914–17*, published in 1962, is a modest corrective to Joyce's account. However, Richards was not a diligent diarist and thus had only fragmentary records of his own. He relied heavily upon *The South Polar Trail* to jog his memory and reproduces many of Joyce's errors, including dates and statistics. Later secondary accounts suffer from limited consultation of the original sources. In the course of my research, I became aware of errors and discrepancies in previous published accounts, which are corrected here without notes. Similarly, discrepancies in the primary sources are reconciled without comment except when the evidence is equivocal.

For this book I have relied mainly on contemporaneous primary sources, including diaries, logs, notes, correspondence, official documents, invoices, photographs, and motion picture footage. Participant interviews and memoirs from later years were valuable as well, although they required careful evaluation. While the memories of the survivors were often startlingly sharp decades later, they also suffered from the effects of time and hindsight. I feel fortunate to have worked with such a wealth of documentary material, and I have hewn closely to the sources. When I describe events and characterize states of mind of the participants, it is drawn from their own records. Spelling and punctuation of quotations are reproduced verbatim.

I consulted the first editions of major works of polar history. If the cited content was identical, the page numbers for editions currently in print are cited unless otherwise noted, since they are more easily accessible to interested readers. The cited edition is indicated by an asterisk (*).

The following abbreviations are used in the notes and bibliography:

AOT   Archives Office of Tasmania (Hobart, Australia)

ATL   Alexander Turnbull Library Manuscripts and Archives Collection, National Library of New Zealand, Te Puna Matauranga (Wellington, New Zealand)

CM   Canterbury Museum Antarctic Manuscripts Collection (Christchurch, New Zealand)

EDP   Ethel Douglas Papers, Private Collection

GL   Guildhall Library (London, England)

GMPM   Greater Manchester Police Museum (Manchester, England)

HBCA   Hudson's Bay Company Archives, Archives of Manitoba (Winnipeg, Canada)

HC   Hocken Collections, Uare Taoka o Häkena, University of Otago (Dunedin, New Zealand)

HW   Henry Worsley, Private Collection

MC   Mawson Collection, South Australian Museum (Adelaide, Australia)

MFP   Mackintosh Family Papers, Private Collection

ML   Mitchell Library, State Library of New South Wales (Sydney, Australia)

NAA   National Archives of Australia (Canberra, Australia)

NLA   National Library of Australia (Canberra, Australia)

NMM   National Maritime Museum (Greenwich, England)

RGS   Royal Geographical Society (London, England)

RPK   Richard and Peggy Krementz, Private Collection

SLV   State Library of Victoria, La Trobe Australian Manuscripts Collection (Melbourne, Australia)

SM   Southland Museum and Art Gallery (Invercargill, New Zealand)

SPRI   Scott Polar Research Institute, University of Cambridge (Cambridge, England)

TNA   The National Archives (the Public Records Office and the Historical Manuscripts Commission) (Kew, England)

US   University of Sydney Archives (Sydney, Australia)

— PREFACE —

p. 2. **There was little threat:** Samoa was formerly held by Germany. In 1914, New Zealand and Australian forces occupied Samoa and New Guinea, the only German territories in the South Pacific.

p. 2. **telegraphist Alfred Goodwin:** New Zealand Post and Telegraph Department, "Awarua Coastal Wireless Station Attendance Book," March 1916, SM.

p. 2. **CT VLB VLB VLB:** *Handbook for Wireless Telegraph Operators*, 1912, and Lionel Hooke, "S.Y. *Aurora* Wireless Log," March 24, 1916, ML.

p. 3. **"Sire":** Lionel Hooke, "S.Y. *Aurora* Wireless Log," March 24, 1916, ML, and Joseph Stenhouse, wireless message to King George V, March 24, 1916, SPRI.

p. 5. **"the need for":** Margery Fisher and James Fisher, *Shackleton and the Antarctic*, p. 397.

p. 5. **"Talking is easy":** Ernest Joyce to Charles Royds, April 7, 1930, SPRI.

— CH. 1: "THAT RESTLESS SPIRIT" —

p. 9. **"Men go out into":** Ernest Shackleton, *The Heart of the Antarctic*, p. 1.

p. 9. **"War in the old":** "Nemo" (pen name of Ernest Shackleton), "The Antarctic Petrel," 1908–9, ATL.

p. 9. **"What is the use":** Winston Churchill, memorandum, January 23, 1914, PRO ADM 1/8368/29, TNA.

p. 9. **By 1913, the North and South Poles:** The question of who reached the North Pole first is still a bone of contention. Frederick Cook declared triumph in 1908 and Robert Peary disputed Cook's claim in 1909. Some observers discount both claims. However, the rivalry did have the effect of shifting the attention of polar explorers to other objectives.

p. 9. **"Enough life and money":** Winston Churchill, Admiralty memorandum minute, February 19, 1914, PRO ADM 1/8368/29, TNA.

p. 9. **"Bold & Napoleonic":** George Curzon, *Geographical Journal*, Vol. 44, No. 1 (1914), pp. 2–3.

p. 10. **"One of the most":** George Curzon to Ernest Shackleton, July 29, 1914, SPRI.

p. 10. **"one of the few":** George Curzon, *Geographical Journal*, Vol. 43, No. 2 (1914), p. 178.

p. 10. **"the exploration of the Antarctic":** as quoted in Alan Gurney, *The Race to the White Continent*, p. 282.

p. 10. **"staring white blank":** George Curzon, *Geographical Journal*, Vol. 44, No. 1 (1914), p. 2.

p. 11. **"I may as well":** Robert F. Scott, *The Voyage of the "Discovery,"* p. 32.

p. 11. **"regarded the expedition":** Hugh Robert Mill, *The Life of Sir Ernest Shackleton*, p. 57.

p. 12. **"the Stone Age":** John King Davis, *High Latitude*, p. 101.

p. 12. **"If he does not":** Albert Armitage, memorandum to Hugh Robert Mill, May 24, 1922, SPRI.

p. 13. **"That man should strive":** Ernest Shackleton to Emily Dorman, August 12, 1898, SPRI. In the first line, Shackleton is quoting Robert Browning's poem "The Statue and the Bust."

p. 14. **"Surely, among all"**: John King Davis, *High Latitude*, p. 101–2.

p. 14. **"The last day out"**: Ernest Shackleton, *Nimrod* diary, January 9, 1909.

p. 15. **"A live donkey is"**: Emily Shackleton to Hugh Mill, August 16, 1922, SPRI.

p. 15. **"the most brilliant incident"**: Roald Amundsen, *The South Pole*, p. 40.

p. 15. **"what the will and energy"**: Roald Amundsen, *The South Pole*, p. 36.

p. 16. **"Land of Hope"**: Douglas Mawson, *The Home of the Blizzard*, p. xvii.

p. 16. **"Beg leave to inform"**: Geoffrey Hattersley-Smith, ed., *The Norwegian with Scott: Tryggve Gran's Antarctic Diary 1910–13*, p. 14.

p. 16. **The gauntlet was thrown:** It was not the first time Amundsen had upstaged British polar strivings. In 1905, Amundsen became the first explorer to navigate the Northwest Passage, the water route joining the Atlantic and Pacific through Arctic Canada. The quest had been the consuming aim of British polar exploration in the nineteenth century.

p. 16. **"Polar exploration is"**: Apsley Cherry-Garrard, *The Worst Journey in the World*, p. vii.

p. 16. **"Great God!"**: Robert F. Scott, *Scott's Last Expedition*, p. 424.

p. 16. **All that remained:** The latitudinal distance from Cape Evans to the South Pole is 742 nautical miles, though Scott's indirect route would have been longer.

p. 17. **"Had we lived"**: Robert F. Scott, *Scott's Last Expedition*, p. 477.

p. 17. **"I feel that another"**: Ernest Shackleton to Emily Shackleton, March 8, 1911, SPRI.

p. 17. **1,500 miles:** Shackleton's estimate was very close to the actual distance, likely within a hundred miles, depending on the precise routing.

p. 18. **£50,000:** In current terms, about $3.62 million.

p. 18. **In her heyday:** David Lindsay, *A Voyage to the Arctic in the Whaler "Aurora,"* p. 219.

p. 18. **he paid £3,200:** In current terms, about $231,000.

p. 18. **No trace:** The story gained momentum with the publication of a 1959 book by Julian L. Watkins, *The World's Hundred Greatest Advertisements*. Searches of 1913–14 issues of the *Times* and other London newspapers have failed to yield the elusive advertisement. Shackleton's personal scrapbook of 1913–22 newspaper clippings contains only the letter to the editor of the *Times*. It is possible that he posted the ad for the *Nimrod* expedition, but as yet it has not been found.

— CH. 2: THE IMPERIAL TRANS-ANTARCTIC EXPEDITION —

p. 19. **"While there is yet"**: *Evening News* (London), December 30, 1913.

p. 19. **"We'll have some British"**: *Manchester Guardian*, December 30, 1913.

p. 19. **"This Expedition calls for"**: Ernest Shackleton to A. E. Cripps, March 31, 1914, SPRI.

p. 20. **"little romantic touches"**: *Evening Standard and St. James Gazette* (London), December 31, 1913.

p. 21. **labeled "Mad, Hopeless, Possible"**: Hugh Robert Mill, *The Life of Sir Ernest Shackleton*, p. 195.

p. 21. **"three sporty girls"**: P. Pegrunel, V. Davey, and E. Webster to Ernest Shackleton, Jan. 11, 1914, SPRI.

p. 21. **"the best navigator"**: Royal Geographical Society, "Report of Conference of a Committee with Sir Ernest Shackleton," March 4, 1914, RGS (abbreviated hereafter as "RGS ITAE Committee Conference Report").

p. 21. **By the age of thirty:** On Mawson's expedition, Davis had served as both master and expedition second-in-command. The typical hierarchy included a captain, the master of the vessel, and a commander of the expedition. Although the commander of a private expedition was also the owner of the ship, the master had full autonomy in matters concerning the ship. The needs of the shore party and the safety of the ship were sometimes at odds, and it took an extraordinary level of judgment and diplomacy to resolve the competing demands in a crisis, as Shackleton knew from experience. Scott eliminated possible conflicts by serving as both captain and expedition commander, as did Amundsen.

p. 21. **"Poor S. works away"**: Frank Wild, *Nimrod* diary, December 19, 1908, SPRI.

p. 21. **"I do not suppose"**: Frank Wild, *Nimrod* diary, January 31, 1909, SPRI.

p. 21. **If funding failed:** The offer was contingent upon financing. Ideally, Shackleton hoped to raise enough to staff a scientific base on the Weddell Sea coast, rather than having the ship sail north. Joyce shared Shackleton's letter with the *Sydney Morning Herald,* which reprinted excerpts in an article on June 29, 1914. He also included the letter in his 1929 book, *The South Polar Trail,* though the text differs. In the newspaper version, Shackleton makes clear that Joyce would not be considered for the transcontinental party; the book version leaves open the possibility that he could be chosen. (See chapter 18.)

p. 22. **"After months of want"**: Ernest Shackleton, *The Heart of the Antarctic,* p. 360–61.

p. 22. **"deep-rooted conviction"**: Colin Bertram, *Arctic and Antarctic: A Prospect of the Polar Regions,* p. 45.

p. 22. **"gay, dashing"**: John King Davis, *High Latitude,* p. 95.

p. 22. **"Poor fellow"**: Eric Marshall, *Nimrod* diary, January 31, 1908.

p. 23. **For command of the Ross Sea:** Eric Marshall to John Kendall, September 15, 1950, SPRI.

p. 23. **"He is a tough nut"**: Dudley Everitt to Judy Tudor, November 9, 1955, RGS.

p. 23. **£60,000 to £100,000:** In current terms, about $4.34 to $7.23 million.

p. 23. **"undignified to take money"**: *Manchester Guardian,* December 30, 1913.

p. 24. **"hidebound and narrow"**: Ernest Shackleton to William S. Bruce, August 20, 1913, SPRI.

p. 24. **"geography and every other"**: Clements Markham to the RGS Council, January 17, 1914, RGS.

p. 24. **A key objection:** When German explorer Wilhelm Filchner first landed and built a hut there in 1912, the ice broke up and rafted out to sea.

p. 25. **Shackleton estimated:** Ernest Shackleton, *Endurance* diary, January 2, 1915, SPRI.

p. 25. **"well commanded"..."I am perfectly aware"**: "RGS ITAE Committee Conference Report," March 4, 1914, RGS.

p. 26. **Colonel Wilfred Beveridge:** Beveridge's diet totaled 5,512 calories per day. *Times* (London) May 30, 1914.

p. 27. **His *Nimrod* attempt:** The diet of Amundsen's polar party was deficient in vitamin C as well, so the seal meat was a useful supplement.

p. 27. **"I have learnt something"**: John King Davis to Ernest Shackleton, March 16, 1914, SPRI.

p. 27. **"modern engines, fine cabins"**: John King Davis, *High Latitude*, p. 236.

p. 27. **Marshall had detected**: Eric Marshall to John Kendall, September 15, 1950, SPRI.

p. 27. **"These polar expeditions"**: Winston Churchill, Admiralty memorandum minute, February 19, 1914, PRO ADM 1/8368/29, TNA. Ultimately, two officers were loaned to the Weddell Sea party.

p. 27. **Æneas Mackintosh, Shackleton finally decided**: Ernest Shackleton, transcript of original cable to Ernest Joyce, March 1913, ATL and *The Sydney Morning Herald*, June 29, 1914.

p. 28. **"jolly good sort"**: Leslie Thomson, diary, January 24, 1915, SPRI.

p. 28. **"in present political conditions"**: Neville Chamberlain to Ernest Shackleton, May 15, 1914, SPRI.

p. 29. **On August 1, however, Macklin**: *Morning Post* (London), July 15, 1914. Shackleton had apparently decided that he needed a physician for the Weddell Sea base as well as his transcontinental party.

p. 30. **"Are you going near"**: *Evening Standard and St. James Gazette* (London), August 1, 1914.

p. 30. **"ship staff stores and"**: Ernest Shackleton, telegram to Admiralty, August 3, 1914, p. 455, PRO ADM 137/51, TNA.

p. 30. **"excursion will be postponed"**: Ernest Shackleton, telegram to Emily Shackleton, August 3, 1914, SPRI.

p. 30. **"Proceed"**: Winston Churchill, telegram to Ernest Shackleton, August 3, 1914, PRO ADM 137/51, TNA.

p. 31. **"I cannot afford"**: Ernest Shackleton to Emily Shackleton, August 17, 1914.

p. 32. **"I am rather tired"**: Ernest Shackleton to Emily Shackleton, August 18, 1914, SPRI.

p. 32. **"I am having a bad time"**: Ernest Shackleton to Janet Stancomb-Wills, August 19, 1914, SPRI.

p. 32. **"I get very tired"**: Ernest Shackleton to Emily Shackleton, August 17, 1914, SPRI.

p. 32. **"I'm not going to"**: *Willesden Chronicle* (Willesden, England), February 16, 1917.

— CH. 3: *AURORA* —

p. 34. **"The ship was leaking"**: Æneas Mackintosh to Frederick White, December 1, 1914, SPRI.

p. 34. **"He had that rare gift"**: John King Davis, *High Latitude*, p. 83.

p. 36. **"a little haphazard"**: Campbell Mackellar to William S. Bruce, April 4, 1916.

p. 36. **"I am still existing"**: Æneas Mackintosh to George Marston, February 27, 1913, SPRI.

p. 37. **"Thrashing seems"**: Joseph Stenhouse, diary, October 3, 1915, SPRI.

p. 38. **Ninnis had enjoyed none**: His uncle on his father's side was Belgrave Ninnis Senior, a wealthy naval surgeon and inspector general of naval hospitals and fleets who had been surgeon to the British Arctic expedition under George Nares (1875–76).

p. 38. **Howard's polar ambitions:** Discharge documents noted "phthisis," a term of the time for tuberculosis or consumption.

p. 39. **"You are looking great":** Stan Hayward to Victor Hayward, January 3, 1915, NMM.

p. 39. **"do anything":** Alexander Stevens, "Report of the Ross Sea Party," SPRI.

p. 40. **"inevitable Woodbine," "For some unknown reason":** "Men of Mark: A. P. Spencer-Smith," *Queen's College Dial* (Cambridge), Lent Term 1907, p. 117.

p. 41. **"a habit of wandering":** Joseph Stenhouse, diary, October 7, 1914, p. 18, SPRI.

p. 41. **"not always striving to avoid":** *Journal of Glasgow University Graduates Association,* November 1953, p. 49.

p. 41. **"the scientist brigade":** Howard Ninnis, diary, September 29, 1914, RGS.

p. 42. **"lavishing attention":** Howard Ninnis, diary, October 22, 1914, RGS.

p. 42. **"Our two 'Launcelots'":** Howard Ninnis, diary, September 28, 1914, RGS.

p. 42. **"He's an absolute 'dear'":** Arnold Spencer-Smith to Charlotte Spencer-Smith, November 8, 1914, CM.

p. 42. **Mackintosh was surprised:** *Sydney Morning Herald,* December 10, 1914.

p. 43. **"She's in a filthy state":** Arnold Spencer-Smith to Charlotte Spencer-Smith, November 8, 1914, CM.

p. 43. **Arguably, it was the ship's:** In addition to whaling and sealing voyages in the Arctic, the *Aurora* had joined the search for Adolphus Greely and the Lady Franklin Bay Expedition in 1884.

p. 43. **"the hundred & one things":** Joseph Stenhouse, diary, November 6, 1914, SPRI.

p. 46. **Mackintosh was stunned:** Æneas Mackintosh, cable to Frederick White, November 20, 1914, quoted in Mackintosh to White, December 1, 1914, SPRI.

p. 46. **"astounded":** Æneas Mackintosh to Frederick White, December 1, 1914, SPRI.

p. 46. **Stenhouse suggested that they raise:** His great-uncle was Andrew Stenhouse, who made his fortune in gypsum mining. Stenhouse Bay, Australia, is named for him.

p. 46. **"My dear Mackintosh"** . . . **"free as gifts":** Ernest Shackleton to Æneas Mackintosh, September 18, 1914, SPRI.

p. 46–47. **"[David] is not given"** . . . **"The ship is held up":** Æneas Mackintosh to Frederick White, December 1, 1914, SPRI.

p. 47. **"extremely capable":** George Barr to Lachlan D. Mackintosh, April 8, 1948, MFP.

p. 47. **Costs soon climbed over £3,000:** This amount was well over the £500 (about $36,000 in current terms) authorized. Referring it to the Navy Office, the yard manager received approval from Sir William Creswell of the Naval Board on November 30 to keep supplying the expedition. Creswell felt that he had no alternative but to approve continued expenditures, given that lives depended on the *Aurora*'s timely sailing. (*The Expenditure Incurred in Connexion with the S.Y. Aurora,* NAA.)

p. 48. **"very energetic":** Alexander Stevens, "Report of the Ross Sea Party," p. 11, SPRI.

p. 48. **"During the stay in Sydney":** Alexander Stevens, "Report of the Ross Sea Party," p. 16, SPRI.

p. 48. **"deplorable circumstances":** Æneas Mackintosh to Frederick White, December 21, 1914, SPRI.

p. 48. **"doing their utmost":** Æneas Mackintosh to Frederick White, December 1, 1914, SPRI.

p. 48. **"Looking extremely happy":** Joseph Stenhouse, diary, December 10, 1914, SPRI.

p. 48. **"was not considered":** Alexander Stevens, "Report of the Ross Sea Party," pp. 7–8, SPRI.

p. 48. **"better class ships":** Alexander Stevens, "Report of the Ross Sea Party," p. 8, SPRI.

p. 49. **"Drink is a curse":** Joseph Stenhouse, diary, December 22, 1914, SPRI.

p. 49. **"Jollification":** Joseph Stenhouse, diary, December 11, 1914, SPRI.

p. 49. **"A rotter":** Edward Wilson, diary, May 22, 1911, SPRI, quoting Bernard Day.

p. 49. **"I absolutely decline":** John King Davis to Ernest Shackleton, March 16, 1914.

p. 49. **Stenhouse advised:** Frank Worsley, notes of conversations with J. R. Stenhouse and E. H. Shackleton, verso letter dated November 17, 1932, HW.

p. 49. **"Many engineers came":** A. H. Larkman, "An Engineer's Antarctic Log (Pt 1)," *New Zealand Engineering,* August 15, 1963, pp. 286–91. The auxiliary two-cylinder steam engine had been supplanted by more powerful triple-expansion engines on merchant ships of the *Aurora*'s size.

p. 50. **salary of £1 per week:** In current terms, about $92. For the sake of comparison, Scott paid his *Terra Nova* scientific staff members £100 to £500 per annum from 1910 to 1913. (J. J. Kinsey Papers, ATL)

p. 50. **"Ship must leave":** Alfred Hutchison and Ernest Perris, cable to Æneas Mackintosh, December 9, 1914, quoted in Æneas Mackintosh to Frederick White, December 21, 1914, SPRI.

p. 50. **"I sighted our crew":** Joseph Stenhouse, diary, December 10, 1914, SPRI.

p. 51. **The crew was not informed:** Although the ship's copy of the articles lists Mackintosh as master, the "Lloyd's Captains Register" shows Stenhouse as master (GL). This information would have been provided by Australian shipping officials to the Board of Trade in Britain, which then passed it on to Lloyd's. That Australian officials listed Stenhouse as master on the second set of articles deposited with the shipping office is borne out by Joseph Kinsey, who held a copy of the document in 1916: "The name of Lieut. J. R. Stenhouse, R.N.R. appears on the *Aurora*'s Register as Master and not A. Mackintosh, as generally understood." (Joseph Kinsey to Robert McNab, July 19, 1916, SLV) The ship's copy of the articles should have been taken to the shipping office for amendment. (ATL)

The reason for the secret change is unknown. The Board of Trade did require sight tests for officers. Australian officials may have objected to Mackintosh serving in command on the grounds of his impairment, and he perhaps acceded, since there was no time for an appeal. The regulations for Royal Naval Reserve Officers do specify that "normal vision is an essential qualification." On these grounds, Mackintosh would have been disqualified from the RNR. (BT 167/138)

p. 51. **"champion roller"**: Arnold Spencer-Smith to Charlotte Spencer-Smith, December 20, 1914, CM.

p. 51. **"with a very lugubrious"**: Joseph Stenhouse, diary, December 19, 1914, SPRI.

p. 51. **"difficult to imagine"**: Richard Richards, *The Ross Sea Shore Party 1914–17*, p. 5.

p. 51. **"Cannot understand"**: Frederick White, cable to Æneas Mackintosh, undated (December 9–21, 1914), SPRI.

p. 52. **"Obligation yours"**: Ernest Perris and Alfred Hutchison, cable to Æneas Mackintosh, December 9, 1914, SPRI.

p. 52. **"My obligations"**: Æneas Mackintosh to Frederick White, December 21, 1914, SPRI.

p. 52. **"of supreme importance"**: Ernest Shackleton to Æneas Mackintosh, SPRI. Shackleton also wrote that if some catastrophe prevented them from laying the depots, his party would carry enough supplies to make the entire journey without them. This unrealistic statement seems intended to allay "the anxiety of feeling that the T. C. Party is absolutely depending on the [Mount Hope] depot" if "some serious accident incapacitated" the Ross Sea party. Shackleton goes on to assert that the Beardmore depot is supremely important, contradicting the reassurance.

p. 52. **"the disorder"**: Richard Richards, *The Ross Sea Shore Party 1914–1917*, p. 5

p. 53. **"nearly took a piece"**: Leslie Thomson, diary, December 23, 1914, SPRI.

p. 53. **"No one cares"**: Joseph Stenhouse, diary, December 23, 1914, SPRI.

— CH. 4: SOUTHING —

p. 54. **"She rides along the waves"**: Æneas Mackintosh, diary, January 2, 1915, SPRI.

p. 54. **"She's a damned oscillating"**: Alfred Larkman, diary, December 26–29, 1914, SPRI.

p. 54. **"Many of our after crowd"**: James Paton, diary, January 2, 1915, HC.

p. 54. **"straight out of *Treasure Island*"**: Arnold Spencer-Smith to Charlotte Spencer-Smith, November 8, 1914, CM.

p. 55. **"the FuFu gang"**: James Paton, diary, January 8, 1915, HC.

p. 55. **"The only disagreeable people"**: Æneas Mackintosh, diary, January 18, 1915, RGS.

p. 55. **For the crew**: The class divide is starkly apparent in public health reports of the time. Recruitment efforts for World War I revealed that two out of three British men were physically unfit for military service. (*Report Upon the Medical Department of the Ministry of National Service*, 1917–19, PRO NATS 1/768, TNA.) Due to poor nutrition, state-schooled children in Britain were on average five inches shorter than those in public schools. (Dennis Winter, *Death's Men: Soldiers of the Great War*, pp. 30–31.)

p. 56. **"An excellent, trustworthy man"**: Clements Markham, "Record of the RGS Antarctic silver medallists," RGS.

p. 56. **"sober, honest"**: Robert F. Scott, reference for Ernest Joyce, November 12, 1905, RGS.

p. 57. **The problem, it was rumored:** Mawson wrote his future wife, Paquita Delprat, telling her that he had fired Joyce on the day before the departure for Antarctica without giving his reason. (December 1, 1911, MC) Morton Moyes, the expedition meteorologist, later told historian A. G. E. Jones that Joyce's excessive drinking and rowdy behavior caused the rift. A two-line letter of reference Mawson wrote for Joyce on the day of the *Aurora*'s departure is most revealing for what it does not say: "On account of the excellent work done by you on Expeditions in the Antarctic I trust that you will be successful in doing well for yourself. Some day I may be able to assist you, and you may be sure I shall not forget the past." (PRO ADM 1/8629/132, TNA) (With thanks to Beau Riffenburgh for pointing out the Mawson letter.)

p. 57. **Shackleton was doubted:** The potential of wireless was dramatically demonstrated in 1912, when two operators aboard the SS *Titanic* transmitted the distress calls that saved hundreds of lives. In fact, Shackleton had testified during the *Titanic* inquiry as an expert on ice navigation. The tragedy inspired the adoption of the Safety of Life at Sea Convention of 1914, mandating wireless on all seagoing passenger vessels and the construction of new coastal receiving stations. The *Endurance* and the *Aurora* were exempt from the requirement.

p. 57. **Shackleton believed that contact:** Shackleton told one associate that he "did not want facilities in his absence." (Douglas Freshfield, interview with Ernest Perris, March 31, 1916, RGS) One can imagine the anguish of concerned family members if negative news reached home and authorities were powerless to act. Then there was the potential financial cost of allowing leaked news of the expedition to jeopardize Shackleton's exclusive deal with the *Daily Chronicle*. His worry was justified: Reports of Amundsen's first crossing of the Northwest Passage galloped ahead of his triumphal return and robbed him of potentially lucrative newspaper deals.

p. 58. **"Everything OK":** Victor Hayward, wireless to Ethel Bridson, December 31, 1914, SPRI.

p. 58. **"We have severed":** Keith Jack, diary, December 31, 1914, SLV.

p. 58. **"rather uncomfortable":** Joseph Stenhouse, diary, January 1, 1915, SPRI.

p. 58. **"Worked out the programme":** Æneas Mackintosh, diary, January 2, 1915, SPRI.

p. 59. **"They look objects of abject pity":** Æneas Mackintosh, diary, January 1, 1915, SPRI.

p. 59. **"a yachting cruise":** Æneas Mackintosh, diary, January 6, 1915, SPRI.

p. 59. **"standing out in prominence":** Joseph Stenhouse, diary, January 7, 1915, SPRI.

p. 60. **"We might with equal chance":** James Clark Ross, *A Voyage of Discovery and Research in the Southern and Antarctic Regions During the Years 1839–1843*, pp. 217–18.

p. 61. **Later in the winter:** The project was ill conceived, based as it was on gathering evidence for the ideas of biologist Ernst Haeckel, which were already out of favor. Haeckel believed that "ontogeny recapitulates phylogeny" in nature—in other words, the embryonic development of an organism reveals the evolutionary origins of the species.

p. 61. **"the worst journey":** Apsley Cherry-Garrard accompanied Dr. Edward Wilson and Birdie Bowers to Cape Crozier in 1911 to study emperor penguin embryos to substantiate the erroneous idea that these flightless birds represent a primitive stage of avian evolution.

p. 61. **"This was regrettable"**: Æneas Mackintosh, diary, January 9, 1915, SPRI.

p. 61. **"We must bustle now"**: Æneas Mackintosh, diary, January 10, 1915, SPRI.

p. 62. **Navigating the pack:** In December 2002, the 399-foot, 13,000-ton U.S. Coast Guard icebreaker *Polar Sea* entered the pack just off Cape Bird, powered by engines generating 60,000 horsepower. The ice was thicker than normal due to a multiyear accumulation. After three weeks of attempting to cut a channel through the nine feet of ice, the icebreaker had gained just three miles, churning through 60,000 gallons of fuel in one twenty-hour period. The *Polar Star* joined forces and it took both ships two weeks to break the ice roughly thirty miles to Hut Point.

p. 62. **"Down below the noise"**: James Paton, diary, January 13, 1915, HC.

p. 62. **"Scurvy is not due"**: John Lachlan Cope, "Medical Report of the Ross Sea Base ITAE," January 1917, SPRI. Cope presciently assumed "vitamines" could be derived from fresh foods or synthesized in tablet form. While the *Aurora* was in Australia, he contacted Burroughs Wellcome about obtaining vitamin extracts, but learned that scientists were nowhere near identifying the biochemistry of these theoretical substances, much less processing them into pills.

p. 62. **"beef, an odoriferous cod fish"**: Frederick Cook, *Through the First Antarctic Night 1898–1899*, p. 234.

p. 64. **"Open water haunts us"**: Æneas Mackintosh, diary, January 15, 1915, SPRI.

p. 65. **"somewhat perturbed"**: Ernest Joyce, *The South Polar Trail*, p. 52.

p. 66. **many still suffered:** The dogs were dosed with extract of male fern (*Dryopteris filix-mas*), a potent remedy for tapeworm (though not for roundworm) known to the ancient Greeks. The active ingredients paralyze the parasites and cause them to detach from the host's intestine, although the head of the tapeworm can sometimes remain attached and regrow. Male fern can be toxic to the patient as well, particularly to weaker dogs.

p. 66. **Purchased from:** Although he ordered huskies from the Hudson's Bay Company in 1914, Shackleton was informed that "only half breeds" could be found for sale. Moreover, an HBC agent cabled that the dogs he had seen were "not in his opinion suitable for Arctic work." Ultimately, one hundred were found, some of which the agent described as "remarkably fine animals," although none had worked in polar conditions. Dog drivers are loath to lose their best animals, so the sellers may well have palmed off inferior dogs on Shackleton.

Northern sledge dogs are known loosely as huskies. There are four main breeds: the malamute, the Samoyed, the Siberian husky, and the Inuit dog, also known as the Canadian Eskimo dog or qimmiq (qimmeq in Greenland). The Inuit dog is native to Arctic Canada, Alaska, and Greenland. It is unclear why Shackleton chose to obtain dogs from the Hudson's Bay Company rather than the Royal Greenland Trading Company, since Amundsen's work with Greenland dogs had demonstrated how well suited they were to Antarctic conditions.

p. 66. **"It will take some time"**: Ernest Joyce, *The South Polar Trail*, p. 54.

p. 67. **Mackintosh plotted a line:** Although the Ross Sea party believed Mount Hope to be at 83°37' S, modern maps show the summit to be at 83°31' S 171°16' E.

p. 67. **During the *Aurora*'s month-long voyage:** Ernest Shackleton to Æneas Mackintosh, September 18, 1914, SPRI. In his account of the expedition, *South,* Shackleton states that he told Mackintosh to lay depots no farther than latitude 80° south in the first season. Joyce seems to echo this when he writes that Shackleton would "expect to find the depôt laid at 80° S in early 1915." (*The South Polar Trail*, p. 60) However, this limited scope is not specified in Shackleton's written instructions, which direct Mackintosh to reach the Beardmore in the first season. Stevens recorded in a report that "Mackintosh hoped to lay depôts to Mount Hope the first summer: indeed, he seemed to regard it as essential that he should." (Stevens, "Report of the Ross Sea Party") Stenhouse, too, wrote to Shackleton that "Mackintosh was very anxious to get all available sledgers away in order, if possible, to junction with you if you came across [in early 1915]." (February 15, 1917, SPRI)

p. 67. **"could not get him":** Ernest Joyce, log, January 24, 1914, RPK.

p. 67. **stores, sledges, and dogs:** Ernest Shackleton to Ernest Joyce, *Sydney Morning Herald*, June 29, 1914. Notably, he deletes this line in the version in his 1929 book to convey the impression that Mackintosh had overstepped his authority. (See chapter 18.)

p. 67. **In response, Mackintosh:** Ernest Shackleton to Æneas Mackintosh, September 18, 1914, SPRI: "All details of organisation and arrangement I leave to your discretion."

p. 67. **Mackintosh told Joyce:** As noted in Chapter 3, Stenhouse was already the master of the ship. The ship's copy of the articles bear the notation that Stenhouse became acting master on January 25, 1915.

p. 67. **"the shore party":** Ernest Shackleton to Ernest Joyce, *Sydney Morning Herald*, June 29, 1914.

p. 68. **Joyce had played:** Shackleton originally intended to establish a base on the Barrier, well east of McMurdo Sound, but yielded to the advice of Joyce and Edgeworth David to sail west. Shackleton listened to the urgings of Joyce over Eric Marshall, who as surgeon outranked Joyce.

p. 68. **"If I had Shacks":** Ernest Joyce, log, January 24, 1914, RPK.

p. 68. **"the silliest damn rot":** Ernest Joyce, log, January 24, 1914, RPK.

p. 68. **"Mack is my Boss":** Ernest Joyce, log, January 24, 1914, RPK.

p. 68. **"Joyce has been considered":** Æneas Mackintosh, diary, January 22, 1915, SPRI.

p. 68. **"Having one eye":** Ernest Joyce, log, January 24, 1914, RPK.

p. 68. **"As he will not":** Ernest Joyce, log, January 24, 1914, RPK.

p. 68. **"There was no show":** Leslie Thomson, diary, January 24, 1915, SPRI.

— CH. 5: THE GREAT BARRIER

p. 69. **Mackintosh charged:** Alexander Stevens, "Report of the Ross Sea Party," p. 23, SPRI.

p. 69. **"There was confusion":** Alexander Stevens, "Report of the Ross Sea Party," p. 22, SPRI.

p. 71. **baptism of frost:** Ernest Shackleton, *The Heart of the Antarctic*, p. 229.

p. 71. **Purchased piecemeal:** Frank Hurley of the *Endurance* indicated in his diary that the Hudson's Bay Company had documented which animals had worked in teams together, but most were separated by Shackleton, some going to the *Endurance* and others to the *Aurora*. (Frank Hurley, diary, April 9, 1915, ML)

p. 71. **"It went just A1":** Irvine Gaze, diary, January 26, 1915, CM.

p. 71. **Pemmican:** The traditional staple of the Antarctic sledging diet was first devised by the Cree Indians for the raw winters of North America. They also mixed dried berries in the concoction.

p. 71. **"crevasses, frostbite":** Ernest Joyce, *The South Polar Trail*, p. 55.

p. 71. **"great capacity as a leader":** Keith Jack, diary, January 25, 1915, SLV.

p. 72. **"splendid sledging mates":** Ernest Joyce, *The South Polar Trail*, p. 59.

p. 72. **"Living a real life":** Irvine Gaze, diary, February 3, 1915, CM.

p. 72. **"My general health":** Irvine Gaze to Æneas Mackintosh, undated, November 1914, SPRI.

p. 73. **"He has excellent":** George Crowther, letter of reference for Keith Jack, September 13, 1904, SPRI.

p. 73. **"Would do any capacity":** Keith Jack to Æneas Mackintosh, December 3, 1914, SPRI.

p. 73. **"almighty great blizzard":** Irvine Gaze, diary, January 25, 1915, CM.

p. 73. **"ripping start":** Irvine Gaze, diary, January 27, 1915, CM.

p. 73. **"It really broke":** Irvine Gaze, diary, January 27, 1915, CM.

p. 74. **Sky and ground:** As a compass is carried closer to the south magnetic pole, the south end of the needle is drawn more strongly downward, necessitating a specially calibrated compass with a nonmagnetic counterweight and a deep housing. Even with such an instrument, the abundance of iron in the rock of Ross Island plays havoc with magnetic compasses.

p. 74. **"We don't know":** Arnold Spencer-Smith, diary, January 26, 1915, SPRI.

p. 74. **"Joyce has just returned":** Irvine Gaze, diary, January 28, 1915, CM.

p. 74. **"The man who bought":** Irvine Gaze, diary, January 27, 1915, CM.

p. 74. **The Australians had already formed:** Richard Richards, "Marooned in the Antarctic" (draft manuscript), MC. Richards believed Mackintosh "did not have the personality" to "organize efficiently before departure from Australia." Due to Mackintosh's discretion, and the fact that the Australians joined very late, they never understood the nature of the financing and organizational problems and assumed that Mackintosh was at fault.

p. 74. **"We all have had":** Keith Jack, diary, January 28, 1915, SLV.

p. 75. **Although the sun circled:** Though the sun is in the sky twenty-four hours each day, it traces a tilted ellipse, and the sun is closer to the horizon in the evening. The difference means less direct rays of sunlight are concentrated on the earth, causing slight cooling and hardening of the snow surface.

p. 75. **"passed Joyce after a struggle":** Ernest Wild, diary, January 28, 1915, ATL.

p. 75. **"a mass of rolling":** Æneas Mackintosh, diary, January 28, 1915, RGS.

p. 75. **"After much beating":** Æneas Mackintosh, diary, January 28, 1915, SPRI.

p. 75. **"the boot, the whip":** Arnold Spencer-Smith, diary, January 29, 1915, SPRI.

p. 75. **"absolute property":** Agreement for employment as a member of the Imperial Trans-Antarctic Expedition, copies signed by various members of the Ross Sea party, SPRI.

p. 76. **"shot & skinned Captain":** Ernest Wild, diary, November 28, 1914, SPRI.

p. 76. **"Relay work no good":** Ernest Wild, diary, January 30, 1915, SPRI.

p. 76. **"someday be Captain":** Lionel Hooke, diary, January 8, 1915, ML.

p. 76. **"the work is cruel":** Arnold Spencer-Smith, diary, January 29, 1915, SPRI.

p. 76. **"A nightmare of Edinburgh"**: Arnold Spencer-Smith, diary, February 6, 1915, SPRI.

p. 76. **"All the old questionings"**: Arnold Spencer-Smith, diary, January 31, 1915, SPRI.

p. 77. **It was also said:** Though Spencer-Smith believed that Oates died on March 17, this date is likely incorrect. The misreading may have arisen because Scott describes the events in a diary entry dated "Friday, Mar. 16 or Saturday 17—Lost track of dates, but the last correct." The entry reads: "[Oates] slept through the night before last, hoping not to wake; but he woke in the morning—yesterday. It was blowing a blizzard . . . He went out into the blizzard and we have not seen him since." Hence, Oates left the tent on March 15 or 16; Scott favored the latter date.

p. 77. **"I am just going outside"**: Robert F. Scott, *Scott's Last Expedition*, p. 462.

p. 77. **"Let us also go"**: Bible, John 11:16 quoted in Arnold Spencer-Smith, diary, February 1, 1915, SPRI.

p. 77. **"I must say I feel"**: Æneas Mackintosh, January 29, 1915, RGS.

p. 78. **"a vast wall surrounding"**: Arnold Spencer-Smith, diary, January 29, 1915, SPRI.

p. 78. **To the west:** The Ross Sea lies beneath the Ross Ice Shelf; Black Island and White Island are remnants of volcanic peaks on the sea floor piercing the ice shelf.

p. 78. **By midday on February 1:** Ernest Joyce, *The South Polar Trail*, p. 59.

p. 79. **"Buck up"**: Æneas Mackintosh, diary, February 1, 1915, RGS.

p. 79. **"The new dog Jock"**: Keith Jack, diary, February 1, 1915, SLV.

p. 79. **"no hurry"**: Irvine Gaze, diary, February 1, 1915, CM.

p. 79. **"b----y rot"**: Irvine Gaze, diary, February 1, 1915, CM.

p. 79. **"knocking hell"**: Irvine Gaze, diary, February 3, 1915, CM.

p. 79. **"hadn't any damned idea"**: Irvine Gaze, diary, February 1, 1915, CM

p. 79. **He knew from experience:** The Ross Ice Shelf is a glacier, up to three thousand feet thick, slowly streaming downhill from the higher elevation of the Antarctic interior to float on the Ross Sea. Like boulders in a stream, rock formations deflect the river of ice. As the flow accelerates and slows, bending and twisting around outcrops, crevasses form along the lines of stress.

p. 80. **"No crevasses as yet"**: Arnold Spencer-Smith, diary, February 1, 1915, SPRI.

p. 80. **likely to crash through:** A man on skis exerts a half pound per square inch of force, while a foot tread is in excess of two pounds per square inch. (Roland Huntford, *Shackleton*, p. 243)

p. 80. **"cruelly hard"**: Arnold Spencer-Smith, diary, February 1, 1915, SPRI.

p. 80. **"proceeding merrily"**: Arnold Spencer-Smith, diary, February 2, 1915, SPRI.

p. 80. **"Of one thing I am sure"**: Arnold Spencer-Smith, diary, February 2, 1915, SPRI.

p. 80. **"vast sea, slightly troubled"**: Arnold Spencer-Smith, diary, February 5, 1915, SPRI.

p. 81. **"None of us get much rest"**: Irvine Gaze, diary, February 3, 1915, CM.

p. 81. **"It was quite painful"**: Arnold Spencer-Smith, diary, February 5, 1915, SPRI.

p. 82. **Mackintosh's dogs fared the worst:** According to Dr. Andrew Bellars, formerly a veterinarian for the British Antarctic Survey, a standard harness would not have been suitable for the Ross Sea party's dogs, since they varied in size and build. Abrasion would have been severe in dogs without husky-type double coats. Also, leather can be inflexible and chafe if poorly designed. Traditional Inuit harnesses were sealskin; BAS used lamp wick, a strong, soft material which tended not to ice and stiffen as readily as leather. Handlers liked to remove the harnesses nightly to dry and put them on incorrectly on occasion so that friction would not aggravate the same spots day after day.

p. 82. **Mackintosh blamed:** Geoff Somers, a member of Will Steger's International Trans-Antarctic Expedition in 1989–90, recalls limiting the days to nine and a half hours and resting every eleventh day on their four-thousand-mile crossing, but these were extremely well fed and conditioned huskies. Experienced dog drivers are divided on the question of physical discipline. Some are opposed to it; others employ it sparingly to stop dangerous behaviors like fighting. In my interviews, dog handlers were overwhelmingly against beating dogs continuously in an attempt to spur them onward.

p. 82. **"that blessed dog":** Irvine Gaze, diary, February 4, 1915, CM.

p. 82. **"Mackintosh would appear":** Keith Jack, diary, February 6, 1915, SLV.

p. 82. **"Let him try":** Irvine Gaze, diary, February 6, 1915, CM.

p. 82. **"We woke to find":** Arnold Spencer-Smith, diary, February 9, 1915, SPRI.

p. 83. **"I had him one side":** Ernest Joyce, log, February 11, 1915, RPK.

p. 83. **"I quite see his point":** Ernest Joyce, log, February 11, 1915, RPK.

p. 83. **"keenly disappointed":** Keith Jack, diary, February 4, 1915, SLV.

p. 83. **"Plans are so easy":** Æneas Mackintosh, diary, February 13, 1915, RGS.

p. 83. **"a rotten slap":** Irvine Gaze, diary, February 11, 1915, CM.

p. 83. **"trust it is best":** Keith Jack, diary, February 11, 1915, SLV.

p. 83. **"We passed Jock's grave":** Arnold Spencer-Smith, diary, February 12, 1915, SPRI.

p. 84. **the fuel tins were leaking:** The same problem had bedeviled Scott on his last expedition. The seals on the cans contracted in the cold and allowed the liquid to leak, a problem he described in his 1913 book, which Mackintosh read. During the chaotic preparation, Mackintosh had little time to modify the containers.

p. 84. **"Wonder what 30 and 40 below":** Irvine Gaze, diary, February 4, 1915, CM.

p. 84. **"Ah, but a man's reach":** Robert Browning, "Andrea del Sarto," p. 402, quoted in Arnold Spencer-Smith, diary, February 16, 1915, SPRI.

p. 84. **The Ross Sea party's tractor:** In May 1914, Shackleton traveled to Finse, Norway, with some members of the Weddell Sea party to test the expedition equipment. The newly designed motor tractors and sledges were plagued with mechanical troubles. Preoccupied with fundraising, Shackleton trusted that the problems would be worked out by the mechanics for each party.

p. 84. **"as a convenience":** Leslie Thomson, diary, January 31, 1915, SPRI.

p. 85. **Dragging a load:** They had no idea that the weight was unprecedented in Antarctic exploration to date. On Shackleton's South Pole attempt, the party had strained under 250 pounds per man; Scott's Cape Crozier party set out with 253 pounds each.

p. 86. **"rank 'criminal' incompetency":** Howard Ninnis, diary, February 14, 1915, RGS.

p. 86. **"You cannot take a very young man"**: Howard Ninnis, diary, February 26, 1915, RGS.

p. 86. **"like visions of Fairy-land"**: Irvine Gaze, diary, February 22, 1915, CM.

p. 86. **"Altogether things"**: Irvine Gaze, diary, February 25, 1915, CM.

p. 86–87. **"By jove we *are* living well"**: Irvine Gaze, diary, February 24, 1915, CM.

p. 87. **"There's a great blizzard"**: Irvine Gaze, diary, February 27, 1915, CM.

p. 87. **"Our good spirit"**: Howard Ninnis, diary, February 21, 1915, RGS.

p. 87. **"The load this party is hauling"**: Keith Jack, diary, February 19, 1915, SLV.

p. 87. **"By what authority"**: Bible, Luke 20:2, quoted in Arnold Spencer-Smith, diary, March 2, 1915, SPRI.

p. 88. **They turned back**: In 1911, Scott and a small group attempted the same journey, following the peninsula to the lower slopes of Mount Erebus, and found the area beyond Hutton Cliffs to be heavily crevassed. They descended just south of Glacier Tongue and continued across the sea ice. In 1958, Roger Mear and others traveled from Hut Point to Cape Evans overland except for short sections on either side of Glacier Tongue. Having experienced the terrain, Mear states that one would need to be very experienced in glacier travel and properly equipped in order to make this journey, neither of which applied to the early expeditioners. Today, the sea ice route is used by the United States and New Zealand Antarctic programs.

p. 88. **"had a devil of a time"**: Irvine Gaze, diary, March 11, 1915, CM.

— CH. 6: EIGHTY DEGREES SOUTH —

p. 89. **"champion of the pack"** . . . **"fiery red eyes"**: Æneas Mackintosh, diary, February 8, 1915, RGS.

p. 90. **Their clothing never completely dried**: Wet clothing loses about 90 percent of its insulation value, and heat loss through conduction is five times as great.

p. 90. **"We call it 'our hour of discontent'"**: Æneas Mackintosh, diary, February 13, 1915, RGS.

p. 90. **"Wild slept like a top"**: Æneas Mackintosh, diary, February 13, 1915, RGS.

p. 90. **"Wild, ever jolly, cheerful"**: Æneas Mackintosh, diary, February 18, 1915, RGS.

p. 91. **"Joyce, a different character"**: Æneas Mackintosh, diary, February 18, 1915, RGS.

p. 91. **"somewhat bombastic"**: Richard Richards to Leslie Quartermain, May 2, 1975, CM.

p. 91. **"an all around athlete"**: Admiral Sir Lewis Baker, letter of reference for Ernest Joyce, RGS.

p. 91. **"But the Code of a Man"**: Robert Service, "The Quitter," quoted in Ernest Joyce, *The South Polar Trail*, p. 70.

p. 91. **"What on earth"**: Æneas Mackintosh, diary, February 15, 1915, RGS.

p. 92. **"What a weird situation"**: Æneas Mackintosh, diary, January 29, 1915, RGS.

p. 92. **"In waking, you build castles"**: Æneas Mackintosh, diary, February 22, 1915, RGS.

p. 92. **"Excitement reigned supreme"**: Ernest Wild, diary, February 20, 1915, ATL.

p. 92. **They dumped their final load**: John King Davis, relief diary, February 9, 1915, SLV and Æneas Mackintosh, plot of course to February 20, 1915, Rocky Mountain Depot, SPRI.

p. 92. **Planned to contain 220 pounds:** Mackintosh's sledging plan for the second season notes that the depot contained 105.492 pounds. They reclaimed some depot rations between February 20 and 24.

p. 92. **They had kept a sharp lookout:** One Ton Depot was reportedly located at latitude 79°28'53" S, longitude 169°22'04" E, although, since the Ross Ice Shelf is constantly moving, the position would have changed by 1915.

p. 92. **The following day:** At depot sites, Mackintosh used the surveyor's instrument to determine their location with the greatest accuracy. For navigating on the fly, Mackintosh noted bearings with a prismatic compass. To take bearings, he stood at the center of an imaginary 360° circle and pointed the needle north. Then, he spotted three prominent landmarks and aligned the compass to determine the angle between each and north in degrees. When the navigator plotted the bearings on a chart as angles, the intersection of the three lines triangulated the position. To find the same point later, the navigator simply sighted the landmarks with the same angle measurements. They left lists of bearings and other information at each depot, usually in a tin tied to a bamboo flag pole.

p. 94. **"Shackleton's party":** Ernest Joyce, log, February 21, 1915, RPK.

p. 94. **"To my sorrow":** Ernest Joyce, *The South Polar Trail*, p. 63.

p. 94. **biscuits were grossly inadequate:** The problem was apparent to Scott when he used Spratt's biscuits on his last expedition: "The dogs are thin as rakes; they are ravenous and very tired," he wrote. "The biscuit alone is not good enough." (Robert F. Scott, *Scott's Last Expedition*, Vol. 1, p. 185) Mackintosh, unaware of Scott's observation, suspected that the dogs actually needed double the daily ration, but his realization was academic. When his party spent more time in the field than anticipated, the supply of dog food was too low to increase the amount.

In winter conditions, a medium-sized (seventy-seven pound) Inuit sledge dog requires between 5,500 and 8,000 calories per day containing 50 to 80 percent fat for endurance work. (Geneviève Montcombroux, *The Canadian Inuit Dog*) American Will Steger and his companions crossed the Antarctic continent by dogsled in 1989–90 and found that his dogs thrived on 5,700 to 8,000 calories per day. Dogs in long-distance races like the Iditarod and Yukon Quest can expend 6,000 to 10,000 calories per day. (Michael Hand and Bruce Novotny, *Pocket Companion to Small Animal Clinical Nutrition*)

p. 95. **Eating only dry biscuits:** Even if the dogs had been given their fill of biscuits, their nutritional needs would not have been satisfied. Overcooking during processing likely also destroyed some essential vitamins and minerals in the raw ingredients. The ideal food was available in abundance: seal meat. As well as being nutritionally complete for dogs, metabolism of the seal fat provides ample water.

p. 95. **"My heart aches":** Ernest Joyce, log, February, 25–26, 1915, RPK.

p. 95. **"What a place":** Æneas Mackintosh, diary, February 25, 1915, RGS.

p. 95. **"Our trouble":** Æneas Mackintosh, diary, February 24, 1915, RGS.

p. 95. **"A rotten miserable time":** Æneas Mackintosh, diary, February 27, 1915, RGS.

p. 95. **"I could not refrain":** Ernest Joyce, log, February 23, 1915, RPK.

p. 96. **The Ross Sea party cut a path:** The cold, arid air of these gravity-fed, high velocity winds—called katabatics—collides with masses of cold, moist air over the Ross Ice Shelf and warmer, moist air from McMurdo Sound. These Antarctic polar depressions, similar to those in the northern hemisphere sometimes called Arctic hurricanes, can produce severe conditions. The Byrd Glacier, scarcely a hundred miles southwest of Minna Bluff, plays a prominent role in generating these mesoscale cyclones, which range between two and two hundred kilometers in diameter.

p. 96. **"It seems hard":** Ernest Joyce, log, February 27, 1915, RPK.

p. 96. **The next day:** The first sunset would have been February 26, but the sun was obscured.

p. 96. **"Pat stopped behind":** Ernest Wild, diary, February 28, 1915, ATL.

p. 96. **"too cold":** Ernest Wild, diary, March 1, 1915, ATL.

p. 96. **"no better reward":** Æneas Mackintosh, diary, March 1, 1915 RGS.

p. 97. **"We went off":** Æneas Mackintosh, diary, March 2, 1915 RGS.

p. 97. **"We shall have to call this":** Ernest Wild, diary, March 2, 1915, ATL.

p. 97. **"On Polar Journeys the dogs":** Ernest Joyce, log, March 6, 1915, RPK.

p. 98. **"Having been without a meal":** Æneas Mackintosh, diary, March 13, 1915, RGS.

p. 98. **full rations:** The diet was slightly more caloric than the *Nimrod* field rations, edging the daily allowance per man upward from thirty-two to thirty-five ounces, but still insufficient. The intensive work of sledging required about twice the modern USDA recommended intake for a typical urban adult male. A. F. Rogers, who with R. J. Sutherland studied the 1957–58 Trans-Antarctic crossing, estimated that man-hauling consumes 10 calories per minute. Dr. Mike Stroud, who monitored participants in two Antarctic man-hauling expeditions in the late 1980s and early 1990s, documented expenditures of 6,000 to 11,650 calories per day. The increased energy expenditure is not only linked to physical work. In one study, cold acclimatization led to an increase in energy requirements of as much as 40 percent (Reed and Lamar, 1997). Calculated with Mackintosh's typical routine of ten hours of sledging with a three-minute "spello" break each hour, the Ross Sea party's expenditure during sledging was on the order of 5,500 to 6,500 calories. On tent-bound days, the men likely burned 1,500 to 2,000 calories per day.

The diet was also inadequate in quality. Deficient in many vitamins and high in protein, the rations were metabolically more difficult to handle than the modern high-carbohydrate diet now recommended for endurance exercise. Protein is harder to digest and requires a higher water intake and more oxygen to metabolize and excrete. The high percentage of fat probably contributed to the lethargy the party reported and the low fiber content of the white flour biscuits promoted constipation and aggravated hemorrhoids.

p. 98. **They were dehydrated:** To conserve fuel, Mackintosh restricted Primus use to one half hour twice a day to melt snow and cook food. It is certain that they could not make enough water to meet their bodies' needs. They were limited to a mug of tea twice a day—containing caffeine, which further hastened water loss—and occasional sips from a shared water skin. Ironically, they were suffering from heat stress on the coldest continent on Earth. The evaporation of sweat normally accounts for about a third of the body's total water loss, but hard physical labor in a cold, dry climate can increase the amount by three-hundredfold (Mark Twight, *Extreme Alpinism*). Twight suggests that heavy exertion may require 600 to 1,200 milliliters per hour in cold, dry environments, especially at altitude. Less fluid meant a drop in blood volume, sharply decreasing the replenishment of oxygen and nutrients in the tissues and the flushing of waste products. Dehydration also increases the susceptibility to frostbite. Consuming snow lowers the core body temperature and accelerates hypothermia.

p. 99. **"a mass of ice"**: Æneas Mackintosh, diary, March 1, 1915, RGS.

p. 99. **"We are 3 old crocks"**: Ernest Joyce, log, March 21, 1915, RPK.

p. 99. **"I ask myself"**: Æneas Mackintosh, diary, March 18, 1915, RGS.

p. 99. **"woe begotten night"**: Ernest Joyce, log, March 25, 1915, RPK.

— CH. 7: HUT POINT —

p. 101. **"not had a rosy time"**: Æneas Mackintosh, diary, March 26, 1915, RGS.

p. 101. **He decided to make:** Joseph Stenhouse to Æneas Mackintosh, March 8, 1915. Stenhouse wrote three other letters to Mackintosh, dated February 3, February 5, and March 11. Mackintosh seems not to have found all of them, given the diary evidence.

p. 101. **"look after the ship"**: Æneas Mackintosh to Joseph Stenhouse, January 24, 1915, SPRI.

p. 102. **"be prepared for all"**: Æneas Mackintosh to Joseph Stenhouse, January 24, 1915, SPRI.

p. 102. **"I cannot describe"**: Victor Hayward, diary, March 24, 1915, NMM.

p. 102. **"raw like steak"**: Ernest Joyce, *The South Polar Trail*, p. 73.

p. 102. **"indescribable"**: Victor Hayward, diary, March 12, 1915, NMM.

p. 102. **"We are still alive"**: Victor Hayward, diary, March 25, 1915, NMM.

p. 102. **"after the great gruelling"**: Ernest Joyce, *The South Polar Trail*, p. 73.

p. 103. **"It really is murder"**: Æneas Mackintosh, diary, April 15, 1915, RGS.

p. 103. **"Believe the taste"**: Keith Jack, diary, March 30, 1915, SLV.

p. 103. **"It was a couple of weeks"**: Ernest Joyce, log, undated entry, March 1915, RPK.

p. 104. **"If the T. C. [transcontinental] Party"**: Ernest Shackleton to Æneas Mackintosh, September 18, 1914, SPRI.

p. 104. **"I have been thinking"**: Victor Hayward, diary, February 27, 1915, NMM.

p. 105. **"very heated arguments"**: Ernest Joyce, log, undated entry, March 1915, RPK.

p. 105. **"Stenhouse knowing"**: Ernest Joyce, log, undated entry, March 1915, RPK.

p. 105. **"Mack is in the wrong"**: Ernest Joyce, log, undated entry, March 1915, RPK.

p. 105. **"shabby treatment"**: Keith Jack, diary, February 4, 1915, SLV.

p. 105. **"Many of the skipper's"**: Keith Jack, diary, February 25, 1915, SLV.

p. 106. **"[Cope] thinks he is"**: Æneas Mackintosh, diary, January 13, 1915, SPRI.

p. 106. **"he had no confidence"**: Alexander Stevens, "Report of the Ross Sea Party," p. 23, SPRI.

p. 106. **"an excellent man"**: Ernest Shackleton to Æneas Mackintosh, September 18, 1914, SPRI.

p. 106. **"one cannot help asking"**: Keith Jack, diary, February 11, 1915, SLV.

p. 106. **"Arguments are rife"**: Æneas Mackintosh, diary, April 4, 1915, RGS.

p. 106. **"evil smelling"**: Keith Jack, diary, May 7, 1915, SLV.

p. 106. **"We are rather up"**: Victor Hayward, diary, April 8, 1915, NMM.

p. 107. **"Here we now are"**: Æneas Mackintosh, diary, March 27, 1915, RGS.

p. 107. **"I am thinking"**: Æneas Mackintosh, diary, April 13, 1915, RGS.

p. 107. the **"queer places"**: Æneas Mackintosh, diary, March 7, 1915, RGS.

p. 107. **"It's too terrible"**: Æneas Mackintosh, diary, March 31, 1915, RGS.

p. 107. **"What a crowd"**: Æneas Mackintosh, diary, April 3, 1915, RGS.

p. 107. **"Little chance of getting"**: Æneas Mackintosh, diary, March 31, 1915, RGS.

p. 107–8. **"But oh! What one pays"**: Æneas Mackintosh, *Nimrod* diary, January 11, 1909, RGS.

p. 108. **"Mackintosh was always the man"**: John King Davis, *High Latitude*, p. 95.

p. 108. **"Late last evening"**: Victor Hayward, diary, March 30, 1915, NMM.

p. 108. **Orcas routinely cruised**: Previous expeditions had experienced frightening brushes with the mammals. In 1909, a *Nimrod* science party was camped on an ice floe. Throughout the night, orcas circled, occasionally bumping the floe from beneath to upset the potential prey into the water. Scientists later established that killer whales do not seek humans as prey and likely mistook the men for seals.

p. 108. **"It seems as though"**: Ernest Joyce, *The South Polar Trail*, p. 76.

p. 108. **"The people who have"**: Ernest Wild, diary, "About May 27," 1915, ATL.

p. 108. **"as though"**: Ernest Joyce, *The South Polar Trail*, p. 76.

p. 108. **"Quite the nicest novel"**: Æneas Mackintosh, diary, February 13, 1915, RGS.

p. 109. **"did not forget"**: Apsley Cherry-Garrard, *The Worst Journey in the World*, p. 302.

p. 109. **"I am very much"**: Robert F. Scott, *Scott's Last Expedition*, pp. 213–14.

p. 109. **"great trouble"**: Cecil Meares interview with Caroline Oates, as quoted in Roland Huntford, *Shackleton*, p. 361.

p. 109. **"cheerful readiness"**: Ernest Shackleton, *The Heart of the Antarctic*, p. 98.

p. 109. **"colonial morals"**: Arnold Spencer-Smith, diary, March 7, 1915, SPRI.

p. 109. **"Cope interjected"**: Keith Jack, diary, May 6, 1915, SLV.

p. 110. **"words warfare"**: Keith Jack, diary, May 6, 1915, SLV.

p. 110. **"The conditions generally here"**: Keith Jack, diary, May 7, 1915, SLV.

p. 110. **"Everybody has had a go"**: Ernest Wild, diary, "About May 27," ATL and SPRI.

p. 110. **"I can't understand"**: Ernest Wild, diary, "About May 27," ATL.

p. 110. **"[Mackintosh] has got some"**: Ernest Wild, diary, "About May 27," ATL.

p. 110. **"I maintain it would"**: Keith Jack, diary, May 25, 1915, SLV.

p. 111. **"Mackintosh probably thinks me"**: Keith Jack, diary, May 25, 1915, SLV.

p. 111. **"like Anarchists"**: Arnold Spencer-Smith, diary, June 2, 1915, CM.

p. 111. **"The hut to us"**: Ernest Joyce, log, June 2, 1915, RPK.

— CH. 8: "AN IDEAL PLACE IN A BLIZZARD" —

p. 112. **"Your first and paramount duty"**: Æneas Mackintosh to Joseph Stenhouse, January 24, 1915, SPRI.

p. 112. **"He is an excellent fellow"**: Æneas Mackintosh, diary, January 13, 1915, RGS.

p. 112. **"the strenuous life"**: Theodore Roosevelt, "The Strenuous Life," speech to the Hamilton Club, Chicago, April 10, 1899.

p. 114. *pukka sahib*: Clarence Mauger to Leslie Quartermain, undated, CM.

p. 114. **"a good man & skipper"**: Clarence Mauger to Leslie Quartermain, July 10, 1963, CM.

p. 114. **"You must not winter"**: Ernest Shackleton to Æneas Mackintosh, September 18, 1914, SPRI.

p. 115. **Exposed features**: Modern meteorologists have studied the weather patterns in the Ross Island region extensively. These phenomena had not been described in Shackleton's day, although the troubles of mariners in McMurdo Sound were well documented.

p. 115. **"unless the most abnormal"**: Ernest Shackleton to Æneas Mackintosh, September 18, 1914, SPRI.

p. 116. **"We've named the Tongue"**: Alfred Larkman, diary, February 14, 1915, SPRI.

p. 116. **"keep the ship from being"**: Ernest Shackleton to Æneas Mackintosh, September 18, 1914, SPRI.

p. 116. **"Well Bo'sun"**: James Paton, diary, February 15, 1915, HC.

p. 117. **"Still, Mr. Stenhouse"**: James Paton, diary, February 15, 1915, HC.

p. 117. **"dodging trouble"**: Leslie Thomson, diary, February 19, 1915, SPRI.

p. 117. **"playing old Harry"**: James Paton, diary, January 21, 1915, HC.

p. 117. **"drop both bower anchors"**: Leslie Thomson, diary, January 30, 1915, SPRI.

p. 118. **"Beggars can't be choosers"**: Leslie Thomson, diary, February 7, 1915, SPRI.

p. 118. **"every prospect"**: Leslie Thomson, diary, February 21, 1915, SPRI.

p. 118. **"Things looked"**: Leslie Thomson, diary, February 21, 1915, SPRI.

p. 118. **"It certainly looks"**: James Paton, diary, February 23, 1915, HC.

p. 118. **"her ribs would crack"**: James Paton, diary, February 24, 1915, HC.

p. 118. **"she rose and fell"**: Leslie Thomson, diary, February 24, 1915, SPRI.

p. 119. **"Luck which seemed"**: Leslie Thomson, diary, February 24, 1915, SPRI.

p. 119. **"It is easy to see now"**: James Paton, diary, February 24, 1915, HC.

p. 119. **"This is the place"**: Leslie Thomson, diary, February 25, 1915, SPRI.

p. 120. **"Very satisfied"**: Leslie Thomson, diary, March 8, 1915, SPRI.

p. 120. **Paton also knew**: These observations were noted on Scott's *Terra Nova* chart. Though it was not published until 1923, members of Shackleton's expedition could have learned about these conditions from Paton and others who sailed aboard the *Terra Nova*.

p. 121. **"She's away!"**: Joseph Stenhouse, diary, March 10, 1915, SPRI.

p. 121 **"nearly upon her beam ends"**: Leslie Thomson, diary, March 10, 1915, SPRI.

p. 122. **The *Aurora* was ranging**: A few hours before, Stenhouse had paid out more anchor cable to secure the ship against the rising seas. At 70 fathoms, the cable stopped, jammed by kinks inside the chain locker. The rule of thumb was five to seven times the water's depth of 25 fathoms, which meant at least 125 fathoms.

p. 122. **"threatened to tear the masts"**: Leslie Thomson, diary, March 12, 1915, SPRI.

p. 122. **"absent shipmates on the Barrier"**: Lionel Hooke, diary, March 13, 1915, ML.

p. 122. **Then, he superseded**: Joseph Stenhouse to Æneas Mackintosh, February 3 and February 11, 1915, SPRI. The first letter suggested that returning parties light a bonfire to signal their arrival at the hut. The second letter stated that Stenhouse would fire rockets at Cape Evans when it appeared that the ice near the cape was safe for crossing. The party that embarked on March 11 had fired a rocket, but it was not seen on the ship. Mackintosh and the members of the Hut Point party do not mention lighting a bonfire or firing rockets in their diaries. Remarkably, Scott's men laid cable and wired telephone service between the huts, but the cable had long since torn away with the seasonal ice. Hooke had rigged telephone wires between the hut and the *Aurora*.

p. 123. **"there were enough"**: *Evening Star* (Dunedin), April 3, 1916.

p. 123. **"not justified"**: Alfred Larkman, diary, March 20, 1915, SPRI.

p. 123. **"My watch on deck"**: Joseph Stenhouse, diary, January 7, 1915, SPRI.

p. 124. **"There is no Sunday"**: James Paton, diary, March 7, 1915, HC.

p. 124. **"with scores"**: Lionel Hooke, diary, January 8, 1915, ML.

p. 124. **"'pore ignorant seamen'"**: Joseph Stenhouse, diary, February 6, 1916, SPRI.

p. 124. **"This strain continually"**: Leslie Thomson, diary, February 18, 1915, SPRI.

p. 124. **"he could not do himself justice"**: Leslie Thomson, diary, January 13, 1915, SPRI.

p. 124. **"From such as these"**: Joseph Stenhouse, diary, February 10, 1916, SPRI.

p. 125. **"rather a nuisance"**: Æneas Mackintosh, diary, January 13, 1915, SPRI.

p. 125. **"ignorant & insolent"**: Alfred Larkman, diary, March 3, 1915, SPRI.

p. 125. **"like coolies"**: Clarence Mauger to Leslie Quartermain, July 10, 1963, CM.

p. 125. **"the ungodly noise"**: Leslie Thomson, diary, March 24, 1915, SPRI.

p. 125. **"respect and almost fear"**: Leslie Thomson, diary, March 25, 1915, SPRI.

p. 125. **"I am afraid that no moorings"**: James Paton, diary, March 31, 1915, HC.

p. 126. **Spells of violent tremors:** The men speculated that the tremors were earthquakes related to the eruptions of Erebus. They were correct, according to volcanologists Philip Kyle and Rick Aster, who have studied Mount Erebus for decades. They reviewed the diary entries and judged that the observed effects were caused by a prolonged swarm of volcanic earthquakes during an exceptionally active period of eruptions for Erebus, although it was not a tectonic earthquake. Infrasonic shaking may have contributed as well. Kyle compared the level of activity to a three-month episode beginning in September 1984, which he documented as the most extensive in the last thirty years. The volcano erupted large volcanic "bombs" a mile into the sky, some over thirty feet across.

p. 126. **"Wintering a ship here"**: James Paton, diary, April 8, 1915, HC.

p. 126. **"seriously believed some misfortune"**: Alexander Stevens, "Report of the Ross Sea Party," p. 25, SPRI.

p. 126. **"taut as any fiddlestring"**: James Paton, diary, April 18, 1915, HC.

p. 126. **"All the cables are now singing"**: Howard Ninnis, diary, April 10, 1915, RGS.

p. 127. **"commence their song"**: Lionel Hooke, diary, May 6, 1915, ML.

p. 127. **"She's away wi' it!"**: Joseph Stenhouse, diary, May 6, 1915, SPRI.

p. 127. **As Thomson arrived on deck:** *Evening Star* (Dunedin), April 3, 1916.

p. 127. **"Fast in the pack and drifting"**: Joseph Stenhouse, diary, May 9, 1915, SPRI.

— CH. 9: MAROONED —

p. 128. **"There was no sign"**: Richard Richards, *The Ross Sea Shore Party 1914–17*, p. 15.

p. 128. **"all snapped like cotton"**: Victor Hayward, diary, June 2, 1915, NMM.

p. 128. **"she had been lost"**: Richard Richards, *The Ross Sea Shore Party 1914–17*, p. 15.

p. 129. **"a knock out blow"**: Æneas Mackintosh, diary, June 20, 1915, SPRI.

p. 129. **"We must put the best face"**: Æneas Mackintosh, diary, June 8, 1915, SPRI.

p. 129. **"If one goes to a place"** . . . **"I pray & hope"**: Æneas Mackintosh, diary, June 20, 1915, SPRI.

p. 129. **"I can imagine"**: Louis Bernacchi, quoting Alfred Lord Tennyson, "Mariana in the South," in *To the South Polar Regions: Expedition of 1898–1900*, p. 135.

p. 130. **"We are ten men"**: Victor Hayward, diary, June 1, 1915, NMM.

p. 130. **"only text-book on the Antarctic"**: John King Davis, *High Latitude*, p. 277.

p. 131. **"excellent both in bulk and taste"**: Æneas Mackintosh, diary, August 18, 1915, SPRI.

p. 131. **"somewhat less fresh"**: Richard Richards, *The Ross Sea Shore Party 1914–17*, p. 17. Jack detected the rank flavor as well. Mackintosh had been unable to obtain pemmican in Australia. Instead, the Ross Sea party used surplus from Scott's last expedition and Mawson's leftovers aboard the *Aurora*. Scott's pemmican, made in 1910 or earlier, had been stored in the Antarctic for four years, freezing and thawing repeatedly. Mawson's pemmican was shipped to the Antarctic in 1911, then returned to Hobart where it stewed in the hold of the *Aurora* for a year through the heat of a summer before the Ross Sea party sailed south.

p. 131. **"grisly and greasy job"**: Arnold Spencer-Smith, diary, January 30, 1915, SPRI.

p. 131. **"So here we are"**: Ernest Wild, diary, June 2, 1915, ATL.

p. 132. **"hard conscientious worker"**: Æneas Mackintosh, diary, June 5, 1915, SPRI.

p. 132. **"doing twice as much"**: Æneas Mackintosh, diary, June 16, 1915, SPRI.

p. 132. **"splendid asset"**: Æneas Mackintosh, diary, July 16, 1915, SPRI.

p. 132. **"All is working smoothly"**: Æneas Mackintosh, diary, June 5, 1915, SPRI.

p. 132. **"a new man"**: Æneas Mackintosh, diary, June 13, 1915, SPRI.

p. 132. **"We are not going to commence"**: Æneas Mackintosh, diary, June 6, 1915, SPRI.

p. 132. **"the local millionaire"**: Æneas Mackintosh, diary, June 22, 1915, SPRI.

p. 132. **"I could not have wished"**: Æneas Mackintosh, diary, June 22, 1915, SPRI.

p. 133. **"the sledging problem"**: Arnold Spencer-Smith, diary, June 25, 1915, CM.

p. 133. **"Working out sledging programme"**: Æneas Mackintosh, diary, June 21, 1915, SPRI.

p. 132. **Only two dogs had survived:** Two dogs put ashore at Cape Evans had perished.

p. 132. **"none of these will be any good"**: Æneas Mackintosh, diary, June 14, 1915, SPRI.

p. 132. **Shackleton had based his plans:** Ernest Shackleton, ITAE diary, January 1, 1915, SPRI.

p. 132. **"Over March is too late"**: Æneas Mackintosh, diary, March 26, 1915, RGS.

p. 132. "Such setbacks—& surprises": Æneas Mackintosh, diary, June 21, 1915, SPRI.

p. 134. "The leadership of men": John King Davis, *High Latitude*, p. 9.

p. 134. "plucky as they come": Richard Richards to Leslie Quartermain, August 4, 1969, CM.

p. 135. "Under the circumstances": Æneas Mackintosh, diary, April 1, 1915, RGS.

p. 135. "Everyone doing as they pleased": Æneas Mackintosh, diary, June 6, 1915, SPRI.

p. 135. "The social problem": Æneas Mackintosh, diary, June 6, 1915, SPRI.

p. 135. "should have been in a better position": Æneas Mackintosh to Ernest Shackleton, 31/ix/15 [sic], SPRI.

p. 135. "I miss the services of an officer": Æneas Mackintosh, diary, June 8, 1915, SPRI.

p. 136. "act as his second": Alexander Stevens, "Report of the Ross Sea Party," p. 29, SPRI.

p. 136. "The loneliness and austerity": John King Davis, *High Latitude*, p. 11.

p. 136. "I miss having another officer": Æneas Mackintosh, diary, June 16, 1915, SPRI.

p. 136. He admitted: Victor Hayward, diary, June 26, 1915, NMM.

p. 137. "mutually & unanimously": Æneas Mackintosh, diary, June 26, 1915, SPRI.

p. 137. "All were agreed": Richard Richards, *The Ross Sea Shore Party 1914–17*, p. 17.

p. 137. "utterly dependent": Richard Richards, *The Ross Sea Shore Party 1914–17*, p. 17.

p. 137. "The job will D.V.": Arnold Spencer-Smith, diary, June 25, 1915, CM. D.V. abbreviates the Latin *Deo volente*.

p. 137. "do their damnedest" . . . "stupendous undertaking": Victor Hayward, diary, June 26, 1915, SPRI.

p. 137. "Wild has made": Æneas Mackintosh, diary, September 28, 1915. SPRI.

p. 138. "One can forgive and forget": Ernest Joyce, *The South Polar Trail*, p. 80.

p. 138. "but was speedily requested to cease": Ernest Joyce, *The South Polar Trail*, p. 83.

p. 138. sennegrass: A kind of Scandinavian hay that polar explorers stuffed in boots to absorb sweat.

p. 138. As the aurora australis streamed: Scientists of the time linked auroral displays to the earth's magnetic field without fully understanding the phenomenon.

p. 139. "general factotum in unpleasant work": Alexander Stevens to Hugh Robert Mill, November 28, 1928, SPRI.

p. 139. "hoodwinked": Æneas Mackintosh, diary, July 2, 1915, SPRI.

p. 139. "all" . . . "under false pretences": Æneas Mackintosh, diary, June 18, 1915, SPRI.

p. 139. "Perhaps he's right": Æneas Mackintosh, diary, June 18, 1915, SPRI.

p. 139. "I cannot fathom him": Keith Jack, diary, June 19, 1915.

p. 139. "Down here": Victor Hayward, diary, February 15, 1915, NMM.

p. 140. "House Cricklewood": Victor Hayward, diary, undated entry, p. 78, NMM.

p. 140. "Spencer-Smith is the finest person": Joseph Stenhouse, diary, October 1, 1914, SPRI.

p. 140. "A cheerful willing soul": Æneas Mackintosh, diary, June 5, 1915, SPRI.

p. 141. **"For peace's sake":** Æneas Mackintosh, diary, August 2, 1915, SPRI.

p. 142. **"a very stupid thing":** Alexander Stevens to Leslie Quartermain, February 2, 1962, CM.

p. 142. **He found one:** In the dry Antarctic environment, fires burn much faster and hotter than in a temperate climate. Nitrocellulose, the basis of dynamite as well as early motion picture film, burns very hot with an intense flame and toxic smoke.

p. 142. **"It might":** Æneas Mackintosh, diary, August 12, 1915, SPRI.

p. 142. **"I did not go out":** Æneas Mackintosh, diary, September 5, 1915, SPRI.

p. 142. **"We saw the Sun":** Ernest Wild, diary, p. 17, ATL.

p. 142. **"The same old sun":** Æneas Mackintosh, diary, August 26, 1915, SPRI.

p. 143. **"intermittent heart":** Medical historian Dr. Joel Howell of the University of Michigan notes that in the parlance of Edwardian physicians, "intermittent heart" indicated occasional skipped beats. At a time when medical debate centered on how much the clinician could learn from instrumental methods of examination, the pulse was thought to reveal much about the patient's health. After 1915, a diagnosis of intermittent heart was considered to be much less meaningful as this approach lost favor. The doctor may have simply detected that Spencer-Smith's heart skipped a beat now and then, which in modern practice is understood as entirely normal. Though it is impossible to be definitive in the absence of more exhaustive data, Dr. Howell considers the probability slim that Spencer-Smith had a serious heart condition.

p. 143. **"at the earliest"** . . . **"the least physically fit":** John Lachlan Cope, "Medical Report of the Ross Sea Base ITAE," January 1917, SPRI.

p. 143. **Tall and rake-thin:** Cope's testing detected traces of sugar in Stevens's urine, which he discounted as not medically significant. However, Cope was not yet a physician. At my request, Dr. John W. Hare of the Joslin Diabetes Center, Boston, reviewed the archival references to Stevens's condition, which included urinary sugar, extreme fatigue, and propensity to abscesses. Described as extremely thin, weak, and debilitated, archival notes and photographs of Stevens suggest he lost weight and energy because of symptomatic diabetes. Dr. Hare believes that Stevens may have had what would now be called insulin-dependent diabetes, although a blood test would have been essential to confirm the diagnosis. Although available at the time, it was cumbersome and infrequently used. In Stevens's case, Dr. Hare considers his diabetes to have been a milder form than the full-blown condition that was then so often fatal for younger patients. The survival rate improved after the availability of insulin in 1922, so Stevens may have been diagnosed and treated after his return to England.

p. 143. **"resistance in liver area":** As Marshall reported in his diary, he suspected Joyce had "myocarditis in a very early stage." Eric Marshall, diary, July 12, 1908, RGS.

p. 143. **"the best is not made":** Æneas Mackintosh, diary, July 29, 1915, SPRI.

p. 144. **"If this letter reaches you":** Æneas Mackintosh to Joseph Stenhouse, August 29, 1915, SPRI.

p. 144. **"head straight out":** Æneas Mackintosh to Joseph Stenhouse, August 29, 1915, SPRI.

— CH. 10: RETURN TO THE BARRIER

p. 145. **"Not a very promising":** Æneas Mackintosh, diary, September 8, 1915, SPRI.

p. 145. **"froze like boards":** Ernest Joyce, *The South Polar Trail*, p. 86.

p. 145. **His plan was:** The diaries and Mackintosh's sledging plan seem to indicate that he was undecided whether to leave depots at both 83° S and Mount Hope. The plan shows a party returning directly from 83° and a dashed line to Mount Hope.

p. 146. **He favored heavier loads:** With the equipment available to the Ross Sea party, the loads were unmanageable. Modern expeditions have pulled much heavier overall loads. For one example, Robert Swan, Gareth Wood, and Roger Mear of the 1985–86 Footsteps of Scott Expedition started with 356 pounds per man. According to Mear, modern techniques and equipment make heavier loads possible. It is far more effective for each man to pull his own load rather than to synchronize pulling. Sleds made from fiberglass or Kevlar, modeled on Laplanders' pulks, are much lighter than the wooden sledges used by the Ross Sea party. Pulk runners are made from low-friction materials like high-density nylon and can be coated with Teflon. Unlike the harnesses of 1914, which concentrated the pulling force at the waist and put greater demands on the legs, newer harnesses shifted the burden to the upper body and allow sledgers to use their body weight to greater advantage.

p. 147. **"No journey ever made":** Robert F. Scott, *The Voyage of the "Discovery,"* Vol. 1, pp. 467–8.

p. 148. **"an unusual element of risk":** Arnold Spencer-Smith to parents, September 30, 1915.

p. 148. **"At present"** . . . **"I am sorry":** Æneas Mackintosh to Ernest Shackleton, 31/ix/15 [sic], SPRI.

p. 149. **Once again, he increased:** According to Mackintosh, the basic gear had been pared down from the 369 pounds carried on each sledge in the first season, before the supplies for the Ross Sea party and the depots were added. (The figure of 182 pounds cited in Joyce's book is incorrect.) For one week, food for three men weighed forty-six pounds; kerosene and denatured alcohol, used to prime the stove, added another seven pounds.

p. 149. **"I don't think":** Ernest Joyce, log, October 10, 1915, RPK.

p. 149. **"Devilish hard pulling":** Keith Jack, diary, October 10, 1915, SLV.

p. 149. **"impossible to carry out":** Victor Hayward, diary, October 10, 1915, NMM.

p. 150. **"We pulled the heaviest sledge":** Ernest Wild, diary, October 11, 1915, SPRI.

p. 150. **"physical farce"** . . . **"to tell the Skipper off":** Ernest Joyce, log, October 11, 1915, RPK.

p. 150. **"We've left the others behind":** Ernest Wild, diary, October 12, 1915, SPRI.

p. 151. **"Skipper has little idea":** Keith Jack, diary, October 9 & 11, 1915, SLV.

p. 151. **Joyce also informed:** None of the party had more than a rudimentary understanding of the forces that created the tormented landscape before them. If glaciology was a science in its adolescence, then the study of the glacial dynamics of Antarctica was in its infancy. Glacier ice behaves at once as a liquid and a solid, flowing at different rates around unyielding rock, like White Island, and shearing. The ice covering the continent is less like a frozen blanket than a dynamic river system, flowing in great streams from the highest altitude of the interior of the continent to the sea. The sledging route crossed the confluence of two such turbulent flows, the McMurdo Ice Shelf and the Ross Ice Shelf.

p. 151. **"not the heart-breaking strain":** Ernest Joyce, *The South Polar Trail,* October 13, 1915, p. 93.

p. 151. **"We were naming them"**: Ernest Joyce, *The South Polar Trail*, p. 95.

p. 152. **"My dear Paton"**: Ernest Joyce to Paton and Bhoys, October 22, 1915, ATL.

p. 153. **"Dear Sir, We leave"**: Ernest Joyce, log, October 26, 1915, RPK. The archival sources conflict on the date of the letter. Joyce states it was dated March 19, Jack records March 13, and a conservator who saw the letter in 1960 noted the date as March 16. None of these dates fit Cherry-Garrard's account of his movements, possibly indicating that the date was only partially legible. However, the description of the depot's location and contents, including motor oil, indicate that the note was likely left at Scott's Petrol Depot, where Cherry-Garrard stopped to pick up dog biscuits for his own journey.

p. 153. **"Rather pathetic"**: Ernest Joyce, log, October 26, 1915, RPK.

p. 154. **"He's down, he's down"**: Keith Jack, diary, October 27, 1915, SLV.

p. 154. **"*Private* I hope you"**: Æneas Mackintosh to Alexander Stevens, October 26, 1915, SPRI.

p. 155. **"invaluable in his advice"**: Æneas Mackintosh to Alexander Stevens, October 26, 1915, SPRI.

p. 155. **"shall have nothing"**: Æneas Mackintosh to Alexander Stevens, October 26, 1915, SPRI.

p. 155. **"has not been a success"**: Æneas Mackintosh to Joseph Stenhouse, October 28, 1915, SPRI.

p. 155. **"have nothing to do"**: Æneas Mackintosh to Joseph Stenhouse, October 28, 1915, SPRI.

p. 155. **"familiar more than you"**: Æneas Mackintosh to Joseph Stenhouse, October 28, 1915, SPRI.

p. 155. **"Seems to be something"**: Irvine Gaze, diary, October 1–26, 1915, CM.

p. 155. **"Trusting this will be plain"**: Æneas Mackintosh to Ernest Joyce, October 28, 1915, SPRI.

p. 155. **"I can't express words"**: Ernest Joyce, log, November 1, 1915, RPK.

p. 156. **"quite hopeless for this task"**: Æneas Mackintosh, diary, February 8, 1915, RGS.

p. 156. **"an unlovely specimen"**: Richard Richards, "Four Dogs" (unpublished manuscript), SPRI.

p. 156. **The party hauled**: In the first season, the outbound journey to the Bluff had taken fourteen days.

p. 156. **"We were going so fast"**: Ernest Joyce, log, November 20, 1915, RPK.

p. 156. **"ripping pace"**: Ernest Joyce, log, November 15, 1915, RPK.

p. 156. **"I for one don't see"**: Irvine Gaze, diary, November 8, 1915, CM.

p. 156. **"white ocean"**: Ernest Joyce, *The South Polar Trail*, p. 110.

p. 156–57. **no spare goggles**: Tinted green or yellow with leather blinders, the glass lenses could not block ultraviolet rays but at least shielded the eyes from the full intensity of the reflected radiation.

p. 157. **"as big as potatoes"**: Ernest Joyce, log, November 19, 1915, RPK.

p. 157. **"Very sick of life"**: Arnold Spencer-Smith, diary, October 12, 1915, SPRI.

p. 157. **"Puzzled and a little anxious"**: Arnold Spencer-Smith, diary, November 24, 1915, SPRI.

p. 158. **"We now must push on"**: Æneas Mackintosh to Ernest Joyce, November 25, 1915, SPRI.

p. 158. **"Letters from Skipper"**: Ernest Joyce, log, November 25, 1915, RPK.

p. 158. **"the mainstay of the work"**: Ernest Joyce, log, November 17, 1915, RPK.

p. 158. **"impetuous and not very tactful"**: Richard Richards to Margery Fisher, March 28, 1957, SPRI.

p. 158. **"the absurdity of his scheme"**: Irvine Gaze, diary, November 25, 1915, CM.

p. 158. **"He'll have to be pretty careful"**: Irvine Gaze, diary, November 25, 1915, CM.

p. 159. **"Richards & I gave"**: Ernest Joyce, log, November 28, 1915, RPK.

p. 159. **"off his head"**: Ernest Joyce, log, November 29, 1915, RPK.

p. 159. **"silly lies"**: Ernest Joyce, log, November 29, 1915, RPK.

p. 159. **"I never in my experience"**: Ernest Joyce, log, November 28, 1915, RPK.

p. 160. **He saw to it:** In this case, the killer was not scurvy. Unlike primates and guinea pigs, dogs synthesize vitamin C and thus are not reliant on dietary sources.

p. 160. **"I was very pleased"**: Æneas Mackintosh to Ernest Joyce, December 4, 1915, SPRI.

p. 160. **"damned heavy"**: Irvine Gaze, diary, December 28, 1915, CM.

p. 161. **"too awful for words"**: Irvine Gaze, diary, December 28, 1915, CM.

p. 161. **"it would be a fine thing"**: Ernest Joyce, log, December 31, 1915, RPK.

p. 161. **"slight hope"**: Richard Richards, diary, summary of December 1915, CM.

p. 161. **"Made all arrangements"**: Ernest Joyce, log, December 31, 1915, RPK.

p. 162. **"We have severed"**: Keith Jack, diary, December 31, 1915, SLV.

p. 162. **"What is everyone doing"**: Keith Jack, diary, December 26, 1915, SLV.

p. 162. **"I often think of the brother"**: Ernest Joyce, log, December 31, 1915, RPK.

p. 162. **"I sincerely hope"**: Ernest Joyce, log, January 2, 1916, RPK.

— CH. 11: MOUNT HOPE —

p. 163. **"Our primus has turned dog"**: Irvine Gaze, diary, January 6, 1916, CM.

p. 163. **"getting too big for his shoes"**: Ernest Joyce, log, December 14, 1915, RPK.

p. 163. **"high and mighty attitude"**: Irvine Gaze, diary, December 14, 1915, CM.

p. 164. **"There was nothing"**: Alexander Stevens to Hugh Mill, November 20, 1928, SPRI.

p. 164. **"I am very pleased"**: Ernest Joyce, log, January 5, 1916, RPK.

p. 164. **"He's a rotter"**: Irvine Gaze, diary, January 6, 1916, CM.

p. 164. **"Out to play DIRT "**: Irvine Gaze, diary, January 6, 1916, CM.

p. 165. **"They gave us news"**: Arnold Spencer-Smith, diary, January 6, 1916, SPRI.

p. 165. **"It's all in the game"**: Arnold Spencer-Smith, diary, flyleaf, SPRI.

p. 165. **"Timeo Danaos"**: Arnold Spencer-Smith, diary, January 6, 1916, SPRI.

p. 165. **"Quibus autem cognitis"**: Arnold Spencer-Smith, diary, January 6, 1916, SPRI.

p. 165. **"It is wonderful the amount"**: Ernest Joyce, log, January 6, 1916, RPK.

p. 165. **"I think I shall have to disobey"**: Ernest Joyce, log, January 8, 1916, RPK.

p. 166. **"Skipper asked me to take over"**: Ernest Joyce, log, January 9, 1916, RPK.

p. 166. **"Parties working harmonious"**: Ernest Joyce, log, January 9, 1916, RPK.

p. 166. **"The Barrier is"**: Ernest Shackleton, *The Heart of the Antarctic*, p. 283.

p. 167. **"the silence was profound"**: Richard Richards, oral history interview, December 2, 1976, NLA.

p. 167. **"endless hours"**: Richard Richards, oral history interview, December 2, 1976, NLA.

p. 167. **"an automatic reaction"**: Richard Richards, oral history interview, December 2, 1976, NLA.

p. 167. **"Skipper & Smith very crocked"**: Victor Hayward, diary, January 13–18, 1916, NMM.

p. 167. **Increasing doses of the cocaine:** Though this may have relieved the pain, the cocaine dilated the pupils and exposed his retinas to more ultraviolet radiation. Thus anesthetized, his eyes were also subject to mechanical injury with Joyce unable to feel the severity of the damage. A similar phenomenon in habitual cocaine abusers, called "crack keratitis," has been noted in modern medical literature.

p. 167. **"Skipper not going it"**: Ernest Joyce, log (transcript), January 15, 1916, ATL.

p. 167. **"2 members of our force"**: Ernest Joyce, log (transcript), January 17, 1916, ATL.

p. 168. **"it would have meant suicide"**: Richard Richards, diary, undated entry, p. 102. CM.

p. 168. **"a malleable character"**: Richard Richards to Leslie Quartermain, August 10, 1966, CM.

p. 168. **"moving spirit"**: Richard Richards to A. G. E. Jones, May 2, 1975, CM.

p. 168. **"I R. W. Richards"**: Æneas Mackintosh, expedition agreement for Richard Richards, January 18, 1916, CM. In fact, Richards did sign the official expedition agreement; this document was later found at the Cape Evans hut (SPRI). An explanation may be that Richards found the unsigned document at the hut later and signed it to affirm the letter signed in the field.

p. 169. **"'Well this bugger Richards'"**: Richard Richards, oral history interview, December 2, 1976, NLA.

p. 169. **"Difficult to pick out Mt. Hope"**: Ernest Joyce, log (transcript), January 19, 1916, ATL.

p. 169. **"The whole country seems"**: Ernest Shackleton, *The Heart of the Antarctic*, p. 295.

p. 170. **"The glacier which we now saw"**: Frank Wild, *Nimrod* diary, December 3, 1908, SPRI.

p. 170. **"Nearly fainted at 11 a.m."**: Arnold Spencer-Smith, diary, January 21, 1916, SPRI.

p. 171. **"A weak character"**: Richard Richards, oral history interview, December 2, 1976, NLA.

p. 171. **"short of restraining him"**: Richard Richards, oral history interview, December 2, 1976, NLA.

p. 171. **"I should be all right," "they rattled off"**: Arnold Spencer-Smith, diary, January 22, 1916, SPRI.

p. 171. **"Nothing doing. Can't see"**: Ernest Wild, diary, January 24, 1916, SPRI.

p. 172. **"A fearful mess"**: Frank Wild, *Nimrod* diary, December 3, 1908, SPRI.

p. 172. **"were going straight"**: Frank Wild, *Nimrod* diary, December 5, 1908, SPRI.

p. 172. **"All around us was such a scene"**: Ernest Joyce, log (transcript), January 26, 1916, ATL.

p. 172. **"We seemed to be"**: Ernest Joyce, *The South Polar Trail*, pp. 136–37.

p. 173. **"a long slope"**: Ernest Shackleton, *Nimrod* diary, December 4, 1916, SPRI.

p. 173. **"climbed the glacier"**: Ernest Joyce, log, January 26, 1916, RPK.

p. 173. **"open road to the South"**: Ernest Shackleton, *The Heart of the Antarctic*, p. 303.

p. 173. **"A most wonderful sight"**: Ernest Joyce, log, January 26, 1916, RPK.

p. 175. **"a place which could not be missed"**: Ernest Joyce, log (transcript), January 26, 1916, ATL. Comparing diary entries and modern charts, the depot was likely placed within a two-mile radius of 83°31'30" S, 171°00' E.

p. 175. **They built a snow cairn:** Shackleton had ordered Mackintosh to erect a series of flagged cairns across the gap, every half mile, to indicate the depot itself, a chore beyond their physical powers at that point.

p. 175. **"As I said"**: Ernest Joyce, log, January 26, 1916, RPK.

p. 175. **"The Skipper telling us"**: Ernest Joyce, log, January 26, 1916, RPK.

p. 175. **"We are now Homeward Bound"**: Ernest Joyce, log, January 27, 1916, RPK.

p. 175. **"Hanging on to the harness"**: Ernest Joyce, log, January 28, 1916, RPK.

— CH. 12: "HOMEWARD BOUND" —

p. 176. **"A year ago to-day"**: Arnold Spencer-Smith, diary, January 25, 1916, SPRI.

p. 176. **"Dreamt that we met Sir Ernest"**: Arnold Spencer-Smith, diary, January 23, 1916, SPRI.

p. 176. **"in case my complaint"**: Arnold Spencer-Smith, diary, January 22, 1916, SPRI.

p. 176. **The prescribed dosage:** One-quarter ounce was the dosage recommended by nutritionist Wilfred Beveridge. The vitamin C content cited here is based on current USDA figures for bottled lime juice. However, it is likely that the Ross Sea party's juice, produced by Rose's, contained even less. Eric Marshall investigated Rose's manufacturing methods and believed that overheating during processing likely destroyed much of the nutrient (Eric Marshall to John Kendall, September 3, 1950, SPRI). Decades later, pioneering scientific studies in which scurvy was induced in test subjects showed that at least ten milligrams per day was needed to keep symptoms at bay (Kenneth Carpenter, *The History of Scurvy and Vitamin C* p. 204). Rapid replenishment at higher levels was shown to cure the disease and prevent death. The current USDA RDA for adult males is ninety milligrams.

p. 177. **Depleted of vitamin C:** The vitamin is essential to maintain collagen, the protein that binds connective tissue fibers like intercellular cement.

p. 177. **"Felt very rotten"**: Arnold Spencer-Smith, diary, January 28, 1916, SPRI.

p. 177. **"It is strange but cheery"**: Arnold Spencer-Smith, diary, January 29, 1916, SPRI.

p. 177. **"It may be scurvy"**: Ernest Joyce, log, January 29, 1916, RPK. In Joyce's book and other versions of his diary, which were later transcripts of this original log, he inserted statements in hindsight indicating that he recognized scurvy symptoms in Mackintosh on January 28, Spencer-Smith on January 29, and Wild, Richards, and himself on February 1.

p. 177. **Mackintosh's party had subsisted:** It has been claimed that Spencer-Smith and Mackintosh disliked the vitamin C–rich seal meat and thus succumbed sooner because they avoided it, but the available evidence contradicts this notion. Spencer-Smith called seal "indigestible," but ate it dutifully, and Mackintosh expressed enthusiasm for it. The real problem for Mackintosh's party was their prolonged reliance on a deficient diet, since the body is unable to store the vitamin and excretes any excess, a process accelerated by cold stress. As of February 1, Mackintosh's party had been living exclusively on the sledging diet for ninety-five days, while it had been fifty days since the last hut visit of Joyce's party.

There is no absolute figure for how long a human being can live on a diet devoid of vitamin C before showing signs of scurvy. On Scott's first expedition, his southern party was away from base for ninety-five days before showing advanced symptoms. In a U.S. study in which scurvy was induced with a diet lacking vitamin C, skin changes were noted in test subjects in the eight- to thirteen-week range and gum changes became apparent after eleven to nineteen weeks. (Kenneth Carpenter, *The History of Scurvy and Vitamin C*, citing Hodges et al., p. 204.)

p. 177–78. **"Where would you be"** . . . **"As regards disaster":** "RGS ITAE Committee Conference Report," March 4, 1914, RGS.

p. 179. **"He seems properly scared"** . . . **"I also told him":** Ernest Joyce, log, January 31, 1916, RPK.

p. 179. **"Skipper is quite lame":** Victor Hayward, diary, January 26–28, 1916, NMM.

p. 179. **"As I said":** Ernest Joyce, log, January 29, 1916, RPK.

p. 179. **"There was nobody in charge":** Richard Richards, oral history interview, December 2, 1976, p. 61, NLA.

p. 179. **"his folly in passing":** Richard Richards, diary, February 28, 1916, CM.

p. 180. **"Now comes one of the trials":** Ernest Joyce, log, January 29, 1916, RPK.

p. 180. **"Turned in wet through":** Ernest Joyce, log, January 31, 1916, RPK.

p. 180. **"Everyone is very kind":** Arnold Spencer-Smith, diary, February 2, 1916, SPRI.

p. 180. **"Wild is a brick":** Ernest Joyce, log (transcript), February 10, 1916, ATL.

p. 180. **"I had a fierce argument":** Arnold Spencer-Smith, diary, February 14, 1916, SPRI.

p. 180. **"A rather more comfortable day":** Arnold Spencer-Smith, diary, February 11, 1916, SPRI.

p. 181. **"Dogs going splendid":** Ernest Joyce, log, February 4, 1916, RPK.

p. 181. **"We shall have to put":** Ernest Joyce, log, February 4, 1916, RPK.

p. 182. **"like ploughing one's way," "Rottener and rottener":** Victor Hayward, diary, February 15–16, 1916, NMM.

p. 182. **"going to pieces":** Richard Richards, diary, undated entry, p. 104, CM.

p. 182. **"The wind—still howling":** Arnold Spencer-Smith, diary, February 18, 1916, SPRI.

p. 183. **"Shall have to make":** Ernest Wild, diary, February 21, 1916, ATL.

p. 183. **"He dosn't howl much":** Ernest Wild, diary, February 17, 1916, ATL.

p. 183. **"Richards, Hayward & I":** Ernest Joyce, log, February 22, 1916, RPK.

p. 183. **"awful beyond words"**: Robert F. Scott, *Scott's Last Expedition*, March 3, 1912, p. 455.

p. 184. **"Oh my hands, my hands"**: Richard Richards, oral history interview, December 2, 1976, NLA.

p. 184–85. **"I'm done, I'm done," "completely gone"**: Richard Richards, diary, February 23, 1916, CM.

p. 185. **"I think he has got scurvy"**: Ernest Joyce, log, February 23, 1916, RPK.

p. 185. **"We wanted Wild"**: Arnold Spencer-Smith, diary, February 23, 1916, SPRI.

p. 185. **"We could only see"**: Richard Richards, diary, February 23, 1916, CM.

p. 185. **ninth day of blizzard**: According to meteorologist David Bromwich of the Byrd Polar Research Center, Ohio State University, a single sustained blizzard of this duration is highly improbable. Rather, a succession of lows likely fueled severe conditions that the men perceived as one event.

p. 186. **Not only scurvy**: The diet was inadequate in many essential nutrients. Especially critical were other water-soluble vitamins not stored in the tissues, notably the B-complex vitamins, which require regular replenishment. Their intake of riboflavin and thiamin was less than half the essential amount. Thiamin was particularly lacking, since the biscuits were made from white flour and the bicarbonate of soda used for leavening destroys the vitamin. The niacin content was less than a third of that required; vitamin B_{12} and folic acid intake were roughly one quarter. Of the vitamins stored in fat tissues, vitamin A and E intakes were low. The manufacture of vitamin D in the skin was impaired due to lack of sunlight and covered skin. The salvaged Scott and Mawson rations they consumed were undoubtedly degraded in nutritional value and more deficient.

p. 186. **Joyce banished**: They had not dressed any of the dog carcasses from the first season to depot the meat.

p. 186. **"If the worst comes"**: Ernest Joyce, log, February 25, 1916, RPK.

p. 186. **"lowered his great head"**: Richard Richards, *The Ross Sea Shore Party 1914–17*, p. 29.

p. 186. **"which seemed to electrify"** . . . **"We all think"**: Ernest Joyce, log, February 26, 1916, RPK.

p. 187. **"This is the longest blizzard"**: Ernest Joyce, log, February 27, 1916, RPK.

p. 187. **"awful—held up here"**: Richard Richards, diary, February 27, 1916, CM.

p. 187. **"dreaded black appearance"**: Richard Richards, diary, February 28, 1916, CM.

p. 187. **"neither able to pull much"**: Richard Richards, diary, February 28, 1916, CM.

p. 187–88. **"good, unselfish fellow"** . . . **"if it's God's will"**: Æneas Mackintosh, open letter, February 28, 1916, transcribed in Irvine Gaze, diary, CM. Certain aspects echo Scott's "Message to the Public," *Scott's Last Expedition*, pp. 476–77.

p. 188. **"Very cold and wet"**: Arnold Spencer-Smith, diary, February 27, 1916, SPRI.

p. 188. **"like a father"**: Ernest Joyce, log (transcript), February 29, 1916, ATL.

p. 188. **"some picnic party"**: Æneas Mackintosh, open letter, undated (February 27, 1916), CM.

p. 188. **"My belly is singing"**: Ernest Wild, diary, February 27, 1916, ATL.

p. 188. **"For God's sake, Richy"**: Richard Richards, oral history interview, December 2, 1976, NLA.

p. 188. **"[The snow] gets through"**: Ernest Joyce, log, March 2, 1916, RPK.

p. 189. **"The hardest damn journey"**: Richard Richards, oral history interview, December 2, 1976, NLA.

p. 189. **"We are all, more or less suffering"**: Victor Hayward, diary, March 1, 1916, NMM.

p. 189. **"a pathetic sight"**: Richard Richards, oral history interview, December 2, 1976, NLA.

p. 189. **"it would be a good job"**: Ernest Joyce, log (transcript), March 4, 1916, ATL.

p. 189. **body temperature:** Hypothermia was not well understood by the medical profession of the time. William Osler's *Principles and Practice of Medicine,* the authoritative text of the day in the United States and Britain, did not recognize hypothermia as a distinct condition with a progression of symptoms requiring treatment. With the paucity of medical knowledge, it is small wonder that polar explorers routinely failed to grasp the dangers of hypothermia. In *Heart of the Antarctic,* Shackleton recorded body temperatures well below normal in his four-man party without apparent alarm. On January 4, 1909, one man's temperature was 95.4° F, while the other three registered at the thermometer's lower limit of 94.2°, indicating they were moderately to severely hypothermic.

p. 189. **"somewhat unbalanced"**: Richard Richards, diary, marginalia, February 29, 1916, CM.

p. 189. **His irrational behavior:** The symptoms described in the diaries, particularly the dilated pupils, suggest that Hayward's temperature had dropped to 89.6° or lower. At this stage, his vital organs were losing the struggle to retain heat, despite the fact that blood flow had been curtailed to the extremities. Below 96° or 97°, the strenuous exercise forced blood into his cold limbs and further lowered his core temperature when the cooled blood returned to his heart. Caught in a downward spiral, Hayward's body produced less heat as his cellular metabolism slowed and his chilled heart became susceptible to arrhythmias. It is impossible to say just how low his temperature dropped. Around 87.8°, blood pressure may be undetectable; below 86°, blood flow to the limbs virtually stops, causing muscle rigidity. Heartbeat and respiration slow, the risk of ventricular fibrillation increases, and loss of consciousness is imminent. Below 80.6°, cardiac function ceases and the sufferer appears clinically dead (author interview with Dr. Andrew Weinberg).

p. 190. **"Am now out of the team"**: Victor Hayward, diary, March 7, 1916, NMM.

p. 190. **As Spencer-Smith declined:** The bloody urine was possibly a consequence of scurvy or the extreme physical exertion, which can lead to a condition called rhabdomyolysis in which muscle fibers are broken down and the cellular contents released into the circulation. Toxins from myoglobin breakdown can cause kidney damage.

p. 190. **"Poor dogs they look"**: Ernest Joyce, log (transcript), March 7, 1916, ATL.

p. 190. **"'O that this too, too solid flesh'"**: Arnold Spencer-Smith, diary, February 2, 1916, SPRI.

p. 190. **Reaching for the medical kit:** The tablets were commonly prescribed for pain relief and diarrhea. Spencer-Smith followed the recommended dosage. Richards wrote in his diary on March 9 that Spencer-Smith had taken four tablets the night of his death, and ten in all were missing from the opium vial. In a 1976 interview, Richards contradicted himself, claiming that only four pills in total were missing from the vial. He also emphasized that Spencer-Smith "didn't poison himself." Richards may have said this to deflect the possible interpretation that Spencer-Smith had committed suicide. However, ten pills would not have exceeded the recommended dosage of up to four per day. Spencer-Smith notes he took two pills on March 5. If he took two each on March 6 and 7, and four on March 8, this would have accounted for the ten tablets and would not have exceeded the recommended amount.

According to Dr. Henry Guly, who has studied the pharmaceuticals in the Burroughs Wellcome expedition medicine kits, the recommended daily dosage likely contained ten milligrams of morphine, a relatively small amount which would have been absorbed more slowly due to the chalk powder in the tablets. Spencer-Smith weighed 194 pounds at the start of the expedition; the modern recommended dose for a man weighing 154 pounds would be fourteen to twenty-eight milligrams every six hours. (The British National Formulary states five to twenty milligrams every four hours.) Still, the pills likely depressed Spencer-Smith's respiration at a time when his organ functions were severely compromised by the effects of nutritional deficiencies, exhaustion, starvation, dehydration, and hypothermia.

p. 190. **"wandering"**: Richard Richards, *The Ross Sea Shore Party 1914–17*, p. 27.

p. 190. **"Spent the afternoon"**: Arnold Spencer-Smith, diary, February 15, 1916, SPRI.

p. 191. **"Smithy had a little joke"**: Ernest Wild, diary, March 10, 1916, ATL. Recalling events of March 9.

p. 191. **"I say Rich"**: Ernest Wild, diary, March 10, 1916, ATL. The description suggests that Spencer-Smith may have succumbed to cardiac arrhythmia related to hypothermia, especially given his inactivity and the extreme cold. However, it must be emphasized that his body was under siege on many fronts. Nutritional deficiencies, electrolyte imbalance, and the opium pills would have affected his cardiac function as well. Also, there was the slight possibility that he suffered from a preexisting heart condition.

p. 191. **"Woke up this morning"**: Ernest Wild, diary, March 10, 1916, ATL.

p. 191. **"as reverently as possible"**: Ernest Wild, diary, March 10, 1916, ATL.

p. 191. **"I am afraid he will share"**: Ernest Joyce, log (transcript), March 9, 1916, ATL.

p. 191. **"the strongest desire to rush"**: Richard Richards, oral history interview, December 2, 1976, NLA.

p. 191. **"a bag of bones"**: Ernest Joyce, *The South Polar Trail*, p. 176.

p. 192. **"How one longs to be out"**: Æneas Mackintosh, diary, March 23, 1915, RGS.

p. 192. **"Told Mackintosh"**: Richard Richards, diary, March 16, 1916, CM.

p. 192. **"a little peculiar"**: Ernest Joyce, log (transcript), February 16, 1916, ATL.

p. 192. **On the morning of March 18**: On arrival at Hut Point on March 14, the party picked up a functional thermometer and recorded actual readings on the next trip.

p. 192. **"As there is no news"**: Ernest Joyce, log, March 11, 1916, RPK.

— CH. 13: "SOME WAY OR OTHER THEY'RE LOST" —

p. 193. **Against all odds**: Richards logged 1,330 nautical miles (1,532 statute miles) in the second season of sledging. Joyce recorded 1,425 nautical miles (1,641 statute);

however, his figures are inconsistent. Richards is likely more accurate. Mackintosh's official navigation book with the second-season records is missing, presumably due to circumstances described later in this chapter. Richards and Joyce independently recorded a total for both seasons of 1,669 nautical miles for Joyce.

p. 193. **"If the T. C. [transcontinental] Party"**: Ernest Shackleton to Æneas Mackintosh, September 18, 1914, SPRI.

p. 193. **"I looked out for Shacks"**: Ernest Wild, diary, April 23, 1916, ATL.

p. 193. **"nothing short of miraculous"**: Victor Hayward, diary, February 27, 1915, NMM.

p. 194. **six novices**: The prevailing view is that inexperienced dog drivers need to begin with two or three dogs and train over a period of weeks or months, building up to a full team once authority is established. The most experienced long-distance dog handlers, such as the Sirius Patrol in Greenland or Arctic Inuit drivers, tend to prefer a maximum of eleven or twelve per sledge. The bigger the team, the greater the probability of fighting and other problems, particularly since Shackleton's dogs had not worked together since birth and the handlers were novices.

p. 194. **the inescapable comparisons**: When the *Endurance* entered the pack ice of the Weddell Sea on January 2, 1915, Shackleton was still trying to balance the logistical equation in his diary. He counted on a team of sixty-nine dogs to provide the power to haul 2,600 pounds of food, fuel, and gear to keep his six-man team alive until they reached the first of the Ross Sea party's depots at Mount Hope. He intended to reach the Ross Sea in 120 days, losing no more than 12 days to weather. The assumptions did not add up. In the first place, he had not factored in the thousands of pounds of dog food necessary before he began slaughtering and feeding the dogs to each other beyond the Pole. It is possible that he expected the motorized transport to carry and depot all of the dog food, as well as other stores, even though the motor sledges had malfunctioned on their testing expedition in Norway in 1914 and the defects had not been worked out before the voyage.

p. 194. **"some way or other"**: Richard Richards, oral history interview, December 2, 1976, NLA.

p. 194. **"I'm afraid she has"**: Ernest Wild, diary, April 23, 1916, SPRI.

p. 194. **The limited provisions**: Thomson had dropped off enough rations to last six men for two months. In addition, there was about five months' worth of old rations from previous expeditions, much of it inedible.

p. 194. **"the life of troglodytes"**: Richard Richards to Leslie Quartermain, November 9, 1960.

p. 194. **"Sealing in the dark"**: Richard Richards, *The Ross Sea Shore Party 1914–17*, p. 17.

p. 195. **"as cheerful as school boys"**: Ernest Joyce, log (transcript), March 11, 1916, ATL.

p. 195. **"To wander more"**: Æneas Mackintosh to Annie Mackintosh, February 28, 1916, MFP.

p. 195. **"1 blue serge suit"**: Victor Hayward, diary, undated entry, p. 79, NMM.

p. 196. **"I don't know why"**: Ernest Joyce, log, May 3–10, 1916, RPK.

p. 196. **The following morning:** The date of these events is given incorrectly by Joyce and Richards as May 7 in their books.

p. 196. **The rocky spur**: When I was on Ross Island in 2002, I often traveled between Hut Point, Cape Evans, and Cape Royds for research, driving a tracked vehicle across the sea ice. The advice from field guides to personnel leaving Hut Point is still the same: Keep an eye on Minna Bluff, and if it disappears, stay off the sea ice.

p. 196. **"I told him he could"**: Ernest Joyce, log, May 3–10, 1916, RPK.

p. 196. **"short of forcibly restraining"**: Richard Richards, *The Ross Sea Shore Party 1914–17*, p. 38.

p. 196. **"After bringing them back"**: Ernest Joyce, log, May 3–10, 1916, RPK.

p. 197. **"I can see these two"**: Richard Richards, *The Ross Sea Shore Party 1914–17*, p. 38.

p. 197. **"If the other two get lost"**: Ernest Wild, diary, May 8, 1916, SPRI.

— CH. 14: "DRIFTING TO GOD KNOWS WHERE" —

p. 198. **"The old ship"**: Joseph Stenhouse, diary, March 8, 1916, SPRI.

p. 198. **"no moorings ever made"**: Frank Worsley, *Endurance*, p. 203.

p. 198. **"Stenhouse endures mental torments"**: Alfred Larkman, diary, June 6, 1915, SPRI.

p. 199. **the receiver was suspiciously silent**: One problem, he suspected, was the aurora australis, the evanescent streamers of light that pulsed in the winter sky. His instincts were correct. It was later learned that auroral displays are caused by the acceleration of electrons along the earth's magnetic field lines, which excites atoms in the atmosphere. The stream of ionized particles causes "polar cap absorption," in which HF and VHF radio waves are absorbed, and LF and VLF radio waves are reflected.

p. 199. **"no use in being despondent"**: Leslie Thomson, diary, May 31, 1915, SPRI.

p. 199. **"We came to the Antarctic"**: James Paton, diary, July 23, 1915, HL.

p. 200. **"perpetual smile"**: Lionel Hooke, diary, January 7, 1915, ML.

p. 200. **"He would be a nasty man"**: Leslie Thomson, diary, May 21, 1915, SPRI.

p. 200. **seizures**: Orrin Devinsky of New York University Medical Center confirmed my suspicion that Shaw's sleepwalking and "fits," as they were described, were a seizure disorder. Reviewing the evidence, he suggested that Shaw suffered from grand mal seizures (now called tonic-clonic seizures) from a sudden surge of electrical activity in his brain. The sleepwalking episodes may have been complex partial seizures or a postseizure state in which he was confused yet able to perform automatic actions, like walking. Sleep deprivation and stress are common aggravating factors. The first onset of Shaw's epilepsy is uncertain; his history of boxing raises the possibility of head trauma. If he had been diagnosed by 1914, Shaw understandably concealed it when he applied, given the stigma associated with epilepsy at the time. Little had changed since *The Seaman's Medical Guide* (1863) had advised against signing sufferers. Knowledge of complex partial seizures or behavioral changes that followed seizures was probably limited to less than 1 percent of the medical community, so Shaw may have gone through life without diagnosis. As the son of a machine shop foreman, he was unlikely to have received the first treatment with a demonstrated antiseizure effect, bromide, which was introduced in the late nineteenth century. Phenobarbital, introduced in 1912, was not in wider use until after World War I.

p. 200. **"punishment for coming"**: Joseph Stenhouse, diary, May 21, 1915, SPRI.

p. 200. **"useless toy"**: Ernest Joyce, *The South Polar Trail*, p. 74.

p. 200. **"a bit of a swanker"**: Ernest Wild, diary, September 1915, p. 18, CM.

p. 200. **his caustic sarcasm**: Clarence Mauger, diary of work started Monday, September 13, 1915, ATL. His defensiveness became the topic of a secretly circulated bit of doggerel:

> *Here we take pleasures & also our meals.*
> *The latter consisting of penguins & seals;*
> *And now and again Old Ninnis here vies,*
> *To cut us in slices with murderous knives.*

p. 200. **"a limestone cave"**: Leslie Thomson, diary, April, 29, 1915, SPRI.

p. 200. **"Unremitting toil"**: Joseph Stenhouse, diary, November 19, 1915, SPRI.

p. 201. **"The first of July!"**: Joseph Stenhouse, diary, July 1, 1915, SPRI.

p. 201. **"had done all any human could"**: Leslie Thomson, diary, May 31, 1915, SPRI.

p. 201. **"The old, old curse"**: Joseph Stenhouse, diary, October 24, 1915, SPRI.

p. 201. **Seven weeks into a voyage**: Stenhouse sailed on the steamship *Gracchus* (British India Steam Navigation Company records, NMM). There is no evidence that he shared this information with Shackleton or Mackintosh.

p. 201. **"This is Hell"**: Joseph Stenhouse, diary, July 1, 1915, SPRI.

p. 201. **"We are all expectation"**: Howard Ninnis, diary, July 13, 1915, RGS.

p. 201. **"Here one sees"**: Joseph Stenhouse, diary, July 13, 1915, SPRI.

p. 202. **like the *Fram***: The ingenious design of Nansen's *Fram* resolved the flaws with a rudder and propeller that could be safely retracted into a well behind the double sternpost, and the sloping bow and stern ensured that the ice would gain no purchase. Under pressure, the *Fram* simply slipped clear of the encroaching pack, in Nansen's words, "as smooth as an eel." (Fridtjof Nansen, *Farthest North*, p. 52)

p. 202. **"I doubt if anyone"**: Howard Ninnis, diary, July 17, 1915, RGS.

p. 202. **"like the collapse"**: Howard Ninnis, diary, July 18, 1915, RGS.

p. 202. **"Our marvellous luck"**: Howard Ninnis, diary, July 21, 1915, RGS.

p. 202. **"slowly, steadily"**: Howard Ninnis, diary, July 21, 1915, SPRI.

p. 203. **"We are nearly 100 miles"**: Lionel Hooke, diary, July 21, 1915, ML.

p. 203. **"Am afraid the ship's back"**: Joseph Stenhouse, diary, July 22, 1915, SPRI.

p. 203. **"God knows"**: Alfred Larkman, diary, July 21–August 1, 1915, SPRI.

p. 204. **"The whole crew are like"**: Lionel Hooke, diary, July 23, 1915, ML.

p. 204. **"the wells and holds"**: James Paton, diary, July 23, 1915, HL.

p. 205. **distress call**: Joseph Stenhouse, wireless to King George V, July 23, 1915, SPRI.

p. 205. **"Still hoping"**: Lionel Hooke, diary, July 26, 1915, ML.

p. 205. **"like Alpine climbing"**: James Paton, diary, August 8, 1915, HL.

p. 205. **Macquarie Island and Awarua**: The station was abandoned later in 1915, when the government recalled the staff to Australia due to the war. Before closing down, the Macquarie Island station had sent messages almost continuously to the *Aurora* during the winter of 1915.

p. 205. **"Hooke dismantled wireless"**: Joseph Stenhouse, diary, September 27, 1915, SPRI.

p. 205. **"I cannot help thinking"**: Joseph Stenhouse, diary, September 21, 1915, SPRI.

p. 205. **he calculated:** Joseph Stenhouse, *Aurora* log, September 22, 1915, SPRI. Stenhouse's best estimate of the ship's position was 69°12' S, 164°32' E. Because of the chronometer problem, the positions noted during the drift were inaccurate. James Wordie calculated the error and published corrected positions in "The Ross Sea Drift of the *Aurora*," *Geographical Journal*, July 1958, p. 222.

The chronometers had not been designed for such extreme conditions and the wrappings of blankets failed to insulate the cases from the cold. For a time, the problem was compounded by the fact that Stenhouse could not see the actual horizon, as it was obscured by pack ice. An artificial horizon is useful in these circumstances, but it was misplaced in the haphazard loading of the ship. On September 3, 1915, it was finally found and put to use.

p. 205. **"This waiting":** Joseph Stenhouse, diary, November 13, 1915, SPRI.

p. 206. **"Feel pretty disgusted":** Lionel Hooke, diary, December 1, 1915, ML.

p. 206. **"Poor beggar":** Joseph Stenhouse, diary, October 27, 1915, SPRI.

p. 206. **"I cannot help remarking":** Joseph Stenhouse, diary, November 15, 1915, SPRI.

p. 206. **"proved himself":** James Paton, diary, July 21, 1915, HL.

p. 207. **"Another month gone":** Joseph Stenhouse, diary, February 1, 1916, SPRI.

p. 207. **"We are too late now":** Joseph Stenhouse, diary, February 10, 1916, SPRI.

p. 207. **"like a racehorse":** James Paton, diary, February 18, 1916, SPRI.

p. 207. **"I feel that we are taking things":** Joseph Stenhouse, diary, February 17, 1916, SPRI.

p. 209. **"And yet it is no good":** Joseph Stenhouse, diary, February 17, 1916, SPRI.

p. 209. **"vast white circle":** James Paton, diary, February 24, 1916, SPRI.

p. 209. **"dangerous customers":** James Paton, diary, March 4, 1916, SPRI.

p. 209. **"Our position, in the midst":** Joseph Stenhouse, diary, February 26, 1916, SPRI.

p. 209. **oakum:** Tarred hemp or manila fibers from unraveled rope.

p. 209. **"The vessel cannot stand":** Joseph Stenhouse, diary, February 28, 1916, SPRI.

p. 209. **"With the lighting of the fires":** Joseph Stenhouse, diary, March 1, 1916, SPRI.

p. 209. **"in the real true blue":** Alfred Larkman, diary, March 14, 1916, SPRI.

p. 209. **the "screaming sixties":** Driven by westerly winds, the Antarctic circumpolar current courses around the earth eastward in an endless clockwise circuit. Between Antarctica and New Zealand, this immensely powerful flow of water channels through a narrowed bottleneck of land. At this chokepoint, and another between the southern tip of South America and the Antarctic peninsula called the Drake Passage, the current can race four times faster than the Gulf Stream, and gales producing fifty-foot waves have been recorded. Few sailors have seen and survived the rogue waves rumored to exceed a hundred feet. (Oceanographer Dr. Elizabeth Hawker kindly reviewed ERS-1 and ERS-2 satellite data on my behalf, which showed maximum recorded wave heights for the region ranging from eight to fifteen meters.)

— CH. 15: "WHEREABOUTS SHACKLETON?" —

p. 211. **"Mackintosh and party"**: Joseph Stenhouse, wireless to Frederick White, March 24, 1916, SPRI.

p. 211. **"Indicate position course speed"**: Senior naval officer, wireless to Joseph Stenhouse, March 24, 1916, SPRI, and Frederick White, wireless to Joseph Stenhouse, March 26, 1916, SPRI.

p. 211. **"whereabouts Shackleton?"**: *New York Times,* wireless to Joseph Stenhouse, March 27, 1916, SPRI.

p. 211. **The news flashed**: With no cablehead at South Georgia, Shackleton had handed letters to a whaling captain in 1914 to carry back to civilization, describing his plans as he prepared to depart South Georgia in December 1914, which the mariner had apparently only recently forwarded to the organizers.

p. 212. **"news of achievements withheld"**: *Mercury* (Leeds), March 25, 1916.

p. 212. **"met Shackleton's Spirit"**: Campbell Mackellar to William S. Bruce, April 4, 1916, SPRI.

p. 212. **"The Shackleton Expedition, Bad News"**: *Daily Chronicle* (London), March 25, 1916.

p. 212. **"Is it too much to ask"**: Emily Shackleton to Andrew Fisher, April 17, 1916, SPRI.

p. 213. **"vessel not under command"**: Alfred Larkman, diary, March 30, 1916, SPRI. Black balls replaced the lights at daybreak.

p. 214. **"War still going strong"**: Alfred Larkman, diary, March 24, 1916, SPRI.

p. 214. **"No news *Endurance*"**: Frederick White, wireless to Joseph Stenhouse, April 1, 1916, 9 pm, SPRI.

p. 214. **"keeper of the conscience"**: "RGS ITAE Committee Conference Report," March 4, 1914, RGS.

p. 215. **"verbal chastiser"**: Hervey Fisher, *From a Tramp's Wallet: A Life of Douglas William Freshfield,* p. 184.

p. 215. **"a great disappointment"**: Douglas Freshfield, notes of interview with Ernest Perris, March 31, 1916, RGS.

p. 215. **Her bunkers were so small**: Ernest Shackleton to Ernest Perris, November 30, 1914, SPRI.

p. 215. **"Shackleton knew perfectly well"**: Douglas Freshfield, notes of interview with Ernest Perris, March 31, 1916, RGS.

p. 215. **"unable to offer"**: Resolution of the RGS Council, April 7, 1916, RGS.

p. 215. **"very few men fitted"**: Hugh Robert Mill to the RGS Council, April 15, 1916, RGS.

p. 215. **"In fact, Davis is the only man"**: Lewis Beaumont to the RGS Council, April 21, 1916, RGS.

p. 215. **"trial by ice"**: Louise Crossley, *Trial by Ice.*

p. 215. **"one year, nine months and eleven days"**: John King Davis, *High Latitude,* p. 238.

p. 216. **"If you are a charlatan"**: John King Davis, *High Latitude,* p. 213.

p. 216. **"There are all these"** . . . **"the ship"**: Douglas Freshfield, interview with John King Davis, April 10, 1916, RGS.

p. 216. **"the ship would be crushed"**: Douglas Freshfield, interview with John King Davis, April 10, 1916, RGS.

p. 216. **"it would be no good"**: "RGS ITAE Committee Conference Report," March 4, 1914, RGS.

p. 216. **"absolute and undivided control"**: John King Davis to Hutchison and Cuff, April 14, 1916, SLV.

p. 216. **"I will have nothing"**: Douglas Freshfield, interview with John King Davis, April 10, 1916, RGS.

p. 217. **"There is no misapprehension"**: Alfred Hutchison and Ernest Perris to John King Davis, April 10, 1916, SLV.

p. 217. **"In my opinion"**: Douglas Freshfield, interview with John King Davis, April 10, 1916, RGS.

p. 217. **Nor could anyone:** The British death toll by the end of the war was 744,702. War Office, *Statistics on the Military Effort of the British Empire during the Great War,* pp. 241–46.

p. 217. **"Shackleton's ideal"**: *Daily Graphic* (London), March 25, 1916.

p. 217. **"Fancy that ridiculous Shackleton"**: Winston Churchill to Clementine Churchill, March 28, 1916, in Martin Gilbert, *Winston S. Churchill,* p. 1468.

p. 218. **"Seeing by the papers"**: Mrs. J. W. Mugridge to RGS, May 24, 1916, RGS.

p. 218. **"Now that we are in touch with civilization"**: Howard Ninnis, diary, April 2, 1916, RGS.

p. 218. **Only as secure:** The London organizers had only arranged to pay monthly allotments to the families of the *Endurance* party and two of the Ross Sea party, Mackintosh and Larkman. Mackintosh had prevailed upon local patrons in New Zealand to advance funds to a few of the neediest members of the Ross Sea party before departure.

p. 219. **D'Anglade volunteered:** He joined the New Zealand Expeditionary Force and shipped out to England in July 1916, where the trail goes cold. His true name and identity are a mystery. Upon joining a previous ship in 1913, the vagabond claimed his identity papers had been lost in a shipwreck and stated that he was twenty-three, both of which the captain doubted. In 1914, he signed on to the *Aurora* as nineteen years of age.

p. 219. **"mentally unsound"**: Howard Ninnis, crew list 1916, April 20, 1916, RGS.

p. 219. **"You are appointed Captain"**: John King Davis, *High Latitude,* p. 247. Davis recalled the date of the cable, May 28, incorrectly in his book. (SLV)

p. 219. **"31st May have arrived"**: Ernest Shackleton, cable, May 31, 1916, RGS.

— CH. 16: PORT CHALMERS —

p. 220. **"What the ice gets"**: Frank Worsley, *Endurance,* p. 4.

p. 220. **"drifting along"**: Frank Hurley, diary, November 7, 1915, ML.

p. 220. **"fast in the Pack & drifting"**: Joseph Stenhouse, diary, May 9, 1915, SPRI.

p. 221. **"I have had a year and a half"**: Ernest Shackleton, cable to Emily Shackleton, June 3, 1916, SPRI.

p. 221. **"There are some failures"**: *Edinburgh Evening Dispatch,* June 2, 1916.

p. 221. **"This mission is one of sheer science"**: *Pall Mall Gazette* (London), March 25, 1916.

p. 221. **"The British public"**: *Outlook* (London), June 3, 1916.

p. 221. **"Congratulate safety"**: RGS, cable to Ernest Shackleton, undated, June 1916, RGS.

p. 222. **Two distinguished scientists**: Masson was a catalyst for Mawson's 1911 expedition and a former professor of Keith Jack.

p. 223. **"exceptionally bad and damaged"**: A. Morrison, "Surveyor's Report, SY *Aurora*," June 28, 1916, SLV.

p. 223. **"many influences at work"**: Joseph Stenhouse to Ernest Shackleton, November 18, 1916. SPRI.

p. 223. **"not all beer and skittles"**: Joseph Kinsey to Edward Saunders, December 7, 1909, ATL.

p. 223. **"a marked difference"**: Joseph Stenhouse to Ernest Shackleton, November 18, 1916, SPRI.

p. 223. **"an interested inspector"**: Joseph Stenhouse to Leonard Tripp, November 14, 1916, ATL.

p. 224. **"Shackleton should perish"**: Notes of conversations with Joseph Stenhouse, Frank Worsley, and Ernest Shackleton, verso letter dated November 17, 1932, HW. Stenhouse reported to Frank Worsley that the comment was made by Thomas Forster Knox, a supporter of Mawson's first expedition. He and Alfred Milsom were the joint mortgagees loaning £700 (in current terms, about $51,000) to the expedition with the *Aurora* as collateral in 1914.

p. 224. **In early May**: Henry Howard to John King Davis, June 7, 1922, SPRI.

p. 224. **"no other course but to accept," "taking unfair advantage"**: *The Expenditure Incurred in Connexion with the S. Y. Aurora*, NAA.

p. 224. **"All that explorers in future will do"**: Douglas Mawson to Andrew Bonar Law, September 14, 1916, MC.

p. 224. **57,000 casualties**: The estimate of the dead was 19,000.

p. 224–25. **by August**: *Report of Marine Department Year 1916–17*.

p. 225. **"disturbed and anxious"**: John King Davis, *High Latitude*, p. 251.

p. 225. **"inexperience and lack of judgement"**: John King Davis to Joseph Kinsey, August 5, 1916, SLV.

p. 225. **"even a slight acquaintance"** . . . **"such a fatal policy"**: John King Davis, *High Latitude*, p. 252. Davis met with Mackintosh and Stenhouse on several occasions in late 1914. There is no record of their conversations, but he presumably discussed navigation issues with them. It also begs the question of why, if Davis judged that Glacier Tongue and Cape Evans were not safe for winter moorings from his experience as *Nimrod*'s first officer, Mackintosh, as second officer, was not similarly impressed by the experience.

p. 225. **"That man obeyed instructions"**: Douglas Freshfield, interview with John King Davis, April 10, 1916, RGS.

p. 225. **In Davis's eyes**: Davis's judgment as a commander had been similarly tested three years before aboard the *Aurora*, and he decided in a difficult situation to disobey orders from Mawson to protect the ship. For firsthand accounts of the events of February 7, 1913, see Douglas Mawson, *The Home of the Blizzard*, and John King Davis, *High Latitude*.

p. 226. **"tactfully explained to him"**: Joseph Kinsey, "Report to the Australian Advisory Committee," August 1916, SLV.

p. 226. **"So keen is he in the work"**: Joseph Kinsey to Robert McNab, July 25, 1916.

p. 226. **"rights of publication"**: W. Graham Greene to Ernest Shackleton, August 2, 1916, SPRI.

p. 226. **"Are you all well?"**: Ernest Shackleton, *South*, p. 242.

p. 226. **"All safe; all well, Boss"**: Hugh Robert Mill, *The Life of Sir Ernest Shackleton*, p. 237.

p. 226. **"I have done it"**: Ernest Shackleton, cable to Emily Shackleton, September 3, 1916, SPRI.

p. 227. **"I cannot settle down"**: Ernest Shackleton to Janet Stancomb-Wills, October 4, 1916, SPRI.

p. 227. **"had no wish to seem"**: John King Davis, *High Latitude*, p. 256.

p. 227. **"absolute and undivided"**: John King Davis, wireless to William Creswell, September 6, 1916, SLV.

p. 227. **"exception to the discourteous manner"**: *Evening Star* (Dunedin), October 5, 1916.

p. 227. **"No man other than one"**: H. Livingstone Tapley to the editor, *Evening Star* (Dunedin), October 7, 1916.

p. 228. **"sign any documents official"**: Joseph Stenhouse to Leonard Tripp, November 14, 1916, ATL.

p. 228. **"unapproachable dignity"**: Frank Wild, memoirs, ML.

p. 228. **"I will not relinquish command"**: Joseph Stenhouse to Leonard Tripp, October 13, 1916, ATL.

p. 228. **"My position here"**: Joseph Stenhouse to Leonard Tripp, October 13, 1916, ATL.

p. 229. **"discourtesy & double dealings"**: Joseph Stenhouse to Leonard Tripp, October 13, 1916, ATL.

p. 229. **"Sir Ernest Shackleton should be informed"**: Ernest Perris to Andrew Bonar Law, October 16, 1916, SLV.

p. 229. **"I am dead tired"**: Ernest Shackleton, cable to Ernest Perris, October 23, 1916, SPRI.

p. 229. **"if necessary put embargo"**: Ernest Shackleton, cable to Leonard Tripp, November 9, 1916, ATL

p. 229. **The minister ordered**: Robert McNab to Leonard Tripp, December 6, 1916, ATL. This action hardened the attitude of the committees in Australia and Britain, as they believed it showed that Shackleton was willing to risk the timely sailing of the *Aurora* and the Ross Sea party's lives for the sake of his own ends.

p. 230. **"a howl against the British"**: Leonard Tripp to Harry Wilson, December 20, 1916, ATL.

p. 230. **"I found him somewhat changed"** . . . **"unable to realize"**: John King Davis, *High Latitude*, pp. 257–60.

p. 230. **It appeared to be**: Davis's insistence on a clear command structure may have stemmed from his experience aboard the *Nimrod* in 1908, when he was present during a dispute between Shackleton and Captain Rupert England over navigation which ultimately ended in England's dismissal. The events are discussed in Davis's *High Latitude*. Shackleton biographers Margery and James Fisher suggested that Kinsey's disenchantment with Shackleton may have partly stemmed from these events.

p. 230. **In fact, Worsley had suggested:** Frank Worsley, *Endurance*, p. 198.

p. 230. **"either childish or criminal":** "Ross Sea Relief Expedition, Report of Australian Advisory Committee," January 31, 1917, SLV.

p. 231. **"The only thing that matters":** John King Davis to Douglas Mawson, December 8, 1916, SPRI.

p. 231. **They also agreed:** Robert McNab to Ernest Shackleton, December 18, 1916, ATL.

p. 231. **"rights as owner to the *Aurora*":** Ernest Shackleton to Robert McNab, December 18, 1916, ATL.

p. 231. **"With every degree of latitude":** John King Davis, *High Latitude*, p. 261.

— CH. 17: RESCUE —

p. 232. **"come home to an old friend":** John King Davis, *High Latitude*, p. 254.

p. 232. **"I just pray that those men":** Ernest Shackleton to Emily Shackleton, December 6, 1916, SPRI.

p. 232–33. **"Aurora is now within 100 miles":** John King Davis, diary, January 9, 1917, SLV.

p. 233. **"It increased our fears":** Howard Ninnis, diary, January 10, 1917, RGS.

p. 233. **"almost more than we":** Howard Ninnis, diary, January 10, 1917, RGS.

p. 234. **"wonderfully white woolens":** Keith Jack, diary, January 10, 1917, SLV.

p. 234. **"any exploring ship ought":** Ernest Joyce, log (transcript), undated entry, ATL.

p. 234. **"A good deal of emotion":** Howard Ninnis, diary, January 10, 1917, RGS.

p. 234. **"I just want to sleep":** Keith Jack, diary, January 10, 1917, SLV.

p. 235. **"Whether they got there":** Ernest Joyce, log, May 3–10, RPK.

p. 235. **"Where is Captain Mackintosh?":** Keith Jack, diary, July 15, 1916, SLV.

p. 235. **"slung their lives away":** Ernest Wild, diary, July 16, 1916, ATL.

p. 235. **"hardships endured":** Keith Jack, diary, July 17, 1916, SLV.

p. 235. **"Hello Richy, I'm not good"** . . . **"off his head":** Richard Richards, oral history interview, December 2, 1976, NLA.

p. 235. **"thieves and murderers":** Richard Richards, oral history interview, December 2, 1976, NLA.

p. 235. **"It seems to me so hard":** Irvine Gaze to Charlotte Spencer-Smith, February 20, 1917, CM.

p. 235. **With Mackintosh gone:** Stevens was plagued with fatigue and abscesses, both of which may have indicated that he had undiagnosed diabetes, contributing to his demoralization. (See notes to chapter 9) He may have spent most of his time in the galley and avoided doing scientific field work as a means of coping with the excessive thirst and hunger caused by the disease without attracting attention.

p. 236. **"If ever persons owed their lives":** Ernest Joyce, log (transcript), February 26, 1916, ATL.

p. 236. **Wild tried the same technique:** The novel was *The Term of His Natural Life* by Marcus Clarke.

p. 236. **"a high degree of conviviality":** Keith Jack, diary, August 20, 1916, SLV.

p. 237. **"He had been my constant companion":** Ernest Joyce, log (transcript), undated (after July 17, 1916), ATL.

p. 237. **"Losses to date"**: Richard Richards, graffiti in Cape Evans hut, August 14, 1916, observed and photographed by the author in 2002. "Shck" presumably abbreviates Shackleton. A numeral "7" to the left of the list may denote the number of survivors at Cape Evans.

p. 237. **"All the same, I felt"**: Ernest Joyce, *The South Polar Trail*, p. 193.

p. 237. **"Come on, the ship's here"**: Richard Richards, oral history interview, December 2, 1976, NLA, and Keith Jack, diary, January 10, 1917, SLV.

p. 238. **"It was not a time for words"**: Keith Jack, diary, January 10, 1917, SLV.

p. 238. **"Ship O"**: Ernest Wild, diary, January 10, 1917, SPRI.

p. 238. **"Joyce old man, more than pleased"**: Ernest Joyce, log (transcript), undated entry, ATL.

p. 238. **"As neither man had any"**: John King Davis, *Report of Voyage by Commander*, ATL.

p. 238. **"a great blow"**: John King Davis to Hugh Robert Mill, February 11, 1917, SPRI.

p. 238. **"little hope in this wilderness"**: Howard Ninnis, diary, January 22, 1917, RGS.

p. 239. **"rather hostile"**: Richard Richards to Leonard Tripp, undated, 1955, ATL.

p. 239. **"playing of chances in equipping"**: The only description of this encounter comes from Joyce in a letter he wrote in 1930 to Sir Charles Royds; Shackleton's diary ceases after December 27. Joyce's disillusionment and anger are well documented, though the details of the exchange cannot be confirmed.

p. 239. **"where [Shackleton's] eyes were"**: Ernest Joyce to Charles Royds, April 7, 1930, SPRI.

p. 239. **"owing to the unfortunate"**: Ernest Joyce, log (transcript), undated entry, ATL.

p. 240. **"amongst the most enjoyable"**: Keith Jack, diary, January 16, 1917, SLV.

p. 240. **"nearly put up with"**: Ernest Joyce, log (transcript), January 1917, undated entry, ATL.

p. 240. **"simply miraculous"**: Ernest Joyce, log (transcript), January 1917, undated entry, ATL.

p. 240. **"ITAE 1914–1917"**: The copper tube was found some distance away in 1947 by Rear Admiral Cruzen of the U.S. Navy's Operation Highjump. The lines from Swinburne's "Atalanta in Calydon: A Tragedy" are: "Things gained are gone, but great things done endure." The misquoted lines from Browning's "Prospice" actually read:

> I was ever a fighter, so—one fight more,
> > The best and the last!
> I would hate that death bandaged my eyes, and forbore,
> > And bade me creep past.
> No! let me taste the whole of it, fare like my peers
> > The heroes of old,
> Bear the brunt, in a minute pay glad life's arrears
> > Of pain, darkness and cold.

p. 241. **"no chance of finding anything"**: John King Davis to Hugh Robert Mill, February 11, 1917, SPRI.

p. 241. **"no doubt that these men"**: John King Davis to Hugh Robert Mill, February 11, 1917, SPRI.

p. 241. **"the harrowing picture"**: John King Davis, *High Latitude*, p. 95.

p. 241. **"so little"**: John King Davis to Hugh Robert Mill, February 11, 1917, SPRI.

p. 241. **"Just why he was so set"**: Richard Richards, *The Ross Sea Shore Party 1914–17*, p. 38.

p. 242. **The nutritional deficiency diseases:** Mackintosh and Spencer-Smith experienced mental symptoms, including hallucinations, which are better explained by deficiencies of the B-complex vitamins. The cardiac arrhythmia noted in Spencer-Smith and Richards can occur with thiamin deficiency (beriberi). Niacin and tryptophan deficiency (pellagra) can cause psychosis and delirium, as well as the skin lesions and gastrointestinal disruption that Cope experienced. The extent of the deficiencies also raises the question of whether Mackintosh and Hayward were fully recovered and if their judgment was literally impaired.

p. 242. **"As these men cannot be alive"**: John King Davis, *Aurora* relief diary, January 17, 1917, SLV.

p. 242. **"utterly dazed and stupefied"** . . . **"the only civilized beings"**: Richard Richards to Leslie Quartermain, April 16, 1963, CM.

p. 242. **"should all be fighting"**: Howard Ninnis, diary, p. 3, RGS.

p. 242. **On January 20:** Shackleton had only allowed his official photographer to shoot events in the Weddell Sea; at least seven members of the Ross Sea party had brought their own cameras and there were numerous prints in the Cape Evans hut. Shackleton realized that Mackintosh had not obtained signed agreements from all of the men to ensure that the members could not publish and jeopardize his newspaper exclusive. He dashed off an agreement in his diary for Jack, which Jack signed. Compared with the others, it was unusual; Shackleton allowed him to keep the scientific results for first publication and promised to increase his salary from £52 to £150 for each year (about $3,600 to $10,500 in current terms) and pay the balance when more funds became available.

p. 242. **"the life and soul"**: Richard Richards, oral history interview, December 2, 1976, NLA.

p. 243. **"There was not a man"**: John King Davis, *High Latitude*, p. 277.

p. 243. **"as a remembrance"**: Inscription, Volume I of *Encyclopedia Britannica*, RGS.

p. 243. **"Do you recall that sweep"**: The poem was inscribed in a copy of Coleridge's "The Rime of the Ancient Mariner" that he presented to Shackleton at the time. Davis was misquoting lines from Robert Service's poem "L'Envoi" as Antarctic explorers often did:

> You may recall that sweep of savage splendor,
> That land that measures each man at his worth,
> And feel in memory, half fierce, half tender,
> The brotherhood of men that know the North.

p. 243. **"This strange voyage"**: John King Davis, *High Latitude*, p. 278.

— CH. 18: "THE MEN THAT DON'T FIT IN" —

p. 244. **"Sir Ernest Shackleton back again"**: *New Zealand Free Lance* (Wellington), February 16, 1917.

p. 244. **"I deeply deplore"**: *Daily Chronicle* (London), February 10, 1917.

p. 245. **"Everyone is very kind"**: Keith Jack, diary, February 11, 1917, SLV.

p. 245. **"We had left the civilized world"**: Richard Richards, *The Ross Sea Shore Party 1914–17*, NLA.

p. 245. **"a man like Churchill"**: Richard Richards, oral history interview, December 2, 1976, NLA.

p. 245. **"His charm and persuasiveness"**: Richard Richards to Margery Fisher, March 28, 1957, SPRI.

p. 245. **On February 12, the citizenry:** The controversy over the relief expedition had received wide press coverage, and New Zealanders had rallied behind Shackleton. An indignant Luke wanted to exclude Davis from the event, but Shackleton refused to countenance the snub and insisted that the *Aurora*'s master be on the dais as well.

p. 245. **"Were positively lionised"**: Keith Jack, diary, February 12, 1917, SLV.

p. 245. **"played the game"**: *Otago Daily Times* (Dunedin), February 12, 1917.

p. 246. **"There are some things"**: Alexander Stevens, "Report of the Ross Sea Party," p. 35, SPRI.

p. 246. **"most of the unpleasantness"**: Alexander Stevens, "Report of the Ross Sea Party," p. 35, SPRI.

p. 246. **"By the way, Stenhouse"**: Arnold Spencer-Smith to Charlotte Spencer-Smith, November 8, 1914, CM.

p. 247. **youngest brother:** *Church Family News* (London), February 1917. His son Martin was killed on September 10, 1916, at Leuze Wood during the Battle of the Somme, and son Charles died on August 3, 1917, at Godewaesvelde near the Belgian border.

p. 247. **"Well, Vic I hope"**: Stan Hayward to Victor Hayward, January 3, 1915. This letter missed the *Aurora*'s departure and was held in port, where it was collected in 1916 for return to Hayward's family.

p. 247. **"On reaching home"**: Alexander Stevens to Hugh Robert Mill, November 20, 1928, SPRI.

p. 247. **"I went out as a geologist"**: Alexander Stevens to Hugh Robert Mill, November 20, 1928, SPRI.

p. 247. **He managed to put Shackleton off:** Keith Jack to Leslie Quartermain, February 22, 1961, CM.

p. 248. **"Mackintosh was a sahib"**: John King Davis to Howard Ninnis, October 12, 1919, HC.

p. 248. **"Obedience is the supreme virtue"**: David Lloyd George, *War Memoirs*, p. 2041.

p. 248. **Their valor:** C. E. W. Bean, *The Official History of Australia in the War 1914–18*, Vol. 5, pp. 30–31. Statistics cited are through March 1918.

p. 248. **"I consider"**: Ernest Shackleton to Ernest Joyce, March 8, 1917 (transcript), ATL.

p. 249. **"I suppose it will be"**: Ernest Joyce to Paton and Bhoys, October 22, 1915, ATL.

p. 249. **"His affairs will be"**: Leonard Tripp to Ernest Shackleton, April 13, 1917, ATL.

p. 249. **The indefatigable Tripp:** Some of the money Tripp raised also went toward paying pressing debts of the Weddell Sea party, including wages. Tripp did not pay any expenses related to the relief expedition.

p. 249. **The members of the Ross Sea party:** Salaries appear to have been calculated according to the agreed weekly rate and time served on the expedition: Stevens received £111 (about $7,800 in current terms); Gaze, £92 ($6,400) and Jack, £100 ($7,000).

p. 249. **With the books balanced:** Jack and Gaze received a bonus of £20 (about $1,400 currently), Richards received £25 ($1,700), and Cope was sent an unspecified amount. A bonus for Wild is not recorded, although the payment of £320 ($22,500), triple the sum he was owed, made him the highest-paid member of the shore party other than Mackintosh and Joyce. Shackleton may have been influenced by the invaluable role Ernest's brother, Frank, played on the *Endurance* expedition. It is unclear whether Stevens received a bonus.

p. 250. **"It seems to me":** Leonard Tripp to Ernest Shackleton, April 25, 1917, ATL.

p. 250. **Above and beyond:** £21,271, or about $1.04 million in current terms. (PRO T1/12337, TNA)

p. 250. **"although we had our little trials":** James Paton, diary, March 17, 1916, HC.

p. 251. **The proceeds were divided:** In March 1917, Tripp noted that the account intended for Mackintosh's widow in London held £538 (about $38,000 in current terms). This amount was over and above Mackintosh's salary, which had already been paid to his wife on a monthly basis through April 1916.

p. 251. **"odour of sanctity":** Ernest Shackleton to Leonard Tripp, July 20, 1917, SPRI.

p. 251. **Their record-setting journey:** Both of Scott's expeditions were plagued by scurvy. Shackleton's *Nimrod* southern party escaped the scourge with the use of pony meat on the march (120 days). Roald Amundsen avoided scurvy as well, although his round-trip journey was only ninety-nine days, compared with two hundred days for the Ross Sea party.

p. 252. **"I am afraid they made":** Ernest Shackleton to Emily Shackleton, June 3, 1916, SPRI.

p. 252. **"a mercy":** Ernest Shackleton to Emily Shackleton, February 27, 1917, SPRI.

p. 252. **"supreme efforts":** Frank Worsley, Notes of conversations with J. R. Stenhouse and E. H. Shackleton, verso letter dated November 17, 1932, HW. The lines refer to the Bible, Romans 13:7, "Render therefore to all their dues: tribute to whom tribute is due; custom to whom custom; fear to whom fear; honor to whom honor."

p. 252. **"This is the story":** *East Kent Times* (Ramsgate, England), July 4, 1917.

p. 252. **"wavered from wild enthusiasm":** *Surrey Comet and General Advertiser* (Kingston Upon Thames, England), May 12, 1917.

p. 252. **"If I have an enemy":** Leonard Tripp to Richard Richards, October 10, 1955, ATL.

p. 252. **"We went at it hammer and tongs":** Orme Masson to John King Davis, March 18, 1917, SLV.

p. 252. **"Have had committee on carpet":** Ernest Shackleton, cable to Leonard Tripp, March 17, 1917, ATL.

p. 252. **"badly organized"** . . . **"the whole motive":** John King Davis to Howard Ninnis, October 12, 1919, HC.

p. 253. **"the tail of his comet":** Howard Ninnis, diary, February 12, 1917, RGS.

p. 253. **word came:** Kavanagh died of his injuries in November 1918.

p. 253. **"even a man":** Leonard Tripp to R. J. Neville, Esq., March 20, 1917, ATL.

p. 253. **Joyce was rejected:** Ernest Joyce to Tannatt William Edgeworth David, September 16, 1919, SPRI. Joyce stated he needed to wear dark glasses for the first eighteen months after his return to civilization to shield his hypersensitive eyes. It is quite possible that he sustained permanent damage from ultraviolet radiation or mechanical injuries sustained while his eyes were numbed with cocaine.

p. 253. **"I don't know what he is up to":** Emily Shackleton and Ernest Shackleton to Leonard Tripp, July 13, 1917, ATL.

p. 253. **"If ever I can assist you":** Ernest Shackleton to Ernest Joyce, March 8, 1917, SPRI.

p. 254. **"Joyce troubling about Albert Medal," "Shut him up":** Ernest Shackleton, cable to Leonard Tripp, July 31, 1917.

p. 254. **"gratitude does not seem to be":** John King Davis to Joseph Kinsey, August 5, 1916, SLV.

p. 254. **"the Great Adventure":** Ernest Shackleton to Ernest Joyce, March 8, 1917, ATL.

p. 254. **"old pal Stennie," "together again":** Frank Worsley to Leonard Tripp, October 26, 1918, ATL.

p. 255. **Joyce was not among:** Shackleton may have helped Joyce obtain these jobs. Guillaume Delprat, Douglas Mawson's father-in-law, was the general manager of Broken Hill. In similar fashion, Joyce landed his sales job through the Commercial Travellers Association, for which Shackleton had lectured in 1917.

p. 255. **"I think"** . . . **"a record":** Ernest Shackleton, *South*, p. 244, 279.

p. 255. **"The ground to be covered," "any sign of the qualities":** Ernest Shackleton, *South*, p. 245.

p. 255. **"though there was a good deal of literature":** Ernest Shackleton, *South*, p. 265.

p. 255. **The executors:** Sir Robert Lucas-Tooth died in 1915.

p. 255. ***South*, the film:** Shot and edited by *Endurance* photographer Frank Hurley, the film was subsequently released in theaters in 1920 as *In the Grip of Polar Ice*. In 1933, a re–edited version with sound, featuring commentary by Frank Worsley, was released under the title *Endurance: The Story of a Glorious Failure*.

p. 255. **"spectacular failure":** *Daily Mail* (London), January 10, 1920.

p. 256. **Cope was planning** . . . **cost of £150,000:** In current terms, about $4.13 million.

p. 256. **"ill-considered in detail":** Francis Younghusband, draft of statement to be read at RGS meeting, February 1920, RGS.

p. 256. **"thoroughly impracticable":** Ernest Shackleton to RGS, February 1920, RGS.

p. 256. **"You have a great story":** Ernest Shackleton to Ernest Joyce, August 8, 1921, SPRI. Two copies of this letter are held by archives; the original has not been found.

p. 256. **"Alaric, King John, Pearls":** Ernest Shackleton, diary, 1914–15, undated endpapers, SPRI.

p. 257. **"double-crossed":** Ernest Joyce to Charles Royds, April 7, 1930, SPRI. The archival evidence on this point is contradictory. Shackleton clearly told Tripp that Joyce should not be permitted to use the film: "Joyce may have photographs not films latter under contract writing him a reference." (Ernest Shackleton, cable to Leonard Tripp, September, 20, 1919, ATL)

Yet Shackleton's August 8, 1921, letter granted Joyce permission to show the film with his lectures. It must be noted that the letter exists only in the form of two typed handbills without Shackleton's signature, which were printed by Joyce's lecture agent, Gerald Christy, who also represented Shackleton. Given Joyce's history of transcribing and revising correspondence, it is an open question whether the letter reads precisely as Shackleton wrote it.

p. 257. **"I am very sorry to hear":** Ernest Joyce to Hugh Robert Mill, March 10, 1924, SPRI.

p. 257. **Joyce asserted:** Joyce told Tripp that Shackleton had promised to pay him a bonus, which Shackleton called "incorrect" (Ernest Shackleton, cable to Leonard Tripp, September 20, 1919, ATL). Expedition records show that Joyce's salary was £250 per year (about $17,500 currently), for a total of £583 from the time he joined the expedition in 1914 until his return in Wellington in 1917. Tripp paid him £636 in cash and covered his hotel bill, meals, and entertainment in Wellington. After Shackleton's death, Joyce insisted that Shackleton had promised him £350 per annum and a bonus for biological specimens, for a grand total of £2,800 *above* the £636 already disbursed to him by Tripp (about $197,000 above the $44,740). There is no written record of Shackleton committing to remuneration on this scale for Joyce. This amount seems exaggerated in light of the fact that Mackintosh was guaranteed £400 per year (about $37,000). (PRO MT 9/997, TNA)

p. 257. **"under the circumstances":** Ernest Joyce to Hugh Robert Mill, March 10, 1924, SPRI.

p. 257. **"calculated to give an incorrect":** Admiralty minute, sheet 3 N 10473/22, January 26, 1923, PRO ADM 1/8629/132, TNA.

p. 258. **But his hunger:** Joyce had returned home from the Antarctic with little improvement in his circumstances, a fact he was ashamed to admit to his widowed mother. He presented her with a duplicate of his Polar Medal which he doctored by filing down the engraved bar that revealed his rank on *Discovery* to be a mere able seaman. In contrast, his elder brother, Joseph, had gone to France as a private in 1914 and returned as a major after a series of battlefield promotions.

p. 258. **"I do not like to write you":** John Quiller Rowett to Leonard Tripp, August 26, 1922, ATL. Joyce was named Ernest Edward Joyce at birth and added his mother's maiden name, Mills, while he was in the Royal Navy.

p. 258. **"out of a job":** Ernest Joyce to Charles Royds, April 7, 1930, SPRI.

p. 258. **"he didn't want him":** Frank Worsley, notes of conversation with J. R. Stenhouse and E. H. Shackleton, verso letter dated November 17, 1932, HW.

p. 258. **"did not want an old fogey":** Ernest Joyce to Charles Royds, April 7, 1930, SPRI.

p. 259. **"I understand that the full story":** Hugh Robert Mill to Alexander Stevens, November 24, 1928, SPRI. One such issue concerned Mackintosh's headlong rush into sledging in early 1915. Had Mackintosh known that Shackleton intended to delay his crossing to late 1915, it is arguable that he would not have hurried into sledging without adequate training. In *The Life of Sir Ernest Shackleton*, Mill states that Shackleton's change of plan "was sent home in time to reach the *Aurora* before she sailed for the Ross Sea" (p. 204); in *The Geographical Journal*, he asserted that there was "no doubt Captain Mackintosh was acquainted" with the plan ("The Position of Sir Ernest Shackleton's Expedition," 47:5, 1916). He also asserted that the new plan would have made "no material difference in the programme of the *Aurora*." His account is contradicted by the RGS meeting minutes in which Ernest Perris told the RGS president, "Shackleton knew perfectly well he could never cross the first season" and that "S's final instruction to Mackintosh . . . Was to be corrected by a long cable which was never sent" (Douglas Freshfield, interview with Ernest Perris, March 31, 1916, RGS). As a longtime RGS associate and a close friend of Shackleton, Mill was in a position to know this and access the correspondence.

p. 259. **"It is true"**: Ernest Joyce, *The South Polar Trail*, pp. 80–81, quoting Ernest Shackleton, *The Heart of the Antarctic*, pp. 5–6.

p. 259. **"I should be sorry"**: Alexander Stevens to Hugh Robert Mill, December 12, 1928, SPRI.

p. 259. **"something steadfast"**: Alexander Stevens to Leslie Quartermain, July 20, 1961, CM.

p. 260. **"great march south"**: John King Davis, draft manuscript of *High Latitude*, SLV.

p. 260. **"astounded"**: Richard Richards to Leslie Quartermain, December 20, 1963, CM.

p. 260. **"The shore party for the Ross Sea"**: *Sydney Morning Herald*, June 29, 1914.

p. 260. **"in charge of all equipment," "I can promise"**: Ernest Joyce, *The South Polar Trail*, p. 28.

p. 260. **"a wonderful performance"**: *Times Literary Supplement* (London), April 11, 1929.

p. 260–61. **"guilelessness of unadorned"**: *Daily Express* (London), April 19, 1929.

p. 261. **"The prose epics of the Antarctic"**: *Evening News* (London), March 1, 1929.

p. 261. **"first-rate"**: *The Geographical Journal*, Vol. 73, No. 6 (1929), pp. 568–69.

p. 261. **"I know you know me"**: Ernest Joyce to Charles Royds, April 7, 1930, SPRI.

p. 261. **"Talking is easy"**: Ernest Joyce to Charles Royds, April 7, 1930, SPRI.

p. 261. **"I've found it"**: Draycot Dell (Ernest Joyce), "Hard-Fist Jemsen," unpublished manuscript, ATL.

p. 261. **"There's a race of men"**: Robert Service, "The Men That Don't Fit In."

p. 262. **"An Interested Observer"**: Ernest Joyce, undated letter, ATL.

p. 262. **"middle-aged, stockily built"**: *Evening News* (London), May 2, 1931, ATL.

p. 262. **"At 64, He Still Wants to Explore"**: *People* (London), May 28, 1939, BL.

p. 263. **On May 2, 1940**: The death certificate records the cause as fibroid myocardial degeneration, congestive heart failure, emphysema, and bronchitis. He had likely been suffering from heart disease for some time.

p. 253. **"Not a rotter"**: Richard Richards to A. G. E. Jones, May 2, 1975, CM.

p. 263. **"plenty of faults"**: Richard Richards to Leslie Quartermain, February 2, 1964, CM.

p. 263. **"After the trouble with my heart"** . . . **"the Pole has been found"**: Richard Richards, oral history interview, December 2, 1976, NLA.

p. 264. **"What is left for exploration?"**: *Daily Mail* (London), March 28, 1933.

p. 264. **"The modern Antarctic"**: Richard Richards, *The Ross Sea Shore Party 1914–17*, p. 3.

p. 265. **"Cape Evans was 'home'"**: Richard Richards, *The Ross Sea Shore Party 1914–17*, p. 10.

p. 265. **"beauty untold"**: Keith Jack, diary, December 4, 1915, SLV.

p. 265. **"I don't wonder"**: Irvine Gaze, diary, February 3, 1915, CM.

p. 265. **"weighing the credits"**: Richard Richards, *The Ross Sea Shore Party 1914–17*, p. 3.

p. 265. **"It was something"**: Richard Richards, oral history interview, December 2, 1976, NLA.

p. 265. **He died at ninety-one:** None of the depots have been found since 1916. British glaciologist Charles Swithinbank, one of the first party to visit the Gateway since the Ross Sea party, looked for the Mount Hope Depot in 1960 during a scientific expedition and found no trace of it.

— EPILOGUE: "THE BROTHERHOOD OF MEN WHO KNOW THE SOUTH" —

p. 269. **George Hubert Wilkins:** A war hero and photographer, Wilkins joined Shackleton's *Quest* expedition and became a pioneer of polar aviation and Arctic submarine exploration.

p. 269. **John Lachlan-Cope:** John Lachlan-Cope did not hyphenate his surname until after he began practicing medicine in 1933. He is thus referred to as "Cope" throughout this book.

p. 270. **"brittle sharpness of intellect"**: *Scottish Geographical Magazine*, Vol. 82 (1966), p. 58.

p. 270. **In 1953, he lost:** If indeed Stevens was diabetic, as some evidence suggests, blindness could have resulted from his condition. Diabetes can cause retinopathy, in which the tiny blood vessels of the retina are damaged and may ultimately progress to blindness.

p. 271. **Casting about for work:** The film was edited by Frank Bickerton, who accompanied Mackintosh on his treasure-hunting expedition to Cocos Island. Bickerton had been a motor mechanic for Mawson's first Antarctic expedition.

p. 272. **"when the war will cease"**: Joseph Stenhouse to Leonard Tripp, November 9, 1917.

p. 275. **"roving disposition"**: M. A. Glenn, Manchester League of Help to W. J. Rule of Mercantile Marine Office, April 8, 1919, PRO BT 99/3286, TNA.

p. 275. **"recompense"**: Editor of *John Bull* to Sidney Webb, December 29, 1924, PRO BT 99/3286, TNA.

p. 275. **"A great explorer"**: John King Davis to Hugh Mill, July 3, 1922, SPRI.

p. 276. **"deepwater sailorman"**: *The Age* (Melbourne), January 8, 1949.

— ACKNOWLEDGMENTS —

p. 283. **"All men dream"**: T. E. Lawrence, *The Seven Pillars of Wisdom*, p. 23.

— NOTES —

p. 287. **"A strange mixture"**: Roland Huntford, *Shackleton*, p. 194.

BIBLIOGRAPHY

The cited edition is indicated by an asterisk (*).

— BOOKS —

Amundsen, Roald. *The South Pole: An Account of the Norwegian Antarctic Expedition in the "Fram" 1910–1912.* Translated by A. G. Chater. London: John Murray, 1912.

Ayres, Philip. *Mawson: A Life.* Melbourne: Melbourne University Press, 1999.

Bagshawe, Thomas W. *Two Men in the Antarctic: An Expedition to Graham Land 1920–22.* Cambridge: Cambridge University Press, 1939.

Ballantyne, Robert. *The World of Ice.* London: Blackie and Son, 1860.

Barnett, Correlli. *The Great War.* London: Penguin Books, 1979.

Bates, L. M. *The Londoner's River.* London: Frederick Muller, 1949.

Bean, Charles Edwin Woodrow. *The Official History of Australia in the War of 1914–1918.* Sydney: Angus & Robertson, 1929.

Bernacchi, Louis. *To the South Polar Regions: Expedition of 1898–1900.* London: Hurst and Blackett, 1901.

Berton, Pierre. *The Arctic Grail.* New York: Viking, 1988.

Bertram, Colin. *Arctic and Antarctic: A Prospect of the Polar Regions.* Cambridge: W. Heffer and Sons, 1957.

Bickel, Lennard. *Shackleton's Forgotten Argonauts.* Melbourne: Macmillan, 1982.

Brittain, Sir Harry. *Happy Pilgrimage.* London: Hutchinson, 1949.

Browning, Robert. *Dramatis Personae.* London: Chapman and Hall, 1864.

Carpenter, Kenneth J. *Beriberi, White Rice, and Vitamin B: A Disease, a Cause, and a Cure.* Berkeley: University of California Press, 2000.

———. *The History of Scurvy and Vitamin C.* Cambridge: Cambridge University Press, 1986.

Cherry-Garrard, Apsley. *The Worst Journey in the World.* London: Constable, 1922.

———. *The Worst Journey in the World.* New York: Carroll & Graf, 1998.*

Church, Ian. *Last Port to Antarctica. Dunedin and Port Chalmers: 100 Years of Polar Service.* Otago: Otago Heritage Books, 1997.

Cook, Frederick. *Through the First Antarctic Night 1898–1899: A Narrative of the Voyage of the Belgica Among Newly Discovered Lands and Over an Unknown Sea about the South Pole.* New York: Doubleday & McClure, 1900.

David, T. W. E., R. Priestley, D. Mawson, T. G. Taylor, E. J. Goddard, J. Murray, and C. Hedley. *British Antarctic Expedition 1907–9 Scientific Reports.* London: William Heinemann, 1914.

Davis, John King. *High Latitude.* Melbourne: Melbourne University Press, 1962.

―――. Trial by Ice: The Antarctic Journals of John King Davis. Edited and introduced by Louise Crossley. Bluntisham, England: Bluntisham Books, Erskine Press, 1997.

―――. With the "Aurora" in the Antarctic, 1911–1914. London: Andrew Melrose, 1919.

Field Manual for the United States Antarctic Program. National Science Foundation, Office of Polar Programs, 2002–2003.

Fisher, Hervey. From a Tramp's Wallet: A Life of Douglas William Freshfield. Norwich, England: Erskine Press, 2002.

Fisher, Margery, and James Fisher. Shackleton and the Antarctic. Boston: Houghton Mifflin Co., 1958.

Flannery, Nancy Robison, ed. The Everlasting Silence: The Love Letters of Paquita Delprat and Douglas Mawson, 1911–1914. Melbourne: Melbourne University Press, 2000.

Funk, Casimir. The Vitamines. Baltimore: Williams & Wilkins, 1922.

Gerlache, Adrian de. Voyage of the Belgica Fifteen Months in the Antarctic. Edited and translated by Maurice Raraty. Norfolk and Huntingdon: The Erskine Press and Bluntisham Books, 1998.

George, David Lloyd. War Memoirs of David Lloyd George. Boston: Little, Brown, and Company, 1933.

Gilbert, Martin. Winston S. Churchill. Volume III, Companion, Part 2, Documents, May 1915–December 1916. London: Heinemann, 1972.

Gurney, Alan. Below the Convergence: Voyages toward Antarctica 1699–1839. New York: W. W. Norton, 1997.

―――. The Race to the White Continent: Voyages to the Antarctic. New York: W. W. Norton, 2000.

Hand, Michael S., and Bruce J. Novotny. Pocket Companion to Small Animal Clinical Nutrition. 4th ed. Topeka, Kan.: Mark Morris Institute Press, 2002.

Handbook for Wireless Telegraph Operators Working Installations Licensed by His Majesty's Postmaster-General. London: His Majesty's Stationery Office, 1912.

Hattersley-Smith, Geoffrey. The Norwegian with Scott: Tryggve Gran's Antarctic Diary 1910–13. Greenwich: National Maritime Museum, 1984.

Headland, R. K. Chronological List of Antarctic Expeditions and Related Historical Events. Cambridge: Scott Polar Research Institute, 1989.

Hillary, Edmund, and Vivian Fuchs. The Crossing of Antarctica: The Commonwealth Trans-Antarctic Expedition 1955–1958. Boston: Little, Brown, 1958.

Huntford, Roland. The Last Place on Earth. New York: Random House, 1999. First published in 1979 under the title Scott and Amundsen.

―――. Nansen. London: Duckworth, 1997.

―――. Shackleton. New York: Atheneum, 1986.

James, Lawrence. The Rise and Fall of the British Empire. London: Abacus, 1994.

Jones, A. G. E. Polar Portraits. Whitby, England: Caedmon of Whitby, 1992.

Joyce, Ernest E. Mills. The South Polar Trail. London: Duckworth, 1929.

Kemp, Peter, ed. The Oxford Companion to Ships and the Sea. Oxford: Oxford University Press, 1988.

Leach, Harry, rev. by William Spooner. The Ship Captain's Medical Guide. London: Simpkin, Marshall, Hamilton, Kent, 1906.

Lindsay, David Moore. A Voyage to the Arctic in the Whaler "Aurora." Boston: Dana Estes, 1911.

The Mariner's Handbook. 7th ed. Taunton, England: Hydrographer of the Navy, 1999.

Markham, Clements. *Antarctic Obsession: A Personal Narrative of the Origins of the British National Antarctic Expedition 1901–1904.* Edited and introduced by Clive Holland. Alburgh, England: Bluntisham Books and the Erskine Press, 1986.

Marriott, Bernadette M., and Sydney J. Carlson, eds., Committee on Military Nutrition Research, Food and Nutrition Board, Institute of Medicine. *Nutritional Needs in Cold and High-Altitude Environments.* Washington: National Academy Press, 1996.

Mawson, Douglas. *The Home of the Blizzard.* New York: St. Martin's Press, 1998.

McKee, Christopher. *Sober Men and True: Sailor Lives in the Royal Navy 1900–1945.* Cambridge: Harvard University Press, 2002.

Mear, Roger, and Robert Swan. *A Walk to the Pole: To the Heart of Antarctica in the Footsteps of Scott.* New York: Crown, 1987.

Mill, Hugh Robert. *The Life of Sir Ernest Shackleton.* London: William Heinemann, 1923.

Mills, William James. *Exploring Polar Frontiers: A Historical Encyclopedia.* Santa Barbara, Calif.: ABC-CLIO, 2003.

Montcombroux, Geneviève. *The Canadian Inuit Dog: Canada's Heritage.* Winnepeg: Whippoorwill, 1997.

Murray, George, ed. *The Antarctic Manual for the Use of the Expedition of 1901.* London: Royal Geographical Society, 1901.

Poulsom, Neville, and Rear Admiral John A. L. Myres. *British Polar Exploration and Research: A Historical and Medallic Record with Biographies 1818–1999.* London: Savannah Publications, 1968.

Osler, William. *The Principles and Practice of Medicine.* 8th ed. New York and London: Appleton and Company, 1912.

Priestley, John Boynton. *The Edwardians.* London: Heinemann, 1970.

Priestley, Raymond. *Antarctic Adventure.* London: T. F. Unwin, 1914.

Quartermain, L. B. *Antarctica's Ten Forgotten Men.* Wellington, New Zealand: Millwood Press, 1981.

Richards, Richard W. *The Ross Sea Shore Party 1914–17.* Special Publication #2. Cambridge, England: Scott Polar Research Institute, 1962.

Riffenburgh, Beau. *Nimrod: Ernest Shackleton and the Extraordinary Story of the 1907–09 British Antarctic Expedition.* London: Bloomsbury, 2004.

Robbins, Keith. *The First World War.* Oxford: Oxford University Press, 1984.

Ross, James Clark. *A Voyage of Discovery and Research in the Southern and Antarctic Regions During the Years 1839–1843.* London: John Murray, 1847.

Roosevelt, Theodore. *The Strenuous Life.* New York: Century, 1901.

Sailing Directions for Antarctica. Washington D.C.: Defense Mapping Agency, Hydrographic and Topographic Center, 1985.

Savours, Ann. *The Voyages of the Discovery: The Illustrated History of Scott's Ship.* London: Chatham, 2001.

Scott, Robert Falcon. *Scott's Last Expedition.* London: Smith, Elder, 1913.

———. *Scott's Last Expedition.* New York: Dodd, Mead, 1929.*

———. *The Voyage of the "Discovery."* London: John Murray, 1905.

Service, Robert W. *Ballads of a Cheechako.* New York: Barse & Hopkins, 1909.

———. *Rhymes of a Rolling Stone.* Toronto: William Briggs, 1912.

———. *The Spell of the Yukon and Other Verses.* New York: Barse & Hopkins, 1907.

Shackleton, Ernest. *South: The Story of Shackleton's Last Expedition 1914–1917.* New York: The Macmillan Co., 1920.

———. *South: The Story of Shackleton's Last Expedition 1914–1917.* New York: Carroll & Graf, 1998.*

———. *The Heart of the Antarctic: Being the Story of the British Antarctic Expedition, 1907–1909.* London: Heinemann, 1909.

———. *The Heart of the Antarctic: Being the Story of the British Antarctic Expedition, 1907–1909.* New York: Carroll & Graf, 1999.*

———, ed. *Aurora Australis.* Cape Royds, Ross Island: Ernest Joyce and Frank Wild at the Sign of the Penguins, 1908.

———, ed. *Aurora Australis.* Shrewsbury, England: Airlife, 1988.

Steger, Will, and John Bowermaster. *Crossing Antarctica.* New York: Alfred A. Knopf, 1992.

Stroud, Mike. *Survival of the Fittest: Understanding Health and Peak Physical Performance.* New York: Random House, 1998.

Swithinbank, Charles. *Forty Years on Ice.* Lewes, Sussex: The Book Guild, 1998.

Thomson, John. *Shackleton's Captain: A Biography of Frank Worsley.* Toronto: Mosaic Press, 1999.

Tiltman, H. Hessell, and T. C. Bridges. *Heroes of Modern Adventure.* Boston: Little Brown, 1927.

———. *More Heroes of Modern Adventure.* London: George G. Harrap, 1929.

Tuchman, Barbara. *The Guns of August.* New York: Ballantine Books, 1962.

———. *The Proud Tower: A Portrait of the World before the War 1890–1914.* New York: The Macmillan Company, 1966.

Twight, Mark F., and James Martin. *Extreme Alpinism.* Seattle: Mountaineers Books, 1999.

University of Cambridge: The Book of Matriculations and Degrees 1901–1912. Cambridge, England: Cambridge University Press, 1915.

Venn, John, and J. A. Venn. *Alumni Cantabrigienses.* Cambridge: Cambridge University Press, 1922–1954.

Walton, D. W. H., ed. *Antarctic Science.* Cambridge, England: Cambridge University Press, 1987.

The War Office. *Statistics of the Military Effort of the British Empire During the Great War 1914–1920.* London: His Majesty's Stationery Office, 1922.

Watkins, Julian L. *The World's Hundred Greatest Advertisements.* New York: Dover, 1959.

Wheeler, Sara. *Cherry: A Life of Apsley Cherry-Garrard.* New York: Random House, 2002.

Winter, Dennis. *Death's Men: Soldiers of the Great War.* London: Penguin Books, 1979.

Worsley, Frank A. *Endurance.* London: P. Allan, 1931.

Yelverton, David. *Antarctica Unveiled.* Denver: University Press of Colorado, 2000.

— JOURNAL AND MAGAZINE ARTICLES —

"Alexander Stevens." *Polar Record* 13 (1966–67): 493.

"Alexander Stevens, M.A., B.Sc., Professor of Geography." *The Journal of Glasgow University Graduates Association,* no. 11 (1953): 49.

Barlow, Frank. "The Telefunken Coastal Wireless Stations." *Break-In: The Journal of the NZART* (1996).

"C. C. Mauger." *Antarctic* 3, no. 9 (1964): 414.

Cherry-Garrard, Apsley. "Review of *South: The Story of Shackleton's Last Expedition 1914–17.*" *The Nation* (1919).

Curzon, G. N. "Address to the Royal Geographical Society by the Rt. Honorable Earl Curzon of Kedleston." *Geographical Journal* 44, no. 1 (1914): 2–3.

———. "Letter from Lord Curon." *Geographical Journal* 43, no. 2 (1915): 178.

David, Tannatt William Edgeworth. "Antarctica and Some of Its Problems." *Geographical Journal* 43, no. 6 (1914): 605–30.

Davis, John King. "The Ross Sea Relief Expedition." *Life* (Jan. 3, 1921): 197–200.

Donaldson, I., T. Campbell, and J. Donaldson. "Energy Requirements of Antarctic Sledge Dogs." *British Journal of Nutrition* 45, no. 95 (1981): 95–98.

Du Baty, Raymond Rallier. "La Voyage de la Curieuse." *La Géographie* 37, no. 1 (1922): 1–38.

Fairhurst, H. "Alexander Stevens." *Scottish Geographical Magazine* 82 (1966): 58.

———. "Emeritus Professor Stevens." *Geography* 51 (1966): 266.

Glenny, M. R. and A. G. E. Jones. "Richard Walter Richards (1893–1985)." *Fram: The Journal of Polar Studies* (1985): 13–16.

Guly, H. R. "Medicine in the Heart of the Antarctic: 1908–2001." *Emergency Medical Journal* 19 (2002): 314–17.

Hardy, Anne. "Straight Back to Barbarism: Antityphoid Inoculation and the Great War, 1914." *Bulletin of the History of Medicine* 74, No. 2 (2000), 265–290.

Harrowfield, David, and Richard McElrea. "'Dick' Richards G. C., Last of Shackleton's Men." *Antarctic* 10, no. 9–10 (1985): 378–82.

"The Home of the Blizzard: An Interesting Account of Wireless Telegraphy and Its Uses in Polar Exploration." *Wireless World* (1916): 779–83.

"'Jimmy' Gaze, Ross Sea Party Survivor." *Antarctic* 8 (1978), no. 6: 209–12.

Jones, A. G. E. "Shackleton's Ross Sea Party 1914–17." *Fram: The Journal of Polar Studies* 1, no. 2 (1984): 429–45.

———. "Tubby: Profile of Ernie Wild." *Polar Record* 18, no. 112 (1976): 43–45.

Kendall, E. J. C. "Scurvy During Some British Polar Expeditions, 1875–1917." *Polar Record* 7 (1955): 467–85.

Larkman, A. H. "An Engineer's Antarctic Log (Pt. I and II)." *New Zealand Engineering* 18, no. 8: 286–91 and no. 9 (1963): 327–28.

Loewe, Fritz. "The Scientific Observations of the Ross Sea Party of the Imperial Trans-Antarctic Expedition 1914–17." *Ohio State University, Institute of Polar Studies*, no. 5 (1963).

———. "On Melting of Fresh-Water Ice in Sea-Water." *Journal of Glaciology* 3 (1961): 1051–52.

"Latest Resources of Science and Invention Enlisted to Equip Shackleton Expedition." *Popular Mechanics* 1914.

"Men of Mark: A. P. Spencer-Smith." *Queens' College Dial*, Lent Term edition 1907.

Mill, H. R. "The Position of Sir Ernest Shackleton's Expedition." *Geographical Journal* 47, no. 5 (1916): 369–76.

———. "The Relief of Shackleton's Ross Sea Party." *Geographical Journal* 49, no. 3 (1917): 218–21.

————. "The Return of Sir Ernest Shackleton." *Geographical Journal* 48, no. 1 (1916): 68–73.

Monaghan, Andrew J., D. Bromwich, Jordan G. Powers, and Kevin W. Manning. "The Climate of the McMurdo, Antarctica Region as Represented by One Year of Forecasts from the Antarctic Mesoscale Prediction System." *Journal of Climate*, 18 (2005): 1174–89.

"Motor Sledges for the Antarctic." *Motor*, Aug. 4, 1914.

Palinkas, L. A., H. L. Reed, and N. Do. "Association between the Polar T3 Syndrome and the Winter-over Syndrome in Antarctica." *Antarctic Journal* 32, no. 5 (Review 1997): 111–14.

Pearson, Michael. "Expedition Huts in Antarctica: 1899–1917." *Polar Record* 28, no. 167 (1992): 261–76.

————. "Sledges and Sledging in Polar Regions." *Polar Record* 31, no. 176 (1995): 3–24.

Reed, H. L., K. R. Reedy, L. A. Palinkas, N. V. Do, N. S. Finney, H. S. Case, H. J. LeMar, J. Wright, and J. Thomas. "Impairment in Cognition and Exercise Performance During Prolonged Antarctic Residence: Effect of Thyroxine Supplementation in Polar T3 Syndrome." *Journal of Clinical Endocrinology and Metabolism* 86: 110–16.

Richards, R. W. "Alexander Stevens: A Tribute by R. W. Richards." *Antarctic* 4, no. 6 (1966): 313.

————. "The Veterans Pass: A. H. Larkman, A Tribute." *Antarctic* 3, no. 3 (Sept. 1962): 127.

Rogers, A. F. "The Death of Chief Petty Officer Evans." *Practitioner* 212 (1974): 570–79.

"The Shackleton Antarctic Expedition." *Scottish Geographical Magazine* 32, no. 5 (1916): 242–47.

"The Shackleton Antarctic Expedition." *Scottish Geographical Magazine* 33, no. 3 (1917): 122–27.

Shackleton, E. H. "The Imperial Trans-Antarctic Expedition, 1914." *Geographical Journal* 43 (1914): pp. 173–78.

Shaughnessy, P. D. "Bird and Mammal Life Recorded during the Antarctic Drift of S. Y. *Aurora*." *Polar Record* 26, no. 159 (1990): 277–80.

"Special Issue Commemorating the Fiftieth Anniversary of the International Glaciological Society." *Journal of Glaciology* (1987).

Taylor, R. J. F. "The Physiology of Sledge Dogs." *Polar Record* 8, no. 55 (1957): 317–21.

————. "The Work Output of Sledge Dogs." *Journal of Physiology* 137, no. 2 (1957): 210–17.

Turner, Herbert Hall. "An Antarctic Log." *The Times Literary Supplement*, April 11, 1929.

"The Veterans Pass: A. K. Jack." *Antarctic* 4, no. 8 (1996): 408.

Watson, Eric. "An Aussie in Airships: The Wartime Experiences of Sir Lionel Hooke." *Australian Society of World War One Aero Historians* (1975–76): p. 66–92.

Weinberg, Andrew D., MD. "Hypothermia Special Situations." *Annals of Emergency Medicine* 22 (1993): 370–77.

Wordie, James Mann. "Review of *The South Polar Trail*." *Geographical Journal* 73, no. 6 (1929): 568–69.

————. "The Ross Sea Drift of the *Aurora* in 1915–1916." *Geographical Journal* 58, no. 3 (1929): 219–24.

Wright, Charles Seymour. "Terrestrial Magnetism and Related Observations. Part 4. The Transmission of Wireless Signals in Relation to Magnetic and Auroral Disturbances." *Australalian Antarctic Expedition 1911–1914, Scientific Reports, Series B*. (1940).

Yelverton, David. "The Bronzes That Never Were: An Unravelling of the Ungazetted Polar Medal Awards." *Miscellany of Honors* 4 (1982): 3–29.

———. "The Naming of Early Polar Medals and Recognition of Duplicates." *Miscellany of Honors* 12 (1998): pp. 15–28.

— UNPUBLISHED SOURCES —

Agreement and Account of Crew, S. Y. *Aurora* (Ship's Articles). ATL.

Antarctic Expedition, E. Shackleton, Correspondence and Reports. NAA.

Army List. TNA.

Aurora Relief Committee Correspondence, 1916. RGS.

Barr, George. Correspondence with L. D. Mackintosh. Private Collection, MFP.

Birth certificates and death certificates. London: General Register Office.

Board of Admiralty and Navy Board Records. TNA.

> Antarctic Expeditions. Loan of officers, stores etc. (ADM 1/8387/222)
> *Discovery* fitting out for Shackleton polar expedition. (ADM 131/83)
> Home Waters Telegrams, 1914. (ADM 137/51)
> Request for services of officers and men of RN. (ADM 1/8368/29)
> Recommendation for the Albert Medal for four members of Shackleton's Trans-Antarctic Expedition. (ADM 1/8629/132) (includes transcript of E. E. Joyce diary)
> Shackleton Relief Advisory Committee Final Report. (ADM 1/8483/60)
> The Shackleton Relief Expedition. (ADM 116/1712)

Board of Trade. Correspondence re: H. Shaw (BT 99/3286). TNA.

British India Steam Navigation Company. Records. NMM.

Cope, John Lachlan. Medical Report of the Ross Sea Base ITAE. SPRI.

Correspondence and Reports Relating to the Shackleton Relief Expedition. SPRI.

Correspondence Relating to the Purchase and Shipping of Sledge Dogs for the Imperial Trans-Antarctic Expedition (A.12/Ft Misc./223, F. 6-44). HBC.

Davis, John King. Diaries and Papers. SLV.

———. Correspondence. HC.

———. Correspondence. RGS.

———. Correspondence. SPRI.

Discovery Expedition Files, RGS/AA. (Including C. Markham, record of RGS Antarctic silver medallists). RGS.

Evans, Frederick Pryce. Log of the *Koonya*. ATL.

Fisher, Margery and James. The Fisher Papers. SPRI.

Gaze, Irvine Owen. Imperial Trans-Antarctic Expedition Diary (1915–1917). CM.

———. Correspondence with Charlotte Spencer-Smith. CM.

———. Correspondence with L. B. Quartermain. CM.

Hayward, Victor George. Diary (1914–1916) and papers. NMM.

His Majesty's Treasury. Allocation of Expenditure between Imperial and Government Funds in Connection with Voyage of the *Aurora* to the Ross Sea. (Shackleton Relief Expedition). (T1/12337). TNA.

Hooke, Lionel. Diary ITAE (1914–16). ML.

———— S. Y. *Aurora* Wireless Log, AWA (Australasia) Limited (1914–16). ML.

Hurley, James Francis (Frank). *Endurance* Diary. ML.

Hutchison, Alfred, and Perris, Ernest. Correspondence with Royal Geographical Society re: Shackleton Relief Expedition. RGS.

Jack, Andrew Keith. Correspondence with L. B. Quartermain. CM.

————. Imperial Trans-Antarctic Expedition Diary, 1914–16. SLV.

Jack, A. K., R. W. Richards, A. Stevens, and A. L. A. Mackintosh, ITAE Observations, Ross Sea Section. SPRI.

Jones, A. G. E. Papers. SPRI.

Joyce, Ernest Edward. Correspondence. RGS.

————. Correspondence with C. W. R. Royds. SPRI.

————. Correspondence with T. W. E. David. US.

————. Log (original). Private Collection. RPK.

————. Log Kept During Imperial Trans-Antarctic Expedition 1915–17 (transcript). ATL.

————. Manuscript of *The South Polar Trail*. ATL.

————. Papers of Ernest E. Joyce. ATL.

————. Sledging correspondence. SPRI.

Kinsey, Joseph J. Papers. ATL.

Larkman, Alfred Herbert. Correspondence with L. B. Quartermain. CM.

————. Diary (1914–16). SPRI.

————. Engine Room Log, 1914–15. SPRI.

Lloyd's Captains Registers. GL.

Lloyd's Register of Shipping. GL.

Mackellar, Campbell. Correspondence with W. S. Bruce. SPRI.

MacKinnon, Alasdair. Ross Sea Party Relief Expedition Diary 1916–1917. RGS.

MacKinnon, F. D. Voyage and Disappearance of the Auxiliary Barque *Aurora*, Arbitrators Report for Underwriters 1927. MC.

Mackintosh, Æneas Lionel Acton. Correspondence. MFP.

————. ITAE Diary (Dec. 31, 1914–Jan. 24, 1915; June 5–Sept. 30, 1915) and expedition papers. SPRI.

————. ITAE Ross Sea Party, Sledging Programme, 1915–16, SPRI.

————. ITAE Sledge Journey Diary (Jan. 25–April 15, 1915). RGS.

————. A Log of the Sledging Parties Giving Bearings, Distance, Course, Sledgemeter Readings and Remarks 26 Jan. 1915–25 March 1915. SPRI.

————. *Nimrod* Diary (1907–1909). RGS.

Marshall, Eric. Correspondence. SPRI.

————. *Nimrod* Diary (1907–1909) (RGS/EMA/6 and 7). RGS.

————. Papers (including Dudley Everitt correspondence) (RGS/EMA/6 and 7). RGS.

Marshall, Hilles. Correspondence. SPRI.

Mauger, Clarence Charles. Correspondence with L. B. Quartermain. CM.

————. Diary of work, started Sept. 13, 1915. ATL.

Mawson, Douglas. Papers. MC.

Medal Rolls Index. TNA.

Mill, Hugh Robert. Papers. SPRI.

Ministry of Transport Records. Seamen Abroad (Code 121): Allotments, Sir Ernest Shackleton's Imperial Trans-Antarctic Expedition 1914–1915 (MT 9/977). TNA.

Morrison, A. Surveyor's Report, S. Y. *Aurora*. SLV.

Moyes, Morton H. Diary, 26 Dec. 1916–7 Feb. 1917, kept while a member of the *Aurora* Antarctic Relief Expedition 1916–17. ML.

Moyes, M. H., A. LeGros, and W. Aylward. Meteorological Log 20 Dec. 1916–9 Feb. 1917. ML.

Moyes, M. H., and A. L. A. Mackintosh. Navigation Book, S. Y. *Aurora*, 16 Dec. 1914–12 Jan. 1915; 21 Dec. 1916–9 Jan. 1917. ML.

New Zealand Post and Telegraph Dept. Awarua Coastal Wireless Station Attendance Book. SM.

Ninnis, Aubrey Howard. Correspondence. SPRI.

——. Diary 1914–1917 (original and transcript) and Papers. (RGS/AHN.) RGS.

——. Log (Private for Ethel Douglas). Private Collection, EDP.

——. Papers. HC.

Paton, James. *Aurora* Diary. HC.

Peninsular & Oriental Steam Navigation Company Records. NMM.

Phillips, Anne. Mackintosh family genealogy. Private Collection, MF.

Priestley, Raymond. Diary (1907–1909). ATL.

Quartermain, Leslie B. Correspondence. CM.

Report of Conference of a Committee of the Royal Geographical Society with Sir Ernest Shackleton, 1914. RGS.

Richards, Richard Walter. Correspondence with L. B. Quartermain. CM.

——. Diary (1915–1916). CM.

——. Marooned in the Antarctic (Draft manuscript of *The Ross Sea Shore Party 1914–17*). MC.

——. Summary Accounts of the Imperial Trans-Antarctic Expedition, Ross Sea Party. SPRI.

——. Verbatim Copy of Diary Kept by R. W. Richards from February 23rd, 1916 to March 19th During the Shackleton Transantarctic Expedition, 1914–17. SPRI.

Royal Geographical Society. Correspondence Relating to John Lachlan Cope Expedition. RGS.

Saunders, Edward R. C. Papers. ATL.

Shackleton, Emily Dorman. Correspondence. SPRI.

Shackleton, Ernest Henry. "The Antarctic Petrel." ATL.

——. Correspondence with E. E. Joyce. ATL.

——. Expedition files (RGS/EHS). RGS.

——. ITAE Diary and Papers (1914–16). SPRI.

——. Relief Diary (1916–17). SPRI.

——. RGS Correspondence re the ITAE. RGS.

——. The Shackleton Collection (including the papers of the Ross Sea party). SPRI.

Shaw, Harold. Manchester Police Records. GMPM.

——. Tasmanian Police Records (POL 324/1/3). AOT.

Smith, Clifford. Spencer-Smith Family Genealogy. Private Collection.

Spence, Captain D. Report on Tug Dunedin (1916), Otago Harbour Board. SPRI.

Spencer-Smith, Arnold Patrick. Correspondence with Charlotte Spencer-Smith. CM.

———. Diary. March 15–Sept. 30, 1915 (original). CM.

———. Sledging Diaries (1915–16) (original). SPRI.

Stenhouse, Joseph Russell. Diaries, Logs, and Papers. SPRI.

Stenhouse, J. R., and A. L. A. Mackintosh. Log of the S. Y. *Aurora* (1914–1916). SPRI.

Stenhouse J. R., et al. Wireless Communications to/from S. Y. *Aurora*, 1914–1917. SPRI.

Stevens, Alexander. Correspondence with L. B. Quartermain. CM.

———. Report by A. Stevens, July 30, 1915. SPRI.

———. Report of the Ross Sea Party. SPRI.

Thomson, Leslie James Felix. Diary. SPRI.

Tripp, Leonard. Papers. ATL.

Wild, Harry Ernest. Diary, Oct. 1914–March 1916 (original). SPRI.

———. Diary, Oct. 1914–March. 1916 (transcript). ATL.

———. Diary, Oct. 1914–March. 1916 (transcript). CM.

Wild, John Robert Francis (Frank). Correspondence. ML.

———. Diary of the Southern Journey, British Antarctic Expedition (*Nimrod*), 1907–09. SPRI.

———. Memoirs (unpublished draft manuscript), 1937. ML.

Wilson, Edward Adrian. *Terra Nova* Diary. SPRI.

Worsley, Frank A. "Notes of conversations with J. R. Stenhouse and E. H. Shackleton," verso and margin of letter dated Nov. 17, 1932. Private Collection. HWC.

— PUBLISHED REPORTS, BROCHURES, AND ACADEMIC WORKS —

"Army Reserver's Report of Operation Highjump 1946–47." Washington, D.C.: War Department, 1947.

The Chief Inspector of Machinery and the Chief Surveyor of Ships. *Report of Marine Department Year 1916–17*. Wellington: The Secretary of the Marine and Inspection of Machinery Department.

Creswell, William. *Ross Sea Relief Expedition: Report of Australian Advisory Committee (to the Prime Minister of Australia)*. Melbourne: Commonwealth Naval Board, 1917.

Davis, John King. *Report of the Proceedings of the Aurora Antarctic Relief Expedition*. Wellington: Government Printer, 1917.

———. *Report of Voyage by Commander 20th Dec., 1916 to 9th Feb., 1917*. Melbourne: Government Printer, 1918.

Lionel Alfred George Hooke, Kt., in Memoriam. Amalgamated Wireless Australasia, 1974.

MacRury, Ian Kenneth. "The Inuit Dog: Its Provenance, Environment and History." MPhil thesis, University of Cambridge, 1991.

Ministry of National Service. *Report upon the Medical Department of the Ministry of National Service*. Kew: The National Archives, Public Records Office.

Parliament of the Commonwealth of Australia, Joint Committee of Public Accounts, *The Expenditure Incurred in Connexion with the S. Y. Aurora, of the Shackleton Expedition, at the Commonwealth Naval Dockyard, Cockatoo Island, Sydney.* Melbourne: The Parliament of the Commonwealth of Australia, 1916.

Shackleton, Ernest Henry. *The Imperial Trans-Antarctic Expedition (Prospectus).* London: Office of the Expedition, 1913.

————. *Sir Ernest Shackleton's Stirring Appeal for Men for the AIF, March 20, 1917.* Sydney: New South Wales Recruiting Committee.

— NEWSPAPERS —

The Age, Melbourne
Birmingham Post, Birmingham, England
The Church Family News, London
The Daily Chronicle, London
The Daily Express, London
The Daily Graphic, London
The Daily Mail, London
The Daily Mirror, London
Daily News and Leader, London
The Daily Telegraph, London
The East Kent Times and District Advertiser, Ramsgate, England
Edinburgh Evening Dispatch, Edinburgh
The Evening News, London
The Evening Standard and St. James Gazette, London
The Evening Star, Dunedin, New Zealand
Evening Sun, London
The Leeds Mercury, Leeds, England
The Manchester Guardian, Manchester, England
The Morning Post, London
The New York Times, New York
New Zealand Free Lance, Wellington, New Zealand
New Zealand Times, Wellington, New Zealand.
The Otago Daily Times, Dunedin, New Zealand
The Outlook, London
The Pall Mall Gazette, London
The People, London
The Southland Times, Invercargill, New Zealand
The Standard, London
The Surrey Comet and General Advertiser, Kingston Upon Thames, England
Sussex Daily News, Brighton, England
The Sydney Morning Herald, Sydney
The Times and *The Times Literary Supplement,* London
The Wanganui Chronicle, Wanganui, New Zealand
The Willesden Chronicle, Willesden, England

— MAGAZINES —

John Bull, London
London Mail, London
The Nation, New York
The Sphere, London

— MAPS AND CHARTS —

"Admiralty South Polar Chart." 1910. With notations by A. L. A. Mackintosh and R. W. Richards, 1915–16 notations. Cambridge: Scott Polar Research Institute.
"Antarctica, Ross Sea–Victoria Land, Franklin Island to McMurdo Sound (29321)." Washinton, D.C.: Defense Mapping Agency Hydrographic Center, 1978.
"British Antarctic Expedition 1910–13 (*Terra Nova*) Map of Erebus Bay–Ross Island." London: Harrison & Sons.
———. "British Antarctic Expedition 1910–13 (*Terra Nova*) Map of Cape Evans (Winter Quarters)." London: Harrison & Sons.
———. "British Antarctic Expedition 1910–13 (*Terra Nova*) Map of Hut Point Peninsula, Ross Island." London: Harrison & Sons.
"Chart to show sledging route and camps 26 Jan 1915–25 March 1915." By A. L. A. Mackintosh. Cambridge, England: Scott Polar Research Institute.
"McMurdo Sound, Antarctica" (ST 57-60). Reston, Va.: U.S. Department of the Interior Geological Survey in cooperation with the National Science Foundation, 1974.
"Mount Discovery" (ST 57-60/10). Reston, Va.: U.S. Department of the Interior Geological Survey in cooperation with the National Science Foundation, 1962.
"Mount Elizabeth" (SU 56-60/15). Reston, Va.: U.S. Department of the Interior Geological Survey in cooperation with the National Science Foundation, 1965.
"A Planning Chart for the Antarctic Region" (4009). Taunton, England: Hydrographer of the Navy, 1994.
"Ross Island" (ST 57-60/6). Reston, Va.: U.S. Department of the Interior Geological Survey in cooperation with the National Science Foundation, 1960; revised 1970.
"Ross Island and Vicinity" (76190-W1-RR-250). Reston, Va.: U.S. Department of the Interior Geological Survey in cooperation with the National Science Foundation, 1970.

— ONLINE SOURCES —

"Comparing the Purchasing Power of Money in Great Britain from 1264 to Any Other Year including the Present." John J. McCusker, Economic History Services, 2001. http://www.eh.net
"What Is Its Relative Value in UK Pounds?" Lawrence H. Officer, Economic History Services, June 30, 2004. http://www.eh.net
"Sunrise/sunset tables." U.S. Naval Observatory, Astronomical Applications Department. http://www.usno.navy.mil
"1901 Census Records." Kew: The National Archives, 1901. http://www.1901census.nationalarchives.gov.uk

— AUDIO AND FILM SOURCES —

Gaze, Irvine. "Interview." Wellington: Radio New Zealand, date unknown.

Hooke, Lionel. Oral history interview. Australian Society of World War One Aero Historians, 1975.

Hurley, James Francis (Frank). *South: Sir Ernest Shackleton's Glorious Epic of the Antarctic,* DVD. Including commentary on the Ross Sea party written and narrated by Kelly Tyler. London: British Film Institute, 2002.

————. *Endurance: The Story of a Glorious Failure.* Including commentary by Frank Worsley. London: British Film Institute Archives, 1933.

Richards, R. W. "Interview with R. W. Richards" by Ros Bowden. *Verbatim.* Sydney: ABC Radio National, 1977.

————. Oral history interview with R. W. Richards (conducted by Lennard Bickel). Canberra: National Library of Australia, 1976.

Stevens, Alexander. "Marooned on a Desert Island." BBC Radio. London: National Sound Archive, date unknown.

Tyler, Kelly, and Sarah Holt, producers/writers. *Shackleton's Voyage of Endurance,* PBS series *Nova,* DVD. Boston: WGBH Boston Video, 2002.

INDEX

Page numbers in *italics* refer to illustrations.

About the author

Historian Kelly Tyler-Lewis travelled for research to Britain, Australia, New Zealand and Antarctica, where she spent two months with the National Science Foundation's Artists and Writers Program. She was a visiting scholar at the Scott Polar Research Institute of the University of Cambridge from 2002 to 2004. She is also an Emmy Award-winning documentary film writer and producer. She lives in Massachusetts.